T0134884

Igor Aizenberg

Complex-Valued Neural Networks with Multi-Valued Neurons

# Studies in Computational Intelligence, Volume 353

**Editor-in-Chief**
Prof. Janusz Kacprzyk
Systems Research Institute
Polish Academy of Sciences
ul. Newelska 6
01-447 Warsaw
Poland
*E-mail:* kacprzyk@ibspan.waw.pl

Igor Aizenberg

# Complex-Valued Neural Networks with Multi-Valued Neurons

 Springer

Igor Aizenberg
Texas A&m University-Texarkana
7101 University Avenue,
Texarkana, TX, 75503
USA
E-mail: igor.aizenberg@tamut.edu

ISBN 978-3-662-50631-8                    ISBN 978-3-642-20353-4 (eBook)

DOI 10.1007/978-3-642-20353-4

Studies in Computational Intelligence          ISSN 1860-949X

© 2011 Springer-Verlag Berlin Heidelberg
Softcover re-print of the Hardcover 1st edition 2011

This work is subject to copyright. All rights are reserved, whether the whole or part
of the material is concerned, specifically the rights of translation, reprinting, reuse
of illustrations, recitation, broadcasting, reproduction on microfilm or in any other
way, and storage in data banks. Duplication of this publication or parts thereof is
permitted only under the provisions of the German Copyright Law of September 9,
1965, in its current version, and permission for use must always be obtained from
Springer. Violations are liable to prosecution under the German Copyright Law.

The use of general descriptive names, registered names, trademarks, etc. in this
publication does not imply, even in the absence of a specific statement, that such
names are exempt from the relevant protective laws and regulations and therefore
free for general use.

*Typeset* & *Cover Design:* Scientific Publishing Services Pvt. Ltd., Chennai, India.

Printed on acid-free paper

9 8 7 6 5 4 3 2 1

springer.com

In memoriam of my father Naum Aizenberg,

founder of complex-valued neural networks

# Acknowledgements

The author would like to express his appreciation and gratitude to many research colleagues from international scientific community who have inspired his thinking and research in the field of complex-valued neural networks.

First of all, the author expresses his great appreciation to the Editor of the Springer book series "Studies in Computational Intelligence" Professor Janusz Kacprzyk (Systems Research Institute of the Polish Academy of Sciences) who encouraged the author to write this book.

The author had a great pleasure and honor and was very happy to collaborate in his research on multi-valued neurons and neural networks based on them with great scientists and great personalities Professor Claudio Moraga (European Centre for Soft Computing, Mieres, Spain and Dortmund University of Technology, Dortmund, Germany), and Professor Jacek M. Zurada (University of Louisville, Kentucky, USA).

The author also appreciates a great opportunity to collaborate with other great scientists Professor Jaakko Astola (Tampere University of Technology, Tampere, Finland) and Professor Joos Vandewalle (Catholic University of Leuven – KU Leuven, Leuven, Belgium). The author also kindly appreciates those great opportunities he had working in KU Leuven (1996-1998) on the invitation of Professor Joos Vandewalle, Tampere University of Technology (1997, 2002, 2005-2006) on the invitation of Professor Jaakko Astola, and Dortmund University of Technology (2003-2005) on the invitation of Professors Claudio Moraga and Brend Reusch.

The author sincerely appreciates great assistance he got from his former and current students and junior colleagues. Particularly, the author would like to express his special thanks to his former students, co-authors in a number of publications, and great programmers whose wonderful contributions to software simulation of complex-valued neural networks with multi-valued neurons is of crucial importance. They are Dr. Dmitriy Paliy (Nokia Research, Helsinki, Finland), Taras Bregin (Jabil Ukraine Ltd, Uzhgorod, Ukraine), and Dr. Constantine Butakoff (University of Pompeu Fabra, Barcelona, Spain).

The author expresses his appreciation and gratitude to his family, to his wife Ella, to his children Zhenia and Alena, and to his mother Svetlana for their continuous support. The author also wants to say many special thanks to his son Zhenia for his kind assistance in preparing some illustrations for the book and in proofing the manuscript.

Some results presented in this book (Sections 3.3, 4.3, 5.2, and 5.3) are based upon work supported by the National Science Foundation under Grant No. 0925080.

# Preface

The use of complex numbers in neural networks is as natural as their use in other engineering areas and in mathematics.

The history of complex numbers shows that although it took a long time for them to be accepted (almost 300 years from the first reference to "imaginary numbers" by Girolamo Cardano in 1545[1] to Leonard Euler's[2] and Carl Friedrich Gauss'[3] works published in 1748 and 1831, respectively), they have become an integral part of mathematics and engineering. It is difficult to imagine today how signal processing, aerodynamics, hydrodynamics, energy science, quantum mechanics, circuit analysis, and many other areas of engineering and science could develop without complex numbers. It is a fundamental mathematical fact that complex numbers are a necessary and absolutely natural part of numerical world. Their necessity clearly follows from the Fundamental Theorem of Algebra, which states that every non-constant single-variable polynomial of degree $n$ with complex coefficients has exactly $n$ complex roots, if each root is counted up to its multiplicity.

Answering a question frequently asked by some "conservative" researches, what one can get using complex-valued neural networks (typical objections are: they have "twice more" parameters, require more computations, etc.), we may say that one may get the same as using the Fourier transform, but not just the Walsh transform in signal processing. There are many engineering problems in the modern world where complex-valued signals and functions of complex variables are involved and where they are unavoidable. Thus, to employ neural networks for their analysis, approximation, etc., the use of complex-valued neural networks is natural. However, even in the analysis of real-valued signals (for example, images or audio signals) one of the most frequently used approaches is frequency domain analysis, which immediately leads us to the complex domain. In fact, analyzing signal properties in the frequency domain, we see that each signal is characterized

---

[1] G. Cardano's work "Arts Magna" ("Great Art or on Algebraic Rules" was published in 1545. For the first time he introduced a notion of "imaginary numbers", however he considered these numbers useless.

[2] L. Euler proved and published in 1744 the relationship between the trigonometric functions and complex exponential function ( $e^{i\varphi} = \cos\varphi + i\sin\varphi$ ). He also suggested to use symbol $i$ for an imaginary unity (the first letter of Latin word *imaginarius*).

[3] C.-F. Gauss gave to complex numbers their commonly used name "complex" and comprehensively described them in his "Memoir to the Royal Society of Göttingen" in 1831. Gauss has also obtained the first mathematically exact proof of algebraic closure of the field of complex numbers in 1799 (this fact was first hypothetically formulated by J. d'Alembert in 1747 and L. Euler in 1751).

by magnitude and phase that carry different information about the signal. To use this information properly, the most appropriate solution is movement to the complex domain because there is no other way to treat properly the phase information. Hence, *one of the most important characteristics of Complex-Valued Neural Networks is the proper treatment of the phase information*. It is important to mention that this phenomenon is important not only in engineering, but also in simulation of biological neurons. In fact, biological neurons when firing generate sequences of spikes (spike trains). The information transmitted by biological neurons to each other is encoded by the frequency of the corresponding spikes while their magnitude is a constant. Since, it is well known that the frequency can be easily transformed to the phase and vice versa, then it should be natural to simulate these processes using a complex-valued neuron.

Complex-Valued Neural Networks (CVNN) is a rapidly growing area. There are different specific types of complex-valued neurons and complex-valued activation functions. But it is important to mention that all Complex-Valued Neurons and Complex-Valued Neural Networks have a couple of very important advantages over their real-valued counterparts. The first one is that they have *much higher functionality*. The second one is their *better plasticity* and *flexibility*: they learn faster and generalize better. The higher functionality means first of all *the ability of a single neuron to learn those input/output mappings that are non-linearly separable in the real domain*. This means the ability to learn them in the initial space without creating higher degree inputs and without moving to the higher dimensional space, respectively. As it will be shown below, such classical non-linearly separable problems as XOR and Parity $n$ are about the simplest that can be learned by a single complex-valued neuron.

It is important to mention that *the first historically known complex-valued activation function was proposed in 1971* (!) by Naum Aizenberg and his co-authors[4]. It was 40 years ago, before the invention of backpropagation by Paul Werbos (1974), before its re-invention and development of the feedforward neural network by David Rumelhart (1986), before the Hopfield neural network was proposed by John Hopfield in 1982. Unfortunately, published only in Russian (although in the most prestigious journal of the former Soviet Union), this seminal paper by Naum Aizenberg and his colleagues and a series of their subsequent publications were not available to the international research community for many years. A problem was that in the former Soviet Union it was strictly prohibited to submit scientific materials abroad and therefore there was no way for Soviet scientists to publish their results in international journals. May be, being wider known, those seminal ideas on complex-valued neurons could stimulate other colleagues to join research in this area much earlier than it really happened (only in 1990s and 2000s). May be, this could help, for example, to widely use neural networks for solving not only binary, but multi-class classification problems as far back as more than 30 years ago… However, the history is as it is, we cannot go to the past and change something there. Let us better concentrate on what we have today.

---

[4] N.N. Aizenberg, Yu. L. Ivaskiv, and D.A. Pospelov, "About one generalization of the threshold function" *Doklady Akademii Nauk SSSR* (*The Reports of the Academy of Sciences of the USSR*), vol. 196, No 6, 1971, pp. 1287-1290 (in Russian).

So what is this book about? First of all, it is not an overview of all known CVNNs. It is devoted to comprehensive observation of one representative of the complex-valued neurons family – the Multi-Valued Neuron (MVN) (and its variation – the Universal Binary Neuron (UBN) ) and MVN-based neural networks. The Multi-Valued Neuron operates with complex-valued weights. Its inputs and output are located on the unit circle and therefore its activation function is a function only of argument (phase) of the weighted sum. It does not depend on the weighted sum magnitude. MVN has important advantages over other neurons: its functionality is higher and its learning is simpler because it is *derivative-free* and it is based on the error-correction rule. A single MVN with a periodic activation function can easily learn those input/output mappings that are non-linearly separable in the real domain (of course, including the most popular examples of them, XOR and Parity $n$). These advantages of MVN become even more important when this neuron is used as a basic one in a feedforward neural network. The Multilayer Neural Network based on Multi-Valued Neurons (MLMVN) is an MVN-based feedforward neural network. Its original backpropagation learning algorithm significantly differs from the one for a multilayer feedforward neural network (MLF)[5]. It is derivative-free and it is based on the error-correction rule as it is for a single MVN. MLMVN significantly outperforms many other techniques (including MLF and many kernel-based and neuro-fuzzy techniques) in terms of learning speed, network complexity and generalization capability.

However, when the reader starts reading this book or when the reader even consider whether to read it, it is also important to understand that this book is not the 2nd edition of the first monograph devoted to multi-valued neurons[6]. Since that monograph was published 11 years ago, and many new results were obtained during this time by the author of this book, his collaborators and other researchers, this book, on the one hand, contains a comprehensive observation of the latest accomplishments and, on the other hand, it also deeply observes a theoretical background behind MVN. This observation is based on the today's view on the MVN place and role in neural networks. It is important that today's understanding is much deeper and comprehensive than it was when the first book was published. Thus, the overlap of this book with the first one is minimal and it is reduced to some basic necessarily definitions, which is just about 5-6% of the content. The most significant part of the book is based on the results obtained by the author independently and in co-authorship with other colleagues and his students. However, contributions made by other research colleagues to MVN-based neural networks are also observed.

---

[5] Often this network based on sigmoidal neurons is also referred to as the multilayer perceptron (MLP) or a "standard backpropagation network". We will use a term MLF throughout this book reserving a term "perceptron" for its initial assignment given by Frank Rosenblatt in his seminal paper F. Rosenblatt, "The Perceptron: A Probabilistic Model for Information Storage and Organization in the Brain, Cornell Aeronautical Laboratory", *Psychological Review*, v65, No. 6, 1958 pp. 386-408.

[6] I. Aizenberg, N. Aizenberg, and J. Vandewalle, *Multi-Valued and Universal Binary Neurons: Theory, Learning, Applications*, Kluwer Academic Publishers, Boston/Dordrecht/London, 2000.

This book is addressed to all people who work in the fascinating field of neural networks. The author believes that it can be especially interesting for those who use neural networks for solving challenging multi-class classification and prediction problems and for those who develop new fundamental theoretical solutions in neural networks. It should be very suitable for Ph.D. and graduate students pursuing their degrees in computational intelligence. It should also be very helpful for those readers who want to extend their view on the whole area of computational intelligence.

The reader is not expected to have some special knowledge to read the book. All readers with basic knowledge of algebra and calculus, and just very basic knowledge of neural networks (or even without having special knowledge in neural networks area) including students can easily understand it.

We avoid using here too deep mathematical considerations (may be except proofs of convergence of the learning algorithms and analysis of that specific separation of an $n$-dimensional space, which is determined by the MVN activation function). However, those readers who do not want or do not need to go to those mathematical details may skip over the corresponding proofs.

In this book, we cover the MVN and MVN-based neural networks theory and consider many of their applications. The most important topics related to multi-valued neurons are covered. Chapter 1 should help the reader to understand why Complex-Valued Neural Networks were introduced. It presents a brief observation of neurons, neural networks, and learning techniques. Since the functionality of real-valued neurons and neural networks is limited, it is natural to consider the complex-valued ones whose functionality is much higher. We also observe in Chapter 1 CVNNs presenting the state of the art in this area. In Chapter 2, the multi-valued neuron is considered in detail along with the basic fundamentals of multiple-valued logic over the field of complex numbers, which is a main theoretical background behind MVN. In Chapter 3, MVN learning algorithms are presented. Chapter 4 is devoted to the multi-valued neural network based on multi-valued neurons, its original derivative-free backpropagation learning algorithm and its applications. In Chapter 5, MVN with a periodic activation function and its binary version, the universal binary neuron, are presented, and it is shown how it is possible to solve non-linearly separable problems using a single neuron without the extension of the feature space. In Chapter 6, some other applications of MVN are considered (solving classification and prediction problems, associative memories). The book is illustrated by many examples of applications.

The author sincerely hopes that this book will provide its readers with new interesting knowledge and will encourage many of them to use MVN and MVN-based neural networks for solving new challenging applied problems. The author will be glad and consider his work successful if more researches will be involved through this book in the really magic world of neural networks and particularly its complex-valued part.

The author also hopes that by this book he can pay a tribute to the founder of Complex-Valued Neural Networks, his teacher, colleague, father and a great personality Naum Aizenberg.

# Contents

# Chapter 1
# Why We Need Complex-Valued Neural Networks?

"Why is my verse so barren of new pride,
So far from variation or quick change?
Why with the time do I not glance aside
To new-found methods and to compounds strange?"

William Shakespeare, Sonnet 76

This chapter is introductory. A brief observation of neurons and neural networks is given in Section 1.1. We explain what is a neuron, what is a neural network, what are linearly separable and non-linearly separable input/output mappings. How a neuron learns is considered in Section 1.2, where Hebbian learning, the perceptron, and the error-correction learning rule are presented. In Section 1.3, we consider a multilayer feedforward neural network and essentials of backpropagation learning. The Hopfield and cellular neural networks are also presented. Complex-valued neural networks, their naturalness and necessity are observed in Section 1.4. It is shown that a single complex-valued neuron can learn non-linearly separable input/output mappings and is much more functional than a single real-valued neuron. Historical observation of complex-valued neural networks and the state of the art in this area are also presented. Some concluding remarks will be given in Section 1.5.

## 1.1 Neurons and Neural Networks: Basic Foundations and Historical View

### 1.1.1 What Is a Neural Network?

As we have clearly mentioned, this book is devoted to complex-valued neural networks, even only to those of them that are based on multi-valued neurons. However, it should not be correct, if we will start immediately from complex-valued neurons and neural networks. To understand, why complex-valued neurons were introduced and to understand that motivation, which was behind their introduction, it is important to observe what a neural network is. It is also important to have a good imagination about those solutions that existed in neural networks that time when the first complex-valued neuron was proposed and about state of the art

I. Aizenberg: Complex-Valued Neural Networks with Multi-Valued Neurons, SCI 353, pp. 1–53.
springerlink.com                    © Springer-Verlag Berlin Heidelberg 2011

in neural networks, to understand why complex-valued neurons are even more important today. It is also important to understand those limitations that are specific for real-valued neurons and neural networks. This will lead us to much clearer understanding of the importance of complex-valued neurons and the necessity of their appearance for overcoming limitations and disadvantages of their real-valued counterparts.

So let us start from the brief historical overview.

What an artificial neural network is? Among different definitions, which the reader can find in many different books, we suggest to use the following given in [1] by Igor Aleksander and Helen Morton, and in [2] by Simon Haykin.

**Definition 1.1.** A *neural network* is a massively parallel distributed processor that has a natural propensity for storing experimental knowledge and making it available for use. It means that: 1) Knowledge is acquired by the network through a learning process; 2) The strength of the interconnections between neurons is implemented by means of the synaptic weights used to store the knowledge.

Let us consider in more detail what stands behind this definition. It is essential that an artificial neural network is a massively parallel distributed processor whose basic processing elements are artificial neurons. The most important property of any artificial neural network and of its basic element, an artificial neuron, is their ability to learn from their environment. *Learning* is defined in [2] as a process by which the free parameters of a neural network (or of a single neuron) are adapted through a continuing process of simulation by the environment in which the network (the neuron) is embedded. This means that both a single artificial neuron and an artificial[1] neural network are *intelligent systems*. They do not perform computations according to the pre-defined externally loaded program, but they *learn* from their environment formed by learning samples that are united in a *learning set*. Once the learning process is completed, they are able to generalize relying on that knowledge, which was obtained during the learning process. The quality of this generalization is completely based on that knowledge, which was obtained during the learning process.

Compared to biological neural networks, artificial neural networks are "neural" in the sense that they have been inspired by neuroscience, but they are not true models of biological or cognitive phenomena. The important conclusion about artificial neural networks, which is done by Jacek Zurada in [3], states that typical neural network architectures are more related to mathematical and/or statistical techniques, such as non-parametric pattern classifiers, clustering algorithms, nonlinear filters, and statistical regression models.

In contrast to algorithmic approaches usually tailored to tasks at hand, neural networks offer a wide palette of versatile modeling techniques applicable to a large class of problems. Here, learning in data-rich environments leads to models of specific tasks. Through learning from specific data with rather general

---

[1] We will omit further the word "artificial" keeping in mind that across this book we have deal with artificial neurons and artificial neural networks. Wherever it will be needed, when a biological neuron will be considered, we will add the word "biological".

neural network architectures, neurocomputing techniques can produce problem-specific solutions [3].

## 1.1.2 The Neuron

We just told that there are many equivalent definitions of a neural network. However, it is quite difficult to find a strict definition of a neuron. We may say that an artificial neuron is on the one hand, an abstract model of a biological neuron, but on the other hand, it is an intelligent information processing element, which can learn and can produce the output in response to its inputs. As a result of learning process, the neuron forms a set of *weights* corresponding to its inputs. Then by weighting summation of the inputs and transformation of the weighted sum of input signals using an *activation* (*transfer*) function it produces the output.

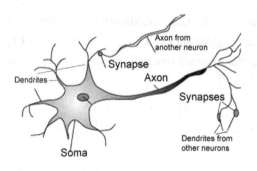

This is really quite similar to what a biological neuron is doing. Let us consider its schematic model (see Fig. 1.1). Indeed, a biological neuron receives input signals thorough its *dendrites* that are connected to *axons* (which transmit output signals) of other neurons via *synapses* where the input signals are being weighted by the synaptic weights. Then the biological neuron performs a weighting summation of inputs in *soma*

**Fig. 1.1** A schematic model of a biological neuron

where it also produces the output, which it transmits to the dendrites of other neurons through the synaptic connections.

The first artificial neuron model was proposed by W. McCulloch and W. Pitts in 1943 [4]. They tried to create a mathematical model of neural information processing as it was considered that time. A common view was that a neuron receives some input signals $x_1, ..., x_n$ that can be excitatory ("1") or inhibitory ("-1"), calculates the weighted sum of inputs $z = w_1 x_1 + ... + w_n x_n$ and then produces the excitatory output ("1") if the weighted sum of inputs exceeds some predetermined threshold value and the inhibitory output ("-1") if it does not. For many years, it is a commonly known fact that a biological neuron is much more sophisticated from the signal processing view point. It is not a discrete binary processing element, its inputs and outputs are continuous, etc. Thus, the McCulloch-Pitts model as a model of a biological neuron is very schematic and it just approaches a basic idea of neural information processing. Nevertheless it is difficult to overestimate the importance of this model. First of all, it is historically the first model of a neuron. Secondly, this model was important for understanding of learning mechanisms that we will consider below. Thirdly, all later neural models are based on the same approach that was in the McCulloch-Pitts model: weighted summation of inputs followed by the transfer function applied to the weighted sum to produce the output.

Let us take a closer look at the McCulloch-Pitts model. As we have mentioned, in this model the neuron is a binary processing element. It receives binary inputs $x_1,...,x_n$ taken their values from the set $\{-1, 1\}$ and produces the binary output belonging to the same set. The weights $w_1,...,w_n$ can be arbitrary real numbers and therefore the weighted sum $z = w_1x_1 +...+ w_nx_n$ can also be an arbitrary real number. The neuron output $f\left(x_1,...,x_n\right)$ is determined as follows:

$$f\left(x_1,...,x_n\right) = \begin{cases} 1, \text{if } z \geq \Theta \\ -1, \text{if } z < \Theta, \end{cases}$$

where $\Theta$ is the pre-determined threshold. The last equation can be transformed if the threshold will be included to the weighted sum as a "free weight" $w_0 = -\Theta$, which is often also called a *bias* and the weighted sum will be transformed accordingly ( $z = w_0 + w_1x_1 +...+ w_nx_n$ ):

$$f\left(x_1,...,x_n\right) = \begin{cases} 1, \text{if } z \geq 0 \\ -1, \text{if } z < 0. \end{cases}$$

This is the same as

$$f\left(x_1,...,x_n\right) = \text{sgn}\left(z\right), \tag{1.1}$$

where *sgn* is a standard sign function, which is equal to 1 when its argument is non-negative and to -1 otherwise (see Fig. 1.2). Thus, function *sgn* in (1.1) is an activation function. It is usually referred to as the *threshold activation function*. The McCulloch-Pitts neuron is also often called the *threshold element* or the *threshold neuron*. [5]. These names were especially popular in 1960s – 1970s.

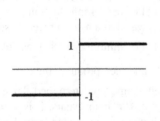

It is important to mention that function *sign* is nonlinear. Hence, the first neuron was a nonlinear processing element. This property is very important. All popular activation functions that are used in neurons are nonlinear. It will

**Fig. 1.2** Threshold Activation Function

not be the overestimation, if we will say that the functionality of a neuron is mainly (if not completely) determined by its *activation function*.

Let us consider now the most general model of a neuron, which is commonly used today (see Fig. 1.3). A neuron has $n$ inputs $x_1,...,x_n$ and weights $w_1,...,w_n$

corresponding to these inputs. It also has a "free weight" (bias) $w_0$, which does not correspond to any input. All together weights form an $(n+1)$-dimensional *weighting vector* $\left(w_0, w_1, ..., w_n\right)$. There is a pre-determined activation function $\varphi(z)$ associated with a neuron. It generates the neuron output limiting it to some reasonable (permissible) range. The neural processing consists of two steps. The first step is the calculation of the weighted sum of neuron inputs

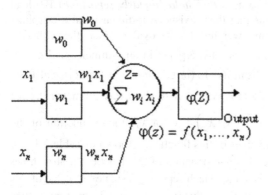

**Fig. 1.3** A general model of a neuron

$z = w_0 + w_1 x_1 + ... + w_n x_n$.
The second step is the calculation of the value of the activation function $\varphi(z)$ for the value $z$ of the weighted sum. This value of the activation function forms the output of the neuron. If input/output mapping is described by some function $f\left(x_1, ..., x_n\right)$, then

$$f\left(x_1, ..., x_n\right) = \varphi(z) = \varphi\left(w_0 + w_1 x_1 + ... + w_n x_n\right). \qquad (1.2)$$

Initially only binary neuron inputs and output were considered. Typically, they were taken from the set $E_2 = \{1, -1\}$ or (rarely) from the set $K_2 = \{0, 1\}^2$. It is important to mention that it is very easy to move from one of these alphabets to another one. For example, if $y \in K_2$ then $x = 1 - 2y \in E_2$, and if $x \in E_2$ then $y = -(x-1)/2 \in K_2$, respectively. Hence, $0 \leftrightarrow 1$, $1 \leftrightarrow -1$. As for the weights, they were taken from the set $\mathbb{R}$ of real numbers. Therefore, the weighted sum in this case is also real and an activation function is a function of a real variable. We may say that mathematically the threshold neuron implements a mapping $f\left(x_1, ..., x_n\right): E_2^n \rightarrow E_2$.

---

### 1.1.3  Linear Separability and Non-linear Separability: XOR Problem

If the neuron performs a mapping $f(x_1,...,x_n): E_2^n \to E_2$, this means that $f(x_1,...,x_n)$ is a Boolean function. If the function *sgn* is used as the activation function of a neuron (thus, if it is the threshold neuron), then this Boolean function is called and commonly referred to as a *threshold (linearly separable)* Boolean function. Linear separability means that there exists an *n*-dimensional hyperplane determined by the corresponding weights (it is evident that the equation $z = w_0 + w_1 x_1 + ... + w_n x_n$ determines a hyperplane in an *n*-dimensional space) and separating 1s of this function from its -1s (or 0s from 1s if the classical Boolean alphabet $K_2 = \{0,1\}$ is used). It is very easy to show this geometrically for *n*=2. Let us consider the function $f(x_1,x_2) = x_1$ or $x_2$, the disjunction of the two Boolean variables. A table of values of this function is shown in Table 1.1.

Fig. 1.4a demonstrates a geometrical interpretation of this function. It also shows what a linear separability is. There is a line, which separates a single "1" value of this function from three "-1" values. It is also clear that there exist infinite amount of such lines. In 1960s study of threshold Boolean functions was very popular.

**Table 1.1** Values of function $f(x_1,x_2) = x_1$ or $x_2$

| $x_1$ | $x_2$ | $f(x_1,x_2) = x_1$ or $x_2$ |
|:---:|:---:|:---:|
| 1 | 1 | 1 |
| 1 | -1 | -1 |
| -1 | 1 | -1 |
| -1 | -1 | -1 |

We can mention at least two comprehensive monographs devoted to this subject [6, 7]. However, the number of threshold or linearly separable Boolean functions is very small. While for *n*=2 there are 14 threshold functions out of 16 and for *n*=3 there are 104 threshold functions out of 256, for *n*=4 there are just about 2000 threshold functions out of 65536. For $n > 4$, the ratio of the number of threshold Boolean functions of *n* variables to $2^{2^n}$ (the number of all Boolean functions of *n* variables) approaches 0.

While threshold Boolean functions can be implemented using a single threshold neuron, other functions that are not threshold cannot. May be the most typical and the most popular example of such a function is XOR problem

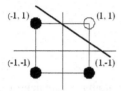

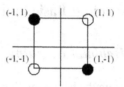

(a) $f(x_1, x_2) = x_1$ or $x_2$ is a linearly separable function. There exists a line, which separates 1 value of this function (a transparent circle) from its -1s (filled circles)

(b) $f(x_1, x_2) = x_1$ xor $x_2$ is a non-linearly separable function. There is no way to find a line, which separates 1s value of this function (transparent circles) from its -1s (filled circles)

**Fig. 1.4**

(the Exclusive OR) $f(x_1, x_2) = x_1$ xor $x_2$, mod 2 sum of the two Boolean variables. This function is non-linearly separable. Let us take a look at the table of values of this function (see Table 1.2) and its graphical representation (see Fig. 1.4b). Geometrically, this problem belongs to the classification of the points in the hypercube, as any problem described by the Boolean function (see Fig. 1.4). Each point in the hypercube is either in class "1" or class "-1". In the case of XOR problem the input patterns (1, 1) and (-1, -1) that are in class "1" are at the opposite corners of the square (2D hypercube). On the other hand, the input patterns (1, -1) and (-1, 1) are also at the opposite corners of the same square, but they are in class "-1". It is clear from this that the function XOR is non-linearly separable, because there is no way to draw a line, which can separate two "1" values of this function from its two "-1" values, which is clearly seen from Fig. 1.4b. Since such a line does not exist, there are no weights using which XOR function can be implemented using a single threshold neuron.

Table 1.2 Values of function $f(x_1, x_2) = x_1$ xor $x_2$

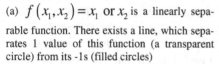

| $x_1$ | $x_2$ | $f(x_1, x_2) = x_1$ xor $x_2$ |
|---|---|---|
| 1 | 1 | 1 |
| 1 | -1 | -1 |
| -1 | 1 | -1 |
| -1 | -1 | 1 |

The existence of non-linearly separable problems was a starting point for neural networks design and likely the XOR problem stimulated creation of the first multilayer neural network. We will consider this network in Section 1.3. However, the most important for us will be the fact that XOR problem can be easily solved using a single complex-valued neuron. We will show this solution in Section 1.4.

## 1.2 Learning: Basic Fundamentals

### 1.2.1 Hebbian Learning

We told from the beginning that the main property of both a single neuron and any neural network is their ability to learn from their environment. How a neuron learns? The first model of the learning process was developed by Donald Hebb in 1949 [8]. He considered how biological neurons learn. As we have already mentioned, biological neurons are connected to each other through synaptic connections: axon of one neuron is connected to dendrites of other ones through synapses (Fig. 1.1). To represent the Hebbian model of learning, which is commonly referred to as *Hebbian learning*, let us cite D. Hebb's fundamental book [8] directly. The idea of the Hebbian learning is as follows ([8], p. 70).

"The general idea is … that any two cells or systems of cells that are repeatedly active at the same time will tend to become 'associated', so that activity in one facilitates activity in the other."

The mechanism of Hebbian learning is the following ([8], p. 63).

"When one cell repeatedly assists in firing another, the axon of the first cell develops synaptic knobs (or enlarges them if they already exist) in contact with the soma of the second cell."

Let us "translate" this idea and mechanism into the language of the threshold neuron. In this language, "1" that is a "positive" signal, means excitation, and "-1" that is a "negative" signal, means inhibition. When the neuron "fires" and produces "1" in its output, this means that weights have to help this neuron to "fire". For example, if the neuron receives a "positive" signal ("1") from some input, then the corresponding weight passing this signal can be obtained by multiplication of the desired output "1" by the input "1" (see Fig. 1.5a). Thus, the weight is equal to 1 and the "positive" input signal will contribute to the positive output of the neuron. Indeed, to produce a "positive" output, according to (1) the weighted sum must be positive. On the contrary, if the neuron "fires", but from some input it receives a "negative" (inhibitory) signal, the corresponding weight has to invert this signal, to make its contribution to the weighted sum and the neuron output positive. Again, the simplest way to achieve this, is to multiply the desired output "1" by the input "-1" (see Fig. 1.5b). The corresponding weight will be equal to -1 and when multiplied by the input, will produce a positive contribution $(-1) \cdot (-1) = 1$ to the weighted sum and output. Respectively, if the neuron does not "fire" and has to produce a "negative" inhibitory output (”-1”), the weights have to help to inhibit the neuron and to produce a negative weighted sum. The weights should be found in the same way: by multiplication of the desired output "-1" by the corresponding input value. If the input is $"-1"$ (inhibitory), then the weight $(-1) \cdot (-1) = 1$ just passes it (see Fig. 1.5c). If the input is "1" (excitatory), the weight $(-1) \cdot 1 = -1$ inverts it (see Fig. 1.5d).

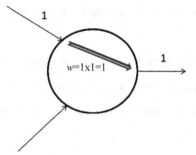

(a) the neuron "fires" and a "firing" input is passed to the output by the positive weight

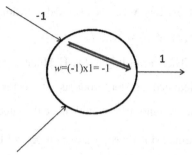

(b) the neuron "fires" and an "inhibitory" input is inverted by the negative weight

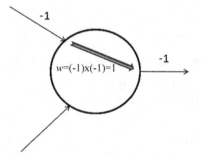

(c) the neuron "inhibits" and an "inhibitory" input is passed to the output by the positive weight

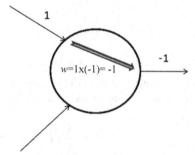

(d) the neuron "inhibits" and a "firing" input is inverted by the negative weight

**Fig. 1.5** Calculation of the weight using the Hebb rule for one of the neuron's inputs and for a single learning sample: the weight is equal to the product of the desired output and input value

To obtain a bias $w_0$, we just need to multiply the desired output by 1, which can be considered as a "virtual constant input" corresponding to this weight.

It is clear that if the threshold neuron has to learn only a single learning sample, then this Hebb rule always produces the weighting vector implementing the corresponding input/output mapping. However, learning from a single learning sample is not interesting, because it is trivial. What about multiple learning samples? In this case, the weights can be found by generalization of the rule for a single learning sample, which we have just described. This generalization leads us to the following representation of the Hebbian learning rule for a single threshold neuron. Let us have $N$ $n$-dimensional learning samples (this means that our neuron has $n$ inputs $x_1, ..., x_n$). Let $\mathbf{f}$ be an $N$-dimensional vector-column[3] of output

---

[3] Here and hereafter we will use a notation $\mathbf{f} = \left( f_1, ..., f_n \right)^T$ for a vector-column, while a notation $F = \left( f_1, ..., f_n \right)$ will be used for a vector-row.

values. Let $\mathbf{x}_1, \ldots, \mathbf{x}_n$ be $N$-dimensional vectors of all possible values of inputs $x_1^j, \ldots, x_n^j, j = 1, \ldots, N$.

Then according to the Hebbian learning rule the weights $w_1, \ldots, w_n$ should be calculated as dot products of vector $\mathbf{f}$ and vectors $\mathbf{x}_1, \ldots, \mathbf{x}_n$, respectively. The weight $w_0$ should be calculated as a dot product of vector $\mathbf{f}$ and an $N$-dimensional vector-constant $\mathbf{x}_0 = (1,1,\ldots,1)^T$ :

$$w_i = (\mathbf{f}, \mathbf{x}_i), i = 0, \ldots, n, \tag{1.3}$$

where $(\mathbf{a}, \mathbf{b}) = a_1 \overline{b_1} + \ldots + a_n \overline{b_n}$ is the dot product of vector-columns $\mathbf{a} = (a_1, \ldots, a_n)^T$ and $\mathbf{b} = (b_1, \ldots, b_n)^T$ in the unitary space ("bar" is a symbol of complex conjugation, in the real space it should simply be ignored).

It can also be suggested to normalize the weights obtained by (1.3):

$$w_i = \frac{1}{N}(\mathbf{f}, \mathbf{x}_i), i = 0, \ldots, n. \tag{1.4}$$

Let us check how rule (1.4) works.

**Example 1.1** Let us learn using this rule the OR problem $f(x_1, x_2) = x_1$ or $x_2$, which is linearly separable and which we have already considered for illustration of the linear separability (Table 1.1, Fig. 1.4a). Let us use rule (1.4) to obtain the weights. From Table 1.1, we have $\mathbf{f} = (1,-1,-1,-1)^T; \mathbf{x}_0 = (1,1,1,1)^T; \mathbf{x}_1 = (1,1,-1,-1)^T; \mathbf{x}_2 = (1,-1,1,-1)^T$. Then, applying Hebbian learning rule (1.4), we obtain the following weights $w_0 = (\mathbf{f}, \mathbf{x}_0) = -0.5; w_1 = (\mathbf{f}, \mathbf{x}_1) = 0.5; w_2 = (\mathbf{f}, \mathbf{x}_2) = 0.5$. Let us now check the results of this learning and apply the weighting vector $W = (-0.5, 0.5, 0.5)$ to all four possible binary inputs of the threshold neuron. The results are summarized in Table 1.3. We see that the weighting vector, which we obtained learning the OR function using the Hebbian learning rule really implements the OR function using the threshold neuron.

The reader for whom neural networks is a new subject may say "Hurrah! It so simple and beautiful!" It is really simple and beautiful, but unfortunately just a minority of all threshold Boolean functions of more than two variables can be learned in this way.

It is also clear that neither of non-threshold Boolean functions can be learned by the threshold neuron using rule (1.4) (non-threshold functions cannot be learned by the threshold neuron at all). By the way, different non-threshold Boolean functions may have the same weighting vectors obtained by rule (1.4). If we apply rule (1.4) to such a non-threshold Boolean function, like XOR (see Table 1.2 and Fig. 1.4b), which is symmetric (self-dual (or odd, in other words) or even), we get the zero weighting vector $(0, ..., 0)$.

**Table 1.3** Threshold neuron implements $f(x_1, x_2) = x_1$ or $x_2$ function with the weighting vector (-0.5, 0.5, 0.5) obtained by Hebbian learning rule (1.4)

| $x_1$ | $x_2$ | $z = w_0 + w_1 x_1 + w_2 x_2$ | $\mathrm{sgn}(z)$ | $f(x_1, x_2) = x_1$ or $x_2$ |
|---|---|---|---|---|
| 1 | 1 | 0.5 | 1 | 1 |
| 1 | -1 | -0.5 | -1 | -1 |
| -1 | 1 | -0.5 | -1 | -1 |
| -1 | -1 | -1.5 | -1 | -1 |

The following natural questions can now be asked by the reader. How those threshold Boolean functions that cannot be learned using the Hebb rule, can be learned? What about multiple-valued and continuous input/output mappings, is it possible to learn them? If the Hebb rule has a limited capability, is it useful? The answer to the first question will be given right in the next Section. Several answers to the second question will be given throughout this book. The third question can be answered right now. The importance of Hebbian learning is very high, and not only because D. Hebb for the first time explained mechanisms of associations developing during the learning process. A vector obtained using the Hebb rule, even if it does not implement the corresponding input/output mapping, can often be a very good first approximation of the weighting vector because it often can "draft" a border between classes when solving pattern recognition and classification problems. In [6] it was suggested to call a vector obtained by (1.4) for a Boolean function the *characteristic vector* of that function. Later the same notion was considered for multiple-valued functions and the Hebb rule was used to learn them using the multi-valued neuron. We will consider this aspect of Hebbian learning later when we will consider multi-valued neurons and their applications (Chapters 2-6).

Using the Hebbian learning it is possible to develop associations between the desired outputs and those inputs that stimulate these outputs. However, the Hebbian learning cannot correct the errors if those weights obtained by the Hibbian rule still do not implement the corresponding input/output mapping. To be able to correct the errors (to adjust the weights in such a way that the error will be minimized or eliminated), it is necessary to use the error-correction learning rule.

## 1.2.2  Perceptron and Error-Correction Learning

The *perceptron* is historically the first artificial neural network. The perceptron was suggested in 1958 by Frank Rosenblatt in [9] as "a hypothetical nervous system", and as the illustration of "some of the fundamental properties of intelligent systems in general".

Today we may say that the Rosenblatt's perceptron as it was defined in the seminal paper [9] is the simplest feedforward neural network (it will be considered in Section 1.3), but in 1958 when F. Rosenblatt published his paper, this today's most popular kind of a neural network was not invented yet. The perceptron was suggested as a network consisted of three types of elements that can simulate recognition of visual patterns (F. Rosenblatt demonstrated the perceptron's ability to recognize English typed letters). The perceptron in its original concept contained three types of units (Fig. 1.6): S-units (sensory) for collecting the input information and recoding it into the form appropriate for A-units (associate units), and R-units (responses). While S-units are just sensors (like eye retina) and R-units are just responsible for reproduction of the information in terms suitable for its understanding, A-units are the neurons, for example the ones with the threshold activation function (later a sigmoid activation was suggested, we will also consider it below). Thus, A-units form a single layer feedforward neural network. All connections among units were usually built at random.

The main idea behind the perceptron was to simulate a process of pattern recognition. At that time when the perceptron concept was suggested, classification was considered only as a binary problem (a two-class classification problem), and the perceptron was primarily used as a binary classifier. Thus, each neuron (each A-unit) performed only input/output mappings $f\left(x_1,...,x_n\right):E_2^n \rightarrow E_2$ (or $f\left(x_1,...,x_n\right):K_2^n \rightarrow K_2$ depending on which Boolean alphabet was used).

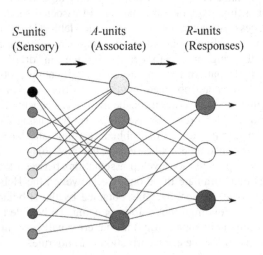

S-units        A-units        R-units
(Sensory)      (Associate)    (Responses)

Later it was suggested to consider a more general case when neuron (perceptron) inputs are real numbers from some bounded set $T \subset \mathbb{R}$ (often the case of $T = \left[0,1\right]$ is considered).

Thus, if $x_i \in T, i = 1,...,n$ a mapping performed by the neuron becomes $f\left(x_1,...,x_n\right):T^n \rightarrow E_2$.

**Fig. 1.6** The Perceptron

One of the main achievements of the perceptron era was the error-correction learning concept first suggested by F. Rosenblatt in [10] and then developed and deeply presented in his monograph [11]. Since in the perceptron all its A-units learn separately and independently, we may consider the error-correction learning rule with regard to a single neuron. We will derive the error-correction learning rule in the same way as it was done by its inventor. We have to mention that this learning rule will be very important for us when we will consider its generalization for complex-valued neurons and neural networks.

Let us consider the threshold neuron with activation function (1.1). Suppose a neuron has to learn some input/output mapping $f(x_1,...,x_n) : E_2^n \to E_2$. This input/output mapping could represent, for example, some binary classification problem. Thus, there are two classes of objects described by $n$-dimensional real-valued vectors. The purpose of the learning process in this case is to train a neuron to classify patterns labeling them as belonging to the first or the second class. Let $d_i \in E_2$ be the desired output for the $i$th learning sample. This means that the input/output mapping has to map a vector $(x_1,...,x_n)$ to some desired output $d$. Suppose we have $N$ learning samples that form a learning set $(x_1^i,...,x_n^i) \to d_i, i = 1,...,N$. Let us have some weighting vector $W = (w_0, w_1,..., w_n)$ (the weights can be generated, for example, by a random number generator). Let $y$ be the actual output of the neuron $y = \text{sgn}(w_0 + w_1 x_1 +...+ w_n x_n)$ and it does not coincide with the desired output $d$. This forms the error

$$\delta = d - y. \tag{1.5}$$

Evidently, the goal of the learning process should be the elimination or minimization of this error through the adjustment of the weights by adding to them the adjustment term $\Delta w$

$$\tilde{w}_i = w_i + \Delta w_i, i = 0,1,...,n. \tag{1.6}$$

We expect that once the weights will be adjusted, our neuron should produce the desired output

$$d = \text{sgn}(\tilde{w}_0 + \tilde{w}_1 x_1 +...+ \tilde{w}_n x_n). \tag{1.7}$$

Taking into account (1.5) and (1.6), (1.7) can be transformed as follows

$$d = \delta + y =$$
$$\text{sgn}((w_0 + \Delta w_0) + (w_1 + \Delta w_1) x_1 +...+ (w_n + \Delta w_n) x_n). \tag{1.8}$$

Then we obtain from (1.8) the following

$$\delta + y =$$
$$\text{sgn}((w_0 + w_1 x_1 +...+ w_n x_n) + (\Delta w_0 + \Delta w_1 x_1 +...+ \Delta w_n x_n)). \tag{1.9}$$

Since the neuron's output is binary and it can be equal only to 1 or -1, according to (1.5) we have the following two cases for the error

$$\delta = \begin{cases} 2, \text{if } d = 1, y = -1 \\ -2, \text{if } d = -1, y = 1. \end{cases} \quad (1.10)$$

Let us consider the first case from (1.10), $d = 1, y = -1, \delta = 2$. Substituting these values to (1.9), we obtain the following

$$\delta + y = 2 - 1 = 1 =$$
$$\text{sgn}\left(\left(w_0 + w_1 x_1 + ... + w_n x_n\right) + \left(\Delta w_0 + \Delta w_1 x_1 + ... + \Delta w_n x_n\right)\right). \quad (1.11)$$

It follows from the last equation that

$$0 \le \left(w_0 + w_1 x_1 + ... + w_n x_n\right) + \left(\Delta w_0 + \Delta w_1 x_1 + ... + \Delta w_n x_n\right),$$

and (since $w_0 + w_1 x_1 + w_n x_n < 0$ because $y = -1$)

$$0 < \left(\Delta w_0 + \Delta w_1 x_1 + ... + \Delta w_n x_n\right),$$
$$\left|w_0 + w_1 x_1 + ... + w_n x_n\right| \le \left|\Delta w_0 + \Delta w_1 x_1 + ... + \Delta w_n x_n\right| \quad (1.12)$$

Let us set

$$\Delta w_0 = \alpha\delta; \Delta w_i = \alpha\delta x_i, i = 1, ..., n, \quad (1.13)$$

where $\alpha > 0$ is some constant, which is called a *learning rate*. Then

$$\Delta w_0 + \Delta w_1 x_1 + ... + \Delta w_n x_n =$$
$$\alpha\delta + \alpha\delta x_1 x_1 + ... + \alpha\delta x_n x_n = \alpha\delta(n+1). \quad (1.14)$$

It is important to mention that $x_i x_i = x_i^2 = 1; i = 1, ..., n$ in (14) because since we consider the threshold neuron with binary inputs, $x_i \in E_2 = \{1, -1\}$. We will see later that it is more difficult to use (1.13) if the neuron inputs are not binary. We also will see later that this difficulty does not exist for the error-correction learning rule for the multi-valued neuron, which will be considered in Section 3.3.

Since $\alpha > 0, \delta = 2 > 0$, then $\alpha\delta(n+1) > 0$ and the 1st inequality from (1.12) holds. However, it is always possible to find a learning rate $\alpha > 0$ such that the 2nd inequality from (1.12) also holds. This means that for the first case in (1.10) the learning rule based on (1.6) and (1.13) guarantees that (1.11) is true and the neuron produces the correct result after the weights are adjusted.

Let us consider the second case in (1.10). $d = -1, y = 1, \delta = -2$. Substituting these values to (1.9), we obtain the following

$$\delta + y = -2 + 1 = -1 =$$
$$\text{sgn}\left(\left(w_0 + w_1 x_1 + ... + w_n x_n\right) + \left(\Delta w_0 + \Delta w_1 x_1 + ... + \Delta w_n x_n\right)\right). \quad (1.15)$$

It follows from the last equation that

$$\left(w_0 + w_1 x_1 + ... + w_n x_n\right) + \left(\Delta w_0 + \Delta w_1 x_1 + ... + \Delta w_n x_n\right) < 0,$$

and (since $w_0 + w_1 x_1 + w_n x_n \geq 0$ because $y = 1$)

$$\left(\Delta w_0 + \Delta w_1 x_1 + ... + \Delta w_n x_n\right) < 0,$$

$$\left|w_0 + w_1 x_1 + ... + w_n x_n\right| \leq \left|\Delta w_0 + \Delta w_1 x_1 + ... + \Delta w_n x_n\right|. \tag{1.16}$$

Let us again use (1.6) and (1.13) to adjust the weights. We again obtain (1.14). Since $\alpha > 0, \delta = -2 < 0$, then $\alpha\delta(n+1) < 0$ and the 1$^{st}$ inequality from (1.16) holds. However, it is always possible to find such learning rate $\alpha > 0$ that the 2$^{nd}$ inequality from (1.16) also holds. This means that for the second case in (1.10) the learning rule based on (1.6) and (1.13) guarantees that (1.15) is true and the neuron produces the correct result after the weights are adjusted. Since for both cases in (1.10) the learning rule based on (1.6) and (1.13) works, then this rule always leads to the desired neuron output after the weights are corrected. We can merge (1.6) and (1.13) into

$$\tilde{w}_0 = w_0 + \alpha\delta;$$

$$\tilde{w}_i = w_i + \alpha\delta x_i, i = 1,...,n, \tag{1.17}$$

where $\delta$ is the error calculated according to (1.5) and $\alpha > 0$ is a learning rate. Equations (1.17) present the *error-correction learning rule*. After the weights are corrected according to (1.17), we obtain for the updated weighted sum the following expression

$$\tilde{z} = \tilde{w}_0 + \tilde{w}_1 x_1 + ... + \tilde{w}_n x_n =$$

$$(w_0 + \alpha\delta) + (w_1 + \alpha\delta x_1)x_1 + ... + (w_1 + \alpha\delta x_n)x_n =$$

$$\underbrace{w_0 + w_1 x_1 + ... + w_n x_n}_{z} + \alpha\delta(n+1) = z + \alpha\delta(n+1). \tag{1.18}$$

Since as we saw, $\delta$ in (1.18) has a sign, which is always opposite to the one of $z$, then it is always possible to choose $\alpha > 0$ such that $\mathrm{sgn}(\tilde{z}) = -\mathrm{sgn}(z)$. If $x_i \in T, i = 1,...,n$, where $T \subset \mathbb{R}$ and $f(x_1,...,x_n):T^n \rightarrow E_2$, then instead of (1.18) we obtain

$$\tilde{z} = \tilde{w}_0 + \tilde{w}_1 x_1 + ... + \tilde{w}_n x_n =$$

$$(w_0 + \alpha\delta) + (w_1 + \alpha\delta x_1)x_1 + ... + (w_1 + \alpha\delta x_n)x_n =$$

$$\underbrace{w_0 + w_1 x_1 + ... + w_n x_n}_{z} + \alpha\delta\left(1 + x_1^2 + ... + x_n^2\right) = \tag{1.19}$$

$$z + \alpha\delta\left(1 + x_1^2 + ... + x_n^2\right).$$

Like in (1.18), $\delta$ in (1.19) has a sign, which is always opposite to the one of $z$. Since $\alpha > 0$ and $1 + x_1^2 + ... + x_n^2 > 0$, it is again possible to choose $\alpha$ such that (1.19) holds. However, it is necessary to be more careful choosing $\alpha$ here than for the binary input case. While in (1.18) $\alpha$ does not depend on the inputs, in (1.19) it does. We will consider later, In Section 3.3, the error-correction learning rule for the multi-valued neuron and we will see that this problem exists there neither for the discrete multiple-valued inputs/output nor for the continuous ones.

## 1.2.3 Learning Algorithm

**Definition 1.2.** A *learning algorithm* is the iterative process of the adjustments of the weights using a learning rule. Suppose we need to learn some learning set containing $N$ learning samples $\left( x_1^i, ..., x_n^i \right) \rightarrow d_i, i = 1, ..., N$. One iteration of the learning process consists of the consecutive checking for all learning samples whether (1.2) holds for the current learning sample. If so, the next learning sample should be checked. If not, the weights should be adjusted according to a learning rule. The initial weights can be chosen randomly. This process should continue either until (1.2) holds for all the learning samples or until some additional criterion is satisfied.

When the learning process is successfully finished, we say that it has *converged* or converged to a weighting vector. Thus, *convergence* of the learning process means its successful completion. No-convergence means that the corresponding input/output mapping cannot be learned.

A *learning iteration* (*learning epoch*) is a pass over all the learning samples $\left( x_1^i, ..., x_n^i \right) \rightarrow d_i, i = 1, ..., N$.

If the learning process continues until (1.2) holds for all the learning samples, we say that the learning process converges with the zero error. If errors for some learning samples are acceptable, as it was mentioned, some additional criterion for stopping the learning process should be used. The most popular additional criterion is the mean square error/root mean square error criterion. In this case, the learning process continues until either of this errors drops below some predetermined acceptable threshold value. This works as follows. Let $\delta_i, i = 1, ..., N$ be the error for the $i$th learning sample. Then the mean square error (MSE) over all learning samples is

$$MSE = \frac{1}{N} \sum_{i=1}^{N} \delta_i^2 , \qquad (1.20)$$

and the root mean square error (RMSE) is

$$RMSE = \sqrt{MSE} = \sqrt{\frac{1}{N} \sum_{i=1}^{N} \delta_i^2} . \qquad (1.21)$$

If either of (1.20) or (1.21) is used, then a learning iteration starts from computation of MSE (RMSE). The learning process should continue until MSE (RMSE) drops below some pre-determined reasonable threshold value.

Another approach to the learning is the error minimization. In this case, the learning algorithm is considered as an optimization problem and it is reduced to the minimization of the error functional. The error is considered as the function of the weights. But in fact, the error is a composite function

$$\delta(W) = \delta(\varphi(z)) = d - \varphi(z) = d - (w_0 + w_1 x_1 + ... + w_n x_n),$$

where $\varphi(z)$ is the activation function. Actually, this approach is the most popular, but since minimization of the error functional using optimization methods requires differentiability of an activation function, it cannot be applied to the threshold neuron whose activation function $\text{sgn}(z)$ is not differentiable. It is widely used for sigmoidal neurons and neural networks based on them. We will observe them in Section 1.3.

Now we have to discuss the convergence of the learning algorithm for a single threshold neuron based on the error-correction rule (1.17). The first proof of the *perceptron convergence theorem* was given by F. Rosenblatt in [10]. It is important to mention that F. Rosenblatt considered only binary inputs. In its most comprehensive form, this convergence theorem states that if the given input/output mapping can be learned and learning samples appear in an arbitrary order, but with a condition that each of them is repeated in the *learning sequence* within some finite time interval, then the learning process converges starting from an arbitrary weighting vector after a finite number of iterations.

This theorem, however, did not clarify the question which input/output mappings can be learned using the perceptron and which cannot.

A more general case of this theorem was considered by A. Novikoff in [12]. He introduced a notion of a *linearly separable set*. The learning set $\left(x_1^i, ..., x_n^i\right) \to d_i, i = 1, ..., N; x_j^i \in T \subset \mathbb{R}, j = 1, ..., n; i = 1, ..., N$ is called linearly separable if there exist a positive constant $s$ and a weighting vector $W$ such that the following condition holds

$$d_i\left(w_0 + w_1 x_1 + ... + w_n x_n\right) > s, i = 1, ..., N.$$

This means that the weighted sum multiplied by the desired output must be greater than some positive constant for all the learning samples. *Novikoff's convergence theorem states that the learning algorithm converges after a finite number of iterations if the learning set is linearly separable.* The idea behind the Novikoff's proof is to show that the assumption that the learning process does not converge after a finite number of iterations contradicts to the linear separability of the learning set. Novikoff showed that the amount of changes to the initial weighting vector is bounded by $\left(2M / s\right)^2$, where $M$ is the maximum norm of an input vector.

Since the norm is always a finite non-negative real number, the number of iterations in the learning algorithm is also finite.

We will see later that this approach used by Novikoff to prove the convergence of the error-correction learning algorithm for the threshold neuron also works to prove the convergence of the learning algorithm for the multi-valued neuron (Section 3.3) and a multilayer neural network based on multi-valued neurons (Chapter 4).

### 1.2.4 Examples of Application of the Learning Algorithm Based on the Error-Correction Rule

Let us consider how learning rule (1.17) can be used to train the threshold neuron using the learning algorithm, which was just defined.

**Example 1.2.** Let us consider again the OR problem $f(x_1, x_2) = x_1$ or $x_2$ (Table 1.1, Fig. 1.4a), which we have already considered above. Our learning set contains four learning samples (see Table 1.1). Let us start the learning process from the weighting vector $W = (1,1,1)$.

Iteration 1.

1) Inputs (1, 1). The weighted sum is equal to $z = 1 + 1 \cdot 1 + 1 \cdot 1 = 3$; $\varphi(z) = \text{sgn}(z) = \text{sgn}(3) = 1$. Since $f(1,1) = 1$, no further correction of the weights is needed.

2) Inputs (1, -1). The weighted sum is equal to $z = 1 + 1 \cdot 1 + 1 \cdot (-1) = 1$; $\varphi(z) = \text{sgn}(z) = \text{sgn}(1) = 1$. Since $f(1,-1) = -1$, we have to correct the weights. According to (1.5) $\delta = -1 - 1 = -2$. Let $\alpha = 1$ in (1.17). Then we have to correct the weights according to (1.17):

$$\tilde{w}_0 = 1 - 2 = -1; \quad \tilde{w}_1 = 1 + (-2) \cdot 1 = -1; \quad \tilde{w}_2 = 1 + (-2) \cdot (-1) = 3.$$

Thus, $\tilde{W} = (-1, -1, 3)$.

The weighted sum after the correction is equal to $z = -1 + (-1) \cdot 1 + 3 \cdot (-1) = -5$; $\varphi(z) = \text{sgn}(z) = \text{sgn}(-5) = -1$. Since $f(1,-1) = -1$, no further correction of the weights is needed.

3) Inputs (-1, 1). The weighted sum is equal to $z = -1 + (-1) \cdot (-1) + 3 \cdot 1 = 3$; $\varphi(z) = \text{sgn}(z) = \text{sgn}(3) = 1$. Since $f(-1,1) = -1$, we have to correct the weights. According to (17) $\delta = -1 - 1 = -2$. Let $\alpha = 1$ in (1.17). Then we have to correct the weights according to (1.17):

$\tilde{w}_0 = -1-2 = -3;\ \tilde{w}_1 = -1+(-2)\cdot(-1) = 1;\ \tilde{w}_2 = 3+(-2)\cdot 1 = 1$.

Thus, $\tilde{W} = (-3,1,1)$.

The weighted sum after the correction is equal to $z = -3+1\cdot(-1)+1\cdot 1 = -3$; $\varphi(z) = \mathrm{sgn}(z) = \mathrm{sgn}(-3) = -1$. Since $f(-1,1) = -1$, no further correction of the weights is needed.

4) Inputs (-1, -1). The weighted sum is equal to $z = -3+1\cdot(-1)+1\cdot(-1) = -5$; $\varphi(z) = \mathrm{sgn}(z) = \mathrm{sgn}(-5) = -1$. Since $f(-1,-1) = -1$, no further correction of the weights is needed.

Iteration 2.

1) Inputs (1, 1). The weighted sum is equal to $z = -3+1\cdot 1+1\cdot 1 = -1$; $\varphi(z) = \mathrm{sgn}(z) = \mathrm{sgn}(-1) = -1$. Since $f(1,1) = 1$, we have to correct the weights. According to (1.17) $\delta = 1-(-1) = 2$. Let $\alpha = 1$ in (1.17). Then we have to correct the weights according to (1.17):

$$\tilde{w}_0 = -3+2 = -1;\ \tilde{w}_1 = 1+2\cdot 1 = 3;\ \tilde{w}_2 = 1+2\cdot 1 = 3.$$

Thus, $\tilde{W} = (-1,3,3)$.

The weighted sum after the correction is equal to $z = -1+3\cdot 1+3\cdot 1 = 5$; $\varphi(z) = \mathrm{sgn}(z) = \mathrm{sgn}(5) = 1$. Since $f(1,1) = 1$, no further correction of the weights is needed.

2) Inputs (1, -1). The weighted sum is equal to $z = -1+3\cdot 1+3\cdot(-1) = -1$; $\varphi(z) = \mathrm{sgn}(z) = \mathrm{sgn}(-1) = -1$. Since $f(1,-1) = -1$, no further correction of the weights is needed.

3) Inputs (-1, 1). The weighted sum is equal to $z = -1+3\cdot(-1)+3\cdot 1 = -1$; $\varphi(z) = \mathrm{sgn}(z) = \mathrm{sgn}(-1) = -1$. Since $f(-1,1) = -1$, no further correction of the weights is needed.

4) Inputs (-1, -1). The weighted sum is equal to $z = -1+3\cdot(-1)+3\cdot(-1) = -7$; $\varphi(z) = \mathrm{sgn}(z) = \mathrm{sgn}(-7) = -1$. Since $f(-1,-1) = -1$, no further correction of the weights is needed.

This means that the iterative process converged after two iterations, there are no errors for all the samples from the learning set, and this learning set presented by the OR function $f(x_1,x_2) = x_1$ or $x_2$ of the two variables is learned. Therefore, the OR function can be implemented with the threshold neuron using the weighting vector $\tilde{W} = (-1,3,3)$ obtained as the result of the learning process.

**Table 1.4** Learning process for the function $f(x_1, x_2) = \overline{x}_1 \,\&\, x_2$

| $x_1$ | $x_2$ | $z = w_0 + w_1 x_1 + w_2 x_2$ | sgn$(z)$ | $f(x_1, x_2) = \overline{x}_1 \,\&\, x_2$ | $\delta$ |
|---|---|---|---|---|---|
| | | \multicolumn{4}{c}{$W = (1,1,1)$} | | | |
| \multicolumn{6}{l}{Iteration 1} | | | | | |
| 1 | 1 | $z = 1 + 1\cdot 1 + 1\cdot 1 = 3$ | 1 | 1 | 0 |
| 1 | -1 | $z = 1 + 1\cdot 1 + 1\cdot(-1) = 1$ | 1 | -1 | -2 |
| | | \multicolumn{4}{c}{$\tilde{w}_0 = 1 - 2 = -1;\ \tilde{w}_1 = 1 + (-2)\cdot 1 = -1;\ \tilde{w}_2 = 1 + (-2)\cdot(-1) = 3$} | | | |
| | | \multicolumn{4}{c}{$W = (-1,-1,3)$} | | | |
| -1 | 1 | $z = -1 + (-1)\cdot(-1) + 3\cdot 1 = 3$ | 1 | 1 | 0 |
| -1 | -1 | $z = -1 + (-1)\cdot(-1) + 3\cdot(-1) = -3$ | -1 | 1 | 2 |
| | | \multicolumn{4}{c}{$\tilde{w}_0 = -1 + 2 = 1;\ \tilde{w}_1 = -1 + 2\cdot(-1) = -3;\ \tilde{w}_2 = -1 + 2\cdot(-1) = -3$} | | | |
| | | \multicolumn{4}{c}{$\widetilde{W} = (1,-3,-3)$} | | | |
| \multicolumn{6}{l}{Iteration 2} | | | | | |
| 1 | 1 | $z = 1 + (-3)\cdot 1 + (-3)\cdot 1 = -5$ | -1 | 1 | 2 |
| | | \multicolumn{4}{c}{$\tilde{w}_0 = 1 + 2 = 3;\ \tilde{w}_1 = -3 + 2\cdot 1 = -1;\ \tilde{w}_2 = -3 + 2\cdot 1 = -1$} | | | |
| | | \multicolumn{4}{c}{$\widetilde{W} = (3,-1,-1)$} | | | |
| 1 | -1 | $z = 3 + (-1)\cdot 1 + (-1)\cdot(-1) = 3$ | 1 | -1 | -2 |
| | | \multicolumn{4}{c}{$\tilde{w}_0 = 3 - 2 = 1;\ \tilde{w}_1 = -1 + (-2)\cdot 1 = -3;\ \tilde{w}_2 = -1 + (-2)\cdot(-1) = 3$} | | | |
| | | \multicolumn{4}{c}{$\widetilde{W} = (1,-3,3)$} | | | |
| -1 | 1 | $z = 1 + (-3)\cdot(-1) + 3\cdot 1 = 7$ | 1 | 1 | 0 |
| -1 | -1 | $z = 1 + (-3)\cdot(-1) + 3\cdot(-1) = 1$ | 1 | 1 | 0 |
| \multicolumn{6}{l}{Iteration 3} | | | | | |
| 1 | 1 | $z = 1 + (-3)\cdot 1 + 3\cdot 1 = 1$ | 1 | 1 | 0 |
| 1 | -1 | $z = 1 + (-3)\cdot 1 + 3\cdot(-1) = -5$ | -1 | -1 | 0 |
| -1 | 1 | $z = 1 + (-3)\cdot(-1) + 3\cdot 1 = 7$ | 1 | 1 | 0 |
| -1 | -1 | $z = 1 + (-3)\cdot(-1) + 3\cdot(-1) = 1$ | 1 | 1 | 0 |

**Example 1.3.** Let us learn the function $f(x_1, x_2) = \overline{x}_1 \,\&\, x_2$ ($\overline{x}$ means the ne-

| # | $x_1$ | $x_2$ | $\overline{x}_1 \,\&\, x_2$ |
|---|---|---|---|
| 1) | 1 | 1 | 1 |
| 2) | 1 | -1 | -1 |
| 3) | -1 | 1 | 1 |
| 4) | -1 | -1 | 1 |

gation of the Boolean variable $x$, in the alphabet $\{1, -1\}$ $\overline{x} = -x$) using the threshold neuron. Table 1.4 shows the function values and the entire learning set containing four input vectors and four values of the function, respectively. We start the learning process from the same weighting vector $W = (1,1,1)$ as in Example 1.2. We hope that so detailed explanations as were given in Example 1.2 will not be needed now. Thus, the iterative process has converged after three iterations, there are no errors for all the elements from the learning set, and our input/output mapping described by the Boolean function $f(x_1, x_2) = \overline{x}_1 \,\&\, x_2$ is implemented on the threshold neuron using the weighting vector $\widetilde{W} = (1, -3, 3)$ obtained as the result of the learning process.

## 1.2.5  Limitation of the Perceptron. Minsky's and Papert's Work

In 1969, M. Minsky and S. Papert published their famous book [13] in which they proved that the perceptron cannot learn non-linearly separable input/output mappings. Particularly, they showed that, for example the XOR problem is unsolvable using the perceptron. Probably from that time the XOR problem is a favorite problem, which is used to demonstrate why we need multilayer neural networks - to learn such problems as XOR. This resulted in a significant decline in interest to neurons and neural networks in 1970s.

We will show later that this problem is the simplest possible problem, which can be solved by a single multi-valued neuron with a periodic activation function (Section 1.4 and Chapter 5).

Thus, a principal limitation of the perceptron is its impossibility to learn non-linearly separable input/output mappings. This limitation causes significant lack of the functionality and reduces a potential area of applications because the most of real-world pattern recognition and classification problems are non-linearly separable.

The next significant limitation of the perceptron is its binary output. Thus, the perceptron can be used neither for solving multi-class classification problems (where the number of classes to be classified is greater than two) nor problems with a continuous output.

In this book, starting from Section 1.4 and thereafter we will show how these limitations can easily be overcome with complex-valued neurons. Non-linearly separable binary problems and multiple-valued problems (including the nonlinearly-separable ones) can be learned using a single multi-valued neuron.

But first, to conclude our observation of neurons and neural networks, let us consider the most popular topologies of neural networks, which were proposed in 1980s and which are now successfully used in complex-valued neural networks.

## 1.3  Neural Networks: Popular Topologies

### 1.3.1  XOR Problem: Solution Using a Feedforward Neural Network

As we have seen, the perceptron cannot learn non-linearly separable problems (input/output mappings). In the third part of his book [11], F. Rosenblatt proposed an idea of multilayer perceptron containing more than one layer of *A*-units (see Fig. 1.6). He projected that this neural network will be more functional and will be able to learn non-linearly separable problems. However, no learning algorithm for this network was proposed that time. In [13], M. Minsky and S. Papert presented their skeptical view on the "multilayer perceptron". They did not hope that it will be more efficient than the classical single layer perceptron, probably because there was still no learning algorithm for a multilayer neural network.

However, the existence of non-linearly separable problems was a great stimulus to develop new solutions. We will see starting from Section 1.4 how easily many of them can be solved using the multi-valued neuron. But first let us again take a historical view. A two-layer neural network containing three neurons in total, which can solve the XOR problem, is described, for example, in [2], where the paper [14] is cited as a source of this solution. We are not sure that this solution was presented for the first time definitely in [14]; most probably it was known earlier. It is quite difficult to discover today who found this solution first. Nevertheless, let us consider it here.

To solve the XOR problem within a "threshold basis" (using the threshold neuron), it is necessary to build a network from threshold neurons. Let us consider a network from three neurons (see Fig. 1.7a). This network contains the input layer, which distributes the input signals $x_1$ and $x_2$, one hidden layer containing Neurons 1 and 2 and one output layer containing a single Neuron 3. This is the simplest possible non-trivial *multilayer feedforward neural network* (MLF). It is the simplest possible network because it contains a minimum amount of layers and neurons to be non-trivial (two layers including one hidden layer and one output layer, two neurons in the hidden layer, and one neuron in the output layer). A network is trivial if it contains just a single hidden neuron and a single output neuron. This network is called *feedforward* because there are no feedback connections there, all signals are transmitted through the network in a strictly feedforward manner.

Let us remind that the function XOR may be presented in the full disjunctive normal form as follows:

$$x_1 \oplus x_2 = x_1 \overline{x}_2 \vee \overline{x}_1 x_2 = f_1(x_1, x_2) \vee f_2(x_1, x_2),$$

where $f_2(x_1, x_2) = \overline{x}_1 x_2$ is that function whose learning and implementation using the threshold neuron was considered in Example 1.3. Let us also remind that learning and implementation of the OR function, which connects functions

$f_1(x_1, x_2)$ and $f_2(x_1, x_2)$ was considered in Example 1.2. Notice that function $f_1(x_1, x_2) = x_1 \overline{x}_2$ may be obtained by changing the order of variables in function $f_2(x_1, x_2) = \overline{x}_1 x_2$. It was shown in [6] that if some Boolean function is threshold, than any function obtained from the first one by the permutation of its variables is also threshold and its weighting vector can be obtained by the permutation of the weights in the weighting vector of the first function corresponding to the permutation of the variables.

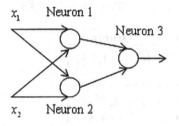

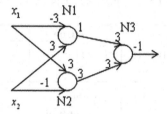

(a) a two layer neural network with two inputs, with one hidden layer containing two neurons, and the output layer containing a single neuron

(b) a two layer neural network with two inputs, with one hidden layer containing two neurons, and the output layer containing a single neuron. The weights that solve the XOR problem are assigned to the neurons

**Fig. 1.7** Simple neural networks

The weight $w_0$ remains unchanged. Therefore a weighting vector $W_{f_1}$ for $f_1(x_1, x_2) = x_1 \overline{x}_2$ may be obtained from the one for $f_2(x_1, x_2) = \overline{x}_1 x_2$ by reordering the weights $w_1$ and $w_2$. Since, as we found in Example 1.3 for function $f_2(x_1, x_2) = \overline{x}_1 x_2$, $W_{f_2} = (1, -3, 3)$, the weighting vector $W_{f_1} = (1, 3, -3)$ implements function $f_2(x_1, x_2) = \overline{x}_1 x_2$ using a single threshold neuron. It is easy to check that this weighting vector gives a correct realization of the function.

This means that if Neuron 1 implements function $f_1(x_1, x_2)$, Neuron 2 implements function $f_2(x_1, x_2)$, and Neuron 3 implements the OR function, then the network presented in Fig. 1.7b implements the XOR function. Let us consider how it works. Thus, Neuron 1 operates with the weighting vector $\tilde{W} = (1, 3, -3)$, Neuron 2 operates with the weighting vector $\tilde{W} = (1, -3, 3)$, and Neuron 3 operates with the weighting vector $\tilde{W} = (-1, 3, 3)$ (see Fig. 1.7b). The network works in the following way. There are no neurons in the input layer. It just distributes the input signals among the hidden layer neurons. The input signals $x_1$ and $x_2$ are accepted in parallel from the input layer by both neurons from the hidden layer (N1 and N2). Their outputs are coming to the corresponding inputs of the single neuron in output layer (N3). The output of this neuron is the output of the entire network. The results are summarized in Table 1.5 ($z$ is the weighted sum of the inputs). For all three neurons their weighted sums and outputs are shown. To be convinced that the network implements definitely the XOR function, its actual values are shown in the last column of Table 1.5.

**Table 1.5** Implementation of the XOR function using a neural network presented in Fig. 1.7b

| Inputs | | Neuron 1 $\tilde{W} = (1,3,-3)$ | | Neuron 2 $\tilde{W} = (1,-3,3)$ | | Neuron 3 $\tilde{W} = (-1,3,3)$ | | $x_1$ xor $x_2$ |
|---|---|---|---|---|---|---|---|---|
| $x_1$ | $x_2$ | $z$ | sgn($z$) output | $z$ | sgn($z$) output | $z$ | sgn($z$) output | |
| 1 | 1 | 1 | 1 | 1 | 1 | 5 | 1 | 1 |
| 1 | -1 | 7 | 1 | -5 | -1 | -1 | -1 | -1 |
| -1 | 1 | -5 | -1 | 7 | 1 | -1 | -1 | -1 |
| -1 | -1 | 1 | 1 | 1 | 1 | 5 | 1 | 1 |

## 1.3.2  Popular Real-Valued Activation Functions

As we see, a multilayer feedforward neural network (MLF) has a higher function-ality compared to the perceptron. It can implement non-linearly separable in-put/output mappings, while the perceptron cannot. Considering in the previous section how MLF may solve the XOR problem, we have not passed this problem through a learning algorithm; we just have synthesized the solution. However, the most wonderful property of MLF is its learning algorithm. MLF was first pro-posed in [15] by D.E. Rumelhart, G.E. Hilton, and R.J. Williams. They also de-scribed in the same paper the backpropagation learning algorithm. It is important to mention that a seminal idea behind the error backpropagation and its use to train a feedforward neural network belongs to Paul Werbos. He developed these ideas in his Harvard Ph. D. dissertation in 1974 and later he included it as a part in his book [16] (Chapters 1-6).

It is also important to mention that starting from mid 1980s, especially from the moment when D. Rumelhart and his co-authors introduced MLF, threshold neu-rons as basic neurons for building neural networks have moved to the background. It became much more interesting to learn and implement using neural networks continuous and multi-valued input/output mappings described by functions $f(x_1,...,x_n):T^n \to T, T \subset \mathbb{R}$, which was impossible using a hard-limited threshold activation function $\text{sgn}(z)$.

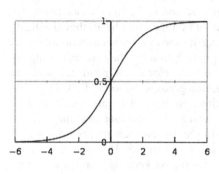
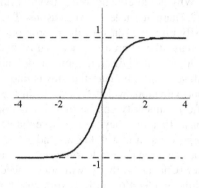

**Fig. 1.8** Logistic function                 **Fig. 1.9** tanh function

Typically, there have been considered $T = [0,1]$ or $T = [-1,1]$. Respectively, new activation functions became very popular from mid 1980s. The most popular of them is a *sigmoid activation function*. It has two forms – the logistic function and the hyperbolic tangent function. Logistic function is as follows

$$\varphi(z) = \frac{1}{1 + e^{-\alpha z}}, \tag{1.22}$$

(see Fig. 1.8), where $\alpha$ is a slope parameter. The curve in Fig. 1.8 got its name "sigmoid" from Pierre François Verhulst (in 1844 or 1845) who studied the population growth described by (1.22). Evidently, the range of function (1.22) is $]0,1[$, the function approaches 0 when $z \to -\infty$ and approaches 1 when $z \to \infty$ (actually, the logistic function approaches its bounds with significantly smaller values of its argument as it is seen from Fig. 1.8). To obtain a sigmoid curve with the range $]-1,1[$, the hyperbolic tangent function

$$\tanh \alpha z = \frac{\sinh \alpha z}{\cosh \alpha z} = \frac{e^{\alpha z} - e^{-\alpha z}}{e^{\alpha z} + e^{-\alpha z}}, \tag{1.23}$$

should be used. The shape of function (1.23) is identical to the one of function (1.22) (see Fig. 1.9) with only distinction that the tanh function cross not the line y=0.5, but the horizontal axis at the origin and it is bounded from the bottom by the line y= -1. $\alpha$ in (1.23) is again a slope parameter and its role is identical to the one in (1.22). It is clear that if $\alpha \to \infty$ in (1.23), then $\tanh \alpha z$ approaches $\text{sgn}(z)$ (compare Fig. 1.2 and Fig. 1.9).

Why definitely sigmoid activation functions (1.22) and (1.23) became so popular? There are at least two reasons. The first reason is that on the one hand, they easily limit the range of the neuron output, but on the other hand, they drastically increase the neuron's functionality making it possible to learn continuous and multiple-valued discrete input/output mappings. Secondly, they are increasing (we will see a little bit later that this is important to develop a computational model for approximation, which follows from the Kolmogorov's theorem [17]). Their specific nonlinearity can be used for approximation of other highly nonlinear functions. Finally, they are differentiable (which is important for the learning purposes; as it is well known and as we will see, the differentiability is critical for that backpropagation learning technique developed in [15, 16]. We will also see later (Chapter 4) that it will not be needed for the backpropagation learning algorithm for a feedforward network based on multi-valued neurons).

Another popular type of an activation function, which is used in real-valued neurons and neural networks, is radial basis function (RBF) first introduced by M.J.D. Powell [18] in 1985. RBF is a real-valued function whose value depends only on the distance from the origin $\varphi(z) = \varphi(\|z\|)$ or on the distance from some pre-determined other point $c$, called a *center*, so that $\varphi(z,c) = \varphi(\|z-c\|)$, where $\| \ \|$ is the norm in the corresponding space. There are different functions that satisfy this property. Perhaps, the most popular of them, which is used in neural networks and machine learning is the Gaussian RBF $\varphi(r) = e^{-\left(\frac{r}{\alpha}\right)^2}$ [2, 5], where $r = z - c$ ($c$ is the corresponding center), $\alpha > 0$ is a parameter.

### 1.3.3   Multilayer Feedforward Neural Network (MLF) and Its Backpropagation Learning

Let us consider in more detail a network with perhaps the most popular topology, namely a multilayer feedforward neural network (MLF), also widely referred to as a multilayer perceptron (MLP) [15, 2]. We will also consider the basic principles of the backpropagation learning algorithm for this network.

Typically, an MLF consists of a set of sensory units (source nodes – analogues of $S$-units in the perceptron) that constitute the *input* layer (which distributes input signals among the first hidden layer neurons), one or more *hidden* layers of neurons, and an *output* layer of neurons. We have already considered a simple example of such a network, which solves the XOR problem. The input signals progresses through the network in a *forward* direction, on a layer-by-layer basis. An important property of MLF is its full connection architecture: the outputs of *all* neurons in a specified layer are connected to the corresponding inputs of *all* neurons of the following layer (for example, the output of a neuron $ij$ (the $i$th neuron from the $j$th layer) is connected to the $i$th input of all neurons from the $j+1^{st}$ layer).

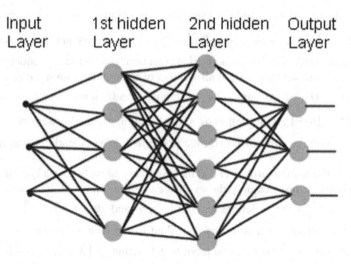

Input          1st hidden          2nd hidden          Output
Layer            Layer                 Layer              Layer

**Fig. 1.10** 3-5-6-3 MLF – Multilayer Feedforward Neural Network. It has 3 inputs, the 1$^{st}$ hidden layer containing 5 neurons, the 2$^{nd}$ hidden layer containing 6 neurons and the output layer containing 3 neurons

This means a full connection between consecutive layers (see Fig. 1.10). To specify a network topology, the notation $n - n_1 - ... - n_i - ... - n_s - n_o$ is used. Here $n$ is the number of network inputs, $n_i, i = 1, ..., s$ is the number of neurons in the $i$th hidden layer, $s$ is the number of hidden layers, and $n_o$ is the number of neurons in the output layer.

This architecture is the result of a "universal approximator" computing model based on the famous Kolmogorov's Theorem [17]. This theorem states the following. There exist fixed (universal) increasing continuous functions $h_{ij}(x)$ on $I = [0,1]$ such that each continuous function of $n$ variables $f(x_1, ..., x_n)$ on $I^n$ can be written in the form

$$f(x_1, ..., x_n) = \sum_{j=1}^{2n+1} g_j \left( \sum_{i=1}^{n} h_i(x_i) \right),$$

(1.24)

where $g_j, j = 1, ..., n$ are some properly chosen continuous functions of one variable.

This result states that any multivariate continuous function can be represented by the superposition of a small number of univariate continuous functions. It is clear that in terms of feedforward neural networks equation (1.24) describes a three layer feedforward neural network whose first two layers contain $n$ and $2n+1$

neurons, respectively, and implement functions $h_i, i = 1,...,n$ and $g_j, j = 1,...,2n+1$, respectively. The output layer of this network contains a single neuron with the linear activation function (its output is equal to the weighted sum; according to (1.20) all weights of the output neuron are equal to 1 except $w_0 = 0$, there are no weighting coefficients in a front of $g_j, j = 1,...,2n+1$). It is well known that a multilayer feedforward neural network is a universal approximator (for the first time this was clearly proven in [19] and [20]).

However, the Kolmogorov's Theorem, being very important, is a typical "existence theorem". It justifies only the existence of the solution. It does not show a mechanism for finding functions $h_i, i = 1,...,n$ and $g_j, j = 1,...,2n+1$. To approach that solution, which exists according to the Kolmogorov's Theorem, a feedforward neural network has to learn that function $f(x_1,...,x_n)$, which we want to approximate. To implement the learning process, the backpropagation learning algorithm was suggested. A problem, which is necessary to solve, implementing the learning process for a feedforward neural network, is finding the hidden neurons errors. While the exact errors of output neurons can be easily calculated as the differences between the desired and actual outputs, for all the hidden neurons their desired outputs are unknown and therefore there is no straightforward way to calculate their errors. But without the errors it is not possible to adjust the weights.

The basic idea behind a backpropagation learning algorithm is sequential propagation of the errors of the neurons from the output layer through all the layers from the "right hand" side to the "left hand" side up to the first hidden layer (see Fig. 1.10), in order to calculate the errors of all other neurons. The heuristic idea is to share the errors of output neurons, which can be calculated because their desired outputs are known (unlike the ones of the hidden neurons), with all the hidden neurons.

Basically, the entire learning process consists of two passes through all the different layers of the network: a forward pass and a backward pass. In the forward pass, the inputs are propagated from the input layer of the network to the first hidden layer and then, layer by layer, output signals from the hidden neurons are propagated to the corresponding inputs of the following layer neurons. Finally, a set of outputs is produced as the actual response of the network. Evidently, during the forward pass the synaptic weights of the network are all fixed. During the backward pass first the errors of all the neurons are calculated and then the weights of all the neurons are all adjusted in accordance with the learning rule. One complete iteration (epoch) of the learning process consists of a forward pass and a backward pass.

Although the error backpropagation algorithm for MLF is well known, we would like to include its derivation here. In our opinion, this is important for the following two reasons. The first reason is to simplify perception of this book for those readers (first of all for students) who are not the experts in neural networks and just make their first steps in this area. The second and even more important reason is to compare this derivation and the backpropagation learning algorithm for MLF with the ones for a complex-valued multilayer feedforward neural network based on multi-valued neurons, which will be considered in Chapter 4. This comparison will be very important for understanding of significant advantages of complex-valued neural networks.

In the derivation of the MLF backpropagation learning algorithm we mostly will follow here [2] and [5].

It is important to mention that the backpropagation learning algorithm is based on the generalization of the error-correction learning rule for the case of MLF. Specifically, the actual response of the network is subtracted from a desired response to produce an error signal. This error signal is then propagated backward through the network, against the direction of synaptic connections – hence the name *"backpropagation"*. The weights are adjusted so as to make the actual output of the network move closer to the desired output. A common property of a major part of real-valued feedforward neural networks is the use of sigmoid activation functions for its neurons. Let us use namely logistic function (1.22).

Let us consider a multilayer neural network with traditional feedforward architecture (see Fig. 1.10), when the outputs of neurons of the input and hidden layers are connected to the corresponding inputs of the neurons from the following layer. Let us suppose that the network contains one input layer, $m$-1 hidden layers and one output layer. We will use here the following notations.

Let

$D_{km}$ - be a desired output of the $k$th neuron from the output ($m$th ) layer

$Y_{km}$ - be the actual output of the $k$th neuron from the output ($m$th) layer.

Then a global error of the network related to the $k$th neuron of the output ($m$th) layer can be calculated as follows:

$$\delta^*_{km} = D_{km} - Y_{km} \text{ - error for the } k\text{th neuron from output } (m\text{th}) \text{ layer.} \quad (1.25)$$

$\delta^*_{km}$ denotes here and further a global error of the network. We have to distinguish it from the local errors $\delta_{km}$ of the particular output neurons because each output neuron contributes to the global error equally with the hidden neurons.

The learning algorithm for the classical MLF is derived from the consideration that the global error of the network in terms of the *mean square error* (MSE) must be minimized. The functional of the error may be defined as follows:

$$E = \frac{1}{N} \sum_{s=1}^{N} E_s , \quad (1.26)$$

where $E$ denotes *MSE*, $N$ is the total number of samples (patterns) in the learning set and $E_s$ denotes the square error of the network for the $s$th pattern;

$$E_s = (D_s - Y_s)^2 = (\delta_s^*)^2, s = 1, ..., N \quad \text{for a single output neuron and}$$

$$E_s = \frac{1}{N_m} \sum_{k=1}^{N_m} (D_{k_s} - Y_{k_s})^2 = \frac{1}{N_m} \sum_{k=1}^{N_m} (\delta_{k_s}^*)^2 ; s = 1, ..., N \quad \text{for } N_m \text{ output neu-}$$

rons. For simplicity, but without loss of generality, we can consider minimization of a *square error* (*SE*) function instead of minimization the MSE function (1.26).

The square error is defined as follows:

$$E = \frac{1}{2} \sum_{k=1}^{N_m} (\delta_{km}^*)^2, \tag{1.27}$$

where $N_m$ indicates the number of output neurons,

$$\delta_{km}^* = D_{k_s} - Y_{k_s}, s = 1, ..., N, \tag{1.28}$$

$m$ is the output layer index, and the factor $\frac{1}{2}$ is used so as to simplify subsequent

derivations resulting from the minimization of $E$. The error function (1.27) is a function of the weights. Indeed, it strictly depends on all the network weights. It is a principal assumption that the error depends not only on the weights of the neurons at the output layer, but on all neurons of the network.

Thus, a problem of learning can be reduced to finding a global minimum of (1.27) as a function of weights. In these terms, this is the optimization problem.

The backpropagation is used to calculate the gradient of the error of the network with respect to the network's modifiable weights. This gradient is then used in a gradient descent algorithm to find such weights that minimize the error. Thus, the minimization of the error function (1.27) (as well, as (1.26) ) is reduced to the search for those weights for all the neurons that ensure a minimal error.

To ensure movement to the global minimum on each iteration, the correction of the weights of all the neurons has to be organized in such a way that each weight $w_i$ has to be corrected by an amount $\Delta w_i$, which must be proportional to the

partial derivative $\dfrac{\partial E}{\partial w_i}$ of the error function $E(W)$ with respect to the weights [2].

For the next analysis, the following notation will be used. Let $w_i^{kj}$ denote the weight corresponding to the $i$th input of the $k$th neuron at the $j$th layer. Furthermore let $z_{kj}$, $y_{kj}$ and $Y_{kj} = y_{kj}(z_{kj})$ represent the weighted sum (of the input signals), the activation function value, and the output value of the $k$th neuron at the $j$th layer, respectively. Let $N_j$ be the number of neurons in the $j$th layer (notice that this means that neurons of the $j+1^{st}$ layer have exactly $N_j$ inputs.) Finally, recall that $x_1, ..., x_n$ denote the inputs to the network (and as such, also the inputs to the neurons of the first layer.)

Then, taking into account that $E(W) = E\big(y(z(W))\big)$ and applying the chain rule for the differentiation, we obtain for the $k^{th}$ neuron at the output ($m^{th}$) layer

$$\frac{\partial E(W)}{\partial w_i^{km}} = \frac{\partial E(W)}{\partial y_{km}} \frac{\partial y_{km}}{\partial z_{km}} \frac{\partial z_{km}}{\partial w_i^{km}}, \quad i = 0, 1, ..., N_{m-1},$$

where

$$\frac{\partial E(W)}{\partial y_{km}} = \frac{\partial}{\partial y_{km}} \left( \frac{1}{2} \sum_k (\delta_{km}^*)^2 \right) = \frac{1}{2} \sum_k \frac{\partial}{\partial y_{km}} (\delta_{km}^*)^2 =$$

$$= \frac{1}{2} \frac{\partial}{\partial y_{km}} (\delta_{km}^*)^2 = \delta_{km}^* \frac{\partial}{\partial y_{km}} (\delta_{km}^*) = \delta_{km}^* \frac{\partial}{\partial y_{km}} \frac{1}{N} \sum_{s=1}^{N} \left( D_{km_s} - Y_{km_s} \right) = -\delta_{km}^*;$$

$$\frac{\partial y_{km}}{\partial z_{km}} = y_{km}'(z_{km}),$$

and

$$\frac{\partial z_{km}}{\partial w_i^{km}} = \frac{\partial}{\partial w_i^{km}} \left( w_0^{km} + w_1^{km} Y_{1,m-1} + ... + w_{N_{m-1}}^{km} Y_{N_{m-1},m-1} \right) = Y_{i,m-1},$$

$i = 0, 1, ..., N_{m-1}.$

Then we obtain the following:

$$\frac{\partial E(W)}{\partial w_i^{km}} = \frac{\partial E(W)}{\partial y_{km}} \frac{\partial y_{km}}{\partial z_{km}} \frac{\partial z_{km}}{\partial w_i^{km}} = -\delta_{km}^* y_{km}' \left( z_{km} \right) Y_{i,m-1}, \quad i = 0, 1, ..., N_{m-1};$$

where $Y_{0,m-1} \equiv 1$. Finally, we obtain now the following

$$\Delta w_i^{km} = -\beta \frac{\partial E(W)}{\partial w_i^{km}} = \begin{cases} \beta \delta_{km}^* y_{km}' \left( z_{km} \right) Y_{i,m-1} & i = 1, ..., N_{m-1} \\ \beta \delta_{km}^* y_{km}' \left( z_{km} \right) & i = 0, \end{cases} \tag{1.29}$$

where $\beta > 0$ is a learning rate.

The part of the rate of change of the square error $E(W)$ with respect to the input weight of a neuron, which is independent of the value of the corresponding input signal to that neuron, is called the *local error* (or simply the error) of that neuron. Accordingly, the local error of the $k$th neuron of the output layer, denoted by $\delta_{km}$, is given by

$$\delta_{km} = y'_{km}\left(z_{km}\right) \cdot \delta_{km}^{*}; k = 1,...,N_m.  \tag{1.30}$$

It is important that we differ local errors of output neurons presented by (1.30) from the global errors of the network presented by (1.28) and taken from the same neurons. Respectively, taking into account (1.30), we can transform (1.29) as follows:

$$\Delta w_i^{km} = -\beta \frac{\partial E(W)}{\partial w_i^{km}} = \begin{cases} \beta \delta_{km} Y_{i,m-1} & i = 1,...,\ N_{m-1} \\ \beta \delta_{km} & i = 0, \end{cases}  \tag{1.31}$$

Let us now find the hidden neurons errors. To find them, we have to backpropagate the output neurons errors (1.30) to the hidden layers. To propagate the output neurons errors to the neurons of all hidden layers, a sequential error backpropagation through the network from the $m$th layer to the $m$-1$^{st}$ one, from the $m$-1$^{st}$ one to the m-2$^{nd}$ one, ..., from the 3$^{rd}$ one to the 2$^{nd}$ one, and from the 2$^{nd}$ one to the 1$^{st}$ one has to be done. When the error is propagated from the layer $j$+1 to the layer $j$, the local error of each neuron of the $j$+1$^{st}$ layer is multiplied by the weight of the path connecting the corresponding input of this neuron at the $j$+1$^{st}$ layer with the corresponding output of the neuron at the $j$th layer. For example, the error $\delta_{i,j+1}$ of the $i^{th}$ neuron at the $j$+1$^{st}$ layer is propagated to the $k$th neuron at the $j$th layer, multiplying $\delta_{i,j+1}$ with $w_k^{i,j+1}$, namely the weight corresponding to the $k$th input of the $i$th neuron at the $j$+1$^{st}$ layer. This analysis leads to the following expression for the error of the $k$th neuron from the $j$th layer:

$$\delta_{kj} = y'_{kj}\left(z_{kj}\right)\sum_{k=1}^{N_{j+1}} \delta_{i,j+1} w_k^{i,j+1}; k = 1,...,N_j.  \tag{1.32}$$

It should be mentioned that equations (1.29)-(1.32) are obtained for the general case, without the connection with some specific activation function. Since we agreed above that we use a logistic function (1.22) in our MLF, a derivative of this function is the following (let us take for simplicity, but without loss of generality, $\alpha = 1$ in (1.22)):

$$y'(z) = \varphi'(z) = \left(\frac{1}{1+e^{-z}}\right)' = \left((1+e^{-z})^{-1}\right)' = -(1+e^{-z})^{-2} \cdot (-e^{-z}) =$$

$$= \frac{e^{-z}}{(1+e^{-z})(1+e^{-z})} = y(z)\frac{e^{-z}}{(1+e^{-z})} = y(z)(1-y(z))$$

because

$$1 - y(z) = 1 - \frac{1}{1+e^{-z}} = \frac{1+e^{-z}-1}{1+e^{-z}} = \frac{e^{-z}}{1+e^{-z}}.$$

Thus $y'(z) = \varphi'(z) = y(z)(1 - y(z))$ and substituting this to (1.32) we obtain the equation for the error of the MLF hidden neurons (the $k$th neuron from the $j$th layer) with the *logistic activation function*:

$$\delta_{kj} = y_{kj}(z_{kj}) \cdot (1 - y_{kj}(z_{kj})) \sum_{i=1}^{N_{j+1}} \delta_{i,j+1} w_k^{i,j+1}; k = 1, ..., N_j . \quad (1.33)$$

Once all the errors are known, (1.30) determine the output neurons errors and (1.33) determine the hidden neurons errors, and all the weights can be easily adjusted by adding the adjusting term $\Delta w$ to the corresponding weight. For the output neurons this term was already derived and it is shown in (1.31). For all the hidden neurons it can be derived in the same way and it is equal for the first hidden layer neurons to

$$\Delta w_i^{k1} = -\beta \frac{\partial E(W)}{\partial w_i^{k1}} = \begin{cases} \beta \delta_{k1} x_i, & i = 1, ..., n \\ \beta \delta_{k1} & i = 0, \end{cases} \quad (1.34)$$

where $x_1, ..., x_n$ are the network inputs, $n$ is the number of them, and $\beta$ is a learning rate. For the rest of hidden neurons

$$\Delta w_i^{kj} = -\beta \frac{\partial E(W)}{\partial w_i^{kj}} = \begin{cases} \beta \delta_{kj} Y_{i,j-1}, & i = 1, ..., N_{m-1} \\ \beta \delta_{kj}, & i = 0, \end{cases} \quad (1.35)$$

$$j = 2, ..., m-1.$$

All the network weights can now be adjusted taking into account (1.31), (1.34) and (1.35) as follows (we consider $N_0 = n$ - the number of "neurons" in the first layer is equal to the number of network inputs, there are $m$-1 hidden layers in the network and the $m$th layer is the output one). Thus, for the $k$th neuron in the $j$th layer we have

$$\tilde{w}_i^{kj} = w_i^{kj} + \Delta w_i^{kj}; i = 0, ..., N_{j-1}; k = 1, ..., N_j; j = 1, ..., m . \quad (1.36)$$

Thus, the derivation and description of the MLF learning algorithm with the error backpropagation is completed. In practice, the learning process should continue either until MSE or RMSE drops below some reasonable pre-defined minimum or until some pre-determined number of learning iterations is exceeded.

It is important to mention that this learning algorithm was really revolutionary. It opened absolutely new opportunities for using neural networks for solving classification and prediction problems that are described by non-linearly separable discrete and continuous functions.

However, we have to point out some specific limitations and disadvantages of this algorithm.

1) The backpropagation learning algorithm for MLF is developed as a method of solving the optimization problem. Its target is to find a global minimum of the error function. As all other such optimization methods, it suffers from a "local minima" phenomenon (see Fig. 1.11).

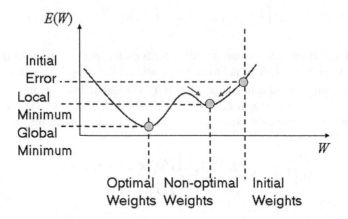

**Fig. 1.11** A "local minima" phenomenon. The learning process may get stuck in a local minimum area. To reach a global minimum, it is necessary to jump over a local minimum using a proper learning rate.

The error function may have many local minima points. A gradient descent method, which is used in the MLF backpropagation learning algorithm, may lead the learning process to the closest local minimum where the learning process may get stuck. This is a serious problem and it has no regular solution. The only method of how to jump over a local minimum is to "play" with the learning rate $\beta$ in (1.31), (1.34), and (1.35) increasing a step of learning. There are many recommendations on how to do that; however all of them are not universal and cannot guarantee that a global minimum of the error function will be reached.

2) Since the MLF backpropagation learning is reduced to solving the optimization problem, an activation function, which is used in MLF neurons, must be differentiable. This is a limitation, because, for example, discrete-valued activation functions cannot be used with this learning algorithm at all, since they are not differentiable. This complicates using MLF as a multi-class classifier and typically it is used just for two-class classification. In this case, the right "half" of the sigmoid activation function is truncated to "1" and the left half to "-1" or 0. For example, for functions (1.22) and (1.23) this means, respectively,

$$\hat{\varphi}_1(z) = \begin{cases} 1, \varphi(z) = \dfrac{1}{1+e^{-\alpha z}} \geq 0.5 \\[3mm] 0, \varphi(z) = \dfrac{1}{1+e^{-\alpha z}} < 0.5, \end{cases}$$

$$\hat{\varphi}_2(z) = \begin{cases} 1, \tanh \alpha z = \dfrac{e^{\alpha z} - e^{-\alpha z}}{e^{\alpha z} + e^{-\alpha z}} \geq 0 \\[4mm] -1, \tanh \alpha z = \dfrac{e^{\alpha z} - e^{-\alpha z}}{e^{\alpha z} + e^{-\alpha z}} < 0. \end{cases}$$

3) Sigmoid functions (1.22) and (1.23) are nonlinear, but their flexibility for approximation of highly nonlinear functions with multiple irregular jumps is limited. Hence, if we need to learn highly nonlinear functions, it is often necessary to extend a network by more hidden neurons.

4) Extension of a network leads to complications during the learning process. The more hidden neurons are in the network, the more is level of heuristics in the backpropagation algorithm. Indeed, the hidden neurons desired outputs and the exact errors are never known. The hidden layer errors can be calculated only on the base of the backpropagation learning algorithm, which is based on the heuristic assumption on the dependence of the error of each neuron on the errors of those neurons to which this neuron is connected. Increasing of the total number of weights in the network leads to complications in solving the optimization problem of the error functional minimization.

These remarks are important for us. When we will consider a backpropagation learning algorithm for the complex-valued multilayer feedforward neural network based on multi-valued neurons (Chapter 4), we will see that this network and its learning algorithm do not suffer from the mentioned disadvantages and limitations.

### 1.3.4 Hopfield Neural Network

In 1982, John Hopfield proposed a fully connected recurrent neural network with feedback links [21]. The *Hopfield Neural Network* is a multiple-loop feedback neural network, which can be used first of all as an associative memory. All the neurons in this network are connected to all other neurons except to themselves that is there are no self-feedbacks in the network (see Fig. 1.12). Thus, the Hopfield network is a *fully connected neural network*. Initially, J. Hopfield proposed to use the binary threshold neurons with activation function (1.1) as the basic ones in this network.

The weight $w_{ij}$ corresponds to the synaptic connection of the $i$th neuron and the $j$th neuron. It is important that in the Hopfield network, for the $i$th and $j$th neurons $w_{ij} = w_{ji}$. Since there is no self-connection, $w_{ii} = 0$. The network works cyclically updating the states of the neurons. The output of the $j$th neuron at cycle $t+1$ is

$$s_j(t+1) = \varphi\left( w_0^j + \sum_{i \neq j} w_{ij} s_i(t) \right). \tag{1.37}$$

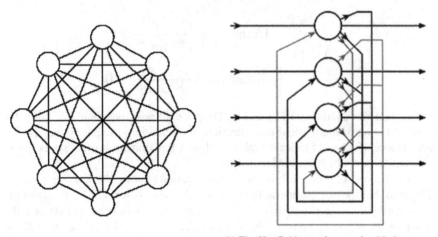

a) The Hopfield neural network with 8 neurons     b) The Hopfield neural network with 4 neurons[4]

**Fig. 1.12** Hopfield Neural Network

A main idea behind the Hopfield net is to use it as the *associative memory* (content-addressable memory). Initially this idea was suggested by Teuvo Kohonen in [22], but D. Hopfield comprehensively developed it in his seminal work [21], which was a great stimulus for the further development of neural networks after a "skeptical period" in 1970s caused by the M. Minsky's and S. Papert's analysis of limited capabilities of the perceptron [13]. The associative memory may learn patterns (for, example, if we want to store $n$ x $m$ images in the associative memory, we should take the $n$ x $m$ Hopfield network whose each neuron learns the intensity values in the corresponding pixels; in this case, there is a one-to-one correspondence between a set of pixels and a set of neurons). The Hebbian learning rule (1.3) or (1.4) can be effectively used for learning. After the learning process is completed, the associative memory may retrieve those patterns, which were learned, even from their fragments or from distorted (noisy or corrupted) patterns. The retrieval process is iterative and recurrent as it is seen from (1.37) ($t$ is the number of cycle-iteration). D. Hopfield showed in [21] that this retrieval process always converges. A set of states of all the neurons on the $t$th cycle is called a *state of the network*. The network state on $t$th cycle is the network input for the $t+1^{st}$ cycle. The network is characterized by its *energy* corresponding to the current state. The energy is determined [21] as

$$E_t = -\frac{1}{2}\sum_i \sum_j w_{ij} s_i(t) s_j(t) + \sum_i w_0^i s_i(t).    \tag{1.38}$$

Updating its states during the retrieval process, the network converges to the local minimum of the energy function (1.38), which is a stable state of the network.

---

[4] This picture is taken from Wikipedia, the free encyclopedia,
http://en.wikipedia.org/wiki/File:Hopfield-net.png

Once the network reaches its stable state, the retrieval process should be stopped. In practical implementation, the retrieval process should continue either until some pre-determined minimum of the energy function (1.38) is reached or until MSE or RMSE between the states on cycle $t$ and $t+1$ drop below some pre-determined minimum.

In [23], D. Hopfield generalized all principles that he developed in [21] for a binary network with threshold neurons for a network with neurons with a continuous monotonic increasing and bounded activation function (for example, a sigmoid function) and with continuous states.

It is important to mention that the Hopfield neural network not only is the first comprehensively developed recurrent neural network. It also stimulated active research in areas of neural networks and dynamical systems in general. It is also worth to mention that the Hopfield network with continuous real-valued neurons suffers from disadvantages and limitations similar to the ones for MLF. For example, it is difficult to use such a network to store gray-scale images with 256 or more gray levels because local minima of the energy function are all located close to the corners of a unitary hypercube. Thus, a stable state of the network tends to a binary state. In Chapter 6, we will observe complex-valued associative memories based on networks with multi-valued neurons that do not suffer from these disadvantages.

## 1.3.5   Cellular Neural Network

The Hopfield neural network as we have seen is a fully connected network. The MLF is a network with full feedforward connections among adjacent layers neurons. We have also seen that the Hopfield network is a recurrent network. It updates its states iteratively until a stable state is reached. In 1988, Leon Chua and Lin Yang proposed another recurrent network with local connections [24] where each neuron is connected just with neurons from its closest neighborhood. They called it the *cellular neural network* (CNN). One of the initial ideas behind this network topology was to use it for image processing purposes. Since the correlation and respectively a mutual dependence between image pixels in any local $n$ x $m$ window is high, the idea was to create a recurrent neural network containing the same amount of neurons as the amount of pixels in an image to be processed. Local connections between the neurons could be used for implementation of various spatial domain filters, edge detectors, etc.

For example, CNN with 3x3 local connections is shown in Fig. 1.13. This network contains $N \times M$ neurons and it is very suitable for processing $N \times M$ images. The output of each neuron is connected to the corresponding inputs of 8 neurons closest to the given neuron (all neurons from a 3x3 neighborhood of a given neuron) and only to them, while outputs of these 8 adjacent neurons are connected to the corresponding inputs of a given neuron and there are only inputs of a given neuron. Unlike the Hopfield net, CNN allows a feedback connection, so each neuron may have one input receiving a signal from its own output. CNN is a recurrent network. Like the Hopfield network, it updates its states iteratively until a stable

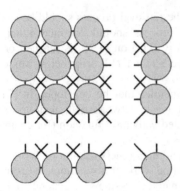

**Fig. 1.13** Cellular Neural Network with 3x3 local connections

state is reached. CNN can be a binary network with the threshold neurons, but it can be based also on neurons with other activation functions, which makes it possible to implement different linear (using a piecewise linear activation function) and nonlinear (using nonlinear activation functions) filters.

Unlike it is in the Hopfield network, weights in CNN are not symmetric. Each neuron is indexed by two indexes-coordinates. The weight corresponding to the $i$th input of the $kj$th neuron is denoted $w_i^{kj}$. As we told, the network works cyclically updating the states of the neurons. The output of the $kj$th neuron at cycle $t+1$ is

$$s_{kj}(t+1) = \varphi\left( w_0^{kj} + \sum_i w_i^{kj} s_{rp}(t) \right);$$

$$k - d + 2 \le r \le k + d - 2, \; j - d + 2 \le p \le j + d - 2,$$

(1.39)

where $\varphi$ is an activation function and $d$ is the closest neighborhood size (for example, for a 3x3 local window $d = 3$). In the CNN community, a very popular topic is mutual influence of a given neuron and those neurons connected to it. This is important for investigation of the stability of the network. In the context of this book, it will be enough for us to consider just equation (1.39), which determines the output of each neuron. The most interesting for us will be CNN based on multi-valued neurons, which can be successfully used as an associative memory (see Chapter 6, Section 6.3), significantly increasing the CNN functionality.

## 1.4  Introduction to Complex-Valued Neurons and Neural Networks

### 1.4.1  Why We Need Them?

We have already mentioned that complex numbers are absolutely natural, as well as real numbers. From this point of view, complex-valued neurons are natural too.

But additionally there are at least three very significant reasons for using complex-valued neurons and neural networks. These reasons are:

1) Unlike a single real-valued neuron, a single complex-valued neuron may learn non-linearly separable problems (a great variety of them) in that initial $n$-dimensional space where they are defined, without any nonlinear projection to a higher dimensional space (very popular kernel-based techniques, and the most popular and powerful of them – the support vector machines (SVM)[5] proposed by Vladimir Vapnik [25, 26] are based on this approach). Thus, a complex-valued neuron is much more functional than a real-valued one.

2) Many real-world problems, especially in signal processing, can be described properly only in the frequency domain where complex numbers are as natural as integer numbers in counting. In the frequency domain, it is essential to treat the amplitude and phase properly. But there is no way to have deal with the phase phenomenon without complex numbers. If we want to analyze any process, in which phase is involved, we should definitely use complex numbers and tools that are suitable for working with them. If we treat the phase as just real numbers belonging to the interval $[0, 2\pi[$ or $[-\pi, \pi[$, then we make a great mistake, because in this way the physical nature of the phase is completely eliminated.

3) Since the functionality of a single complex-valued neuron is higher than the one of a single real-valued neuron, the functionality of complex-valued neural networks is also higher than the functionality of their real-valued counterparts. A smaller complex-valued neural network can learn faster and generalize better than a real-valued neural network. This is true for feedforward complex-valued networks and for Hopfield-like complex-valued networks. More functional neurons connected into a network ensure that this network also is more functional than its real-valued counterpart. We will see below (Chapter 4) that, for example, a feedforward multilayer neural network with multi-valued neurons (MLMVN) completely outperforms MLF. Even smaller MLMVN learns faster and generalizes better than larger MLF. Moreover, there are many problems, which MLF is not able to solve successfully, while MLMVN can. We will also see that a Hopfield-like neural network with multi-valued neurons can store much more patterns and has better retrieval rate as an associative memory, than a classical Hopfield network (Chapter 6, Section 6.3). Moreover, we will also see that just partially connected neural network with multi-valued neurons can also be used as a very powerful associative memory.

However, it is important for better understanding of the foregoing Chapters, to consider right now the first two of three mentioned reasons in more detail.

---

[5] While we presented in detail the most important classical neural network techniques, we do not present here in detail the SVM essentials. We believe that the interested reader can easily find many sources where SVM are described in detail. This book is devoted to complex-valued neural networks, but at least so far no complex-valued SVM were considered. However, we will compare a number of CVNN techniques presented in this book with SVM in terms of number of parameters they employ and generalization capability.

## 1.4.2  Higher Functionality

We have briefly observed what a neuron is, and what a neural network is. We have also observed not all, but the most important turning-points in real-valued artificial neurons and neural networks. We have mentioned several times that real-valued neurons and real-valued neural networks have some specific limitations. May be the most important of these limitations is impossibility of a single real-valued neuron to learn non-linearly separable input/output mappings in that initial linear $n$-dimensional space where the corresponding input/output mapping is defined. The classical example of such a problem, which cannot be learned by a single real-valued neuron due to its non-linear separability, is XOR as we have seen.

**Table 1.6** Threshold neuron implements $f(x_1, x_2) = x_1$ xor $x_2$ function with the weighting vector (0, 1, 1, 2) in 3-dimensional space $(x_1, x_2, x_1 x_2)$

| $x_1$ | $x_2$ | $x_1 x_2$ | $z = w_0 + w_1 x_1 + w_2 x_2 + w_3 x_1 x_2$ | $\text{sgn}(z)$ | $f(x_1, x_2) = x_1$ xor $x_2$ |
|---|---|---|---|---|---|
| 1 | 1 | 1 | 4 | 1 | 1 |
| 1 | -1 | -1 | -2 | -1 | -1 |
| -1 | 1 | -1 | -2 | -1 | -1 |
| -1 | -1 | 1 | 0 | 1 | -1 |

The reader may notice that the XOR problem can be learned using a single real-valued threshold neuron if the initial 2-dimensional space $E_2^2$ where it is defined, will be nonlinearly extended to the 3-dimensional space by adding to the two inputs $x_1$ and $x_2$ a nonlinear (quadratic) third input $x_1 x_2$, which is determined by the product of the two initial inputs [27]. Indeed, let us consider the space $(x_1, x_2, x_1 x_2)$, which is obtained from $E_2^2$ by adding a quadratic term and, for example, the weighting vector $W = (0, 1, 1, 2)$[6]. This solution is shown in Table 1.6.

Actually, this solution confirms the Cover's theorem [28] on the separability of patterns, which states that a pattern classification problem is more likely to be linearly separable in a high dimensional feature space when nonlinearly projected into a high dimensional space. In fact, all kernel-based machine learning techniques including SVM are based on this approach. If some problem is non-linearly separable in that initial $n$-dimensional space where it is defined (for example, some classification problem described by some $n$ features), it can be projected nonlinearly into a higher dimensional space where it becomes linearly separable. We have to understand that any feedforward neural network is also doing the same. It extends the

---

[6] We could also use here the weighting vector $W = (0, 0, 0, 1)$.

initial space (if it contains more neurons in any hidden layer than network inputs) or at least transforms it nonlinearly into another space. When we considered in Section 1.3 solution of the XOR problem using MLF (see Fig. 1.7 and Table 1.5) containing three neurons and two layers, we nonlinearly projected initial space $(x_1, x_2)$ into another (functional) space $(f_1(x_1), f_2(x_2))$ using the first layer neurons where the problem becames linearly separable using the third (output) neuron. This transformation is in fact nonlinear because it is implemented through a nonlinear activation function of neurons.

Nevertheless, is it possible to learn non-linearly separable problems using a single neuron without the extension or transformation of the initial space? The answer is "Yes!" It is just necessary to move to the complex domain!

In all neurons and neural networks that we have considered so far weights and inputs are real and weighted sums are real, respectively. Let us consider now complex-valued weights. Thus, weights can be arbitrary complex numbers. Inputs and outputs will still be real. Moreover, lest us consider even a narrow case of binary inputs and outputs. So, our input/output mapping is described by the function $f(x_1, ..., x_n) : E_2^n \to E_2$, which is a Boolean function. However, since our weights are complex ($w_i \in \mathbb{C}$, $i = 0, 1, ..., n$) and inputs are real $x_i \in E_2 = \{1, -1\}$, a weighted sum is definitely complex $w_0 + w_1 x_1 + ... + w_n x_n = z \in \mathbb{C}$. This means, that an activation function must be a function from $\mathbb{C}$ to $E_2$. Let us define the following activation function

$$\varphi(z) = \begin{cases} 1, & \text{if } 0 \le \arg z < \pi/2 \text{ or } \pi \le \arg z < 3\pi/2 \\ -1, & \text{if } \pi/2 \le \arg z < \pi \text{ or } 3\pi/2 \le \arg z < 2\pi, \end{cases} \tag{1.40}$$

where $\arg z$ is the argument of the complex number $z$ in the range $[0, 2\pi[$. Evidently $\varphi(z)$ maps $\mathbb{C}$ to $E_2$, so $\varphi(z) : \mathbb{C} \to E_2$. Activation function (1.40) divides the complex plane into 4 sectors (see Fig. 1.14) that coincide with the quarters of the complex plane formed by its separation with real and imaginary axes. Depending on $\arg z$, $\varphi(z)$ is equal to 1 in the 0$^{\text{th}}$ and the 2$^{\text{nd}}$ sectors (the 1$^{\text{st}}$ and the 3$^{\text{rd}}$ quarters) and to -1 in the 1$^{\text{st}}$ and the 3$^{\text{rd}}$ sectors (the 2$^{\text{nd}}$ and the 4$^{\text{th}}$ quarters).

Let us return to the most popular classical example of non-linearly separable problem – XOR. Let us show that a single neuron with the activation function (1.40) can easily implement the non-linearly separable XOR function without any extension of the original

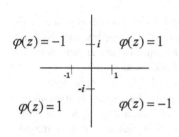

**Fig. 1.14** Activation function (1.40)

**Table 1.7** A complex-valued neuron with the activation function (1.40) implements the $f\left(x_1, x_2\right) = x_1 \text{ xor } x_2$ function with the weighting vector $(0, i, 1)$ in the original 2-dimensional space

| $x_1$ | $x_2$ | $z = w_0 + w_1 x_1 + w_2 x_2$ | $\arg(z)$ | $\varphi(z)$ | $f\left(x_1, x_2\right) = x_1 \text{ xor } x_2$ |
|-------|-------|-------------------------------|-----------|--------------|-------------------------------------------------|
| 1 | 1 | $i+1$ | $\pi/4$ | 1 | 1 |
| 1 | -1 | $i-1$ | $3\pi/4$ | -1 | -1 |
| -1 | 1 | $-i+1$ | $5\pi/4$ | -1 | -1 |
| -1 | -1 | $-i-1$ | $7\pi/4$ | 1 | -1 |

2-dimensional space. Let us take the weighting vector $W = \left(0, i, 1\right)$ ($i$ is an imaginary unity). The results are shown in Table 1.7.

These results shows that the XOR problem, which was for many years, on the one hand, a stumbling block in neurons theory [13] and, on the other hand, was a main argument for necessity of neural networks due to a limited functionality of a single neuron, can in fact be easily solved using a single neuron! But what is the most important – this is a single neuron with the complex-valued weights! This solution was for the first time shown by the author of this book in 1985 [29] and then it was deeply theoretically justified by him in [30].

The ability of a single neuron with complex-valued weights to solve non-linearly separable problems like XOR clearly shows that a single complex-valued neuron has a higher functionality than a single real-valued neuron. This is a crucial point!

We will show later (Chapter 5) why those problems that are non-linearly separable in the space $\mathbb{R}^n$ (or its subspace) can be linearly separable in the space $\mathbb{C}^n$ or its subspace. We will see there that problems like XOR and Parity $n$ ($n$-input XOR or mod 2 sum of $n$ variables) are likely the simplest non-linearly separable problems that can be learned by a single complex-valued neuron. We will also show that activation function (1.40) is a particular case of the 2-valued periodic activation function, which determines a universal binary neuron (UBN), which in turn is a particular case of the multi-valued neuron with a periodic activation function.

### 1.4.3  Importance of Phase and Its Proper Treatment

We have already mentioned that there are many engineering problems in the modern world where complex-valued signals and functions of complex variables are involved and where they are unavoidable. Thus, to employ neural networks for their analysis the use of complex-valued neural networks is natural.

However, even in the analysis of real-valued signals (for example, images or audio signals) one of the most efficient approaches is the frequency domain analysis, which immediately involves complex numbers. In fact, analyzing signal properties in the frequency domain, we see that each signal is characterized by magnitude and phase that carry different information about the signal. A fundamental result showing the crucial importance of phase and its proper treatment was presented in 1981 by Alan Oppenheim and Jae Lim [31]. They have considered, particularly, the importance of phase in images. They have shown that the information about all the edges, shapes, and, respectively, about all the objects located in an image, is completely contained in phase. Magnitude contains just the information about the contrast, about contribution of certain frequencies in the formation of an image, about the noisy component in the image, but not about what is located there. Thus, *phase is much more informative and important for image understanding and interpretation and for image recognition*, respectively.

(a) Original image "Lena"                      (b) Original image "Airplane"

(c) Image obtained by taking the inverse Fourier transform from the synthesized spectrum (magnitude of the "Airplane" original spectrum and phase of the "Lena" original spectrum)

(d) Image obtained by taking the inverse Fourier transform from the synthesized spectrum (magnitude of the "Lena" original spectrum and phase of the "Airplane" original spectrum)

**Fig. 1.15** The importance of phase

These properties can be easily confirmed by the experiments that are illustrated in Fig. 1.15. Let us take two well known test images[7] "Lena" (Fig. 1.15a) and

---

[7] These test images have been downloaded from the University of Sothern California test image database "The USC-SIPI Image Database", http://sipi.usc.edu/database/

"Airplane" (Fig. 1.15b). Let us take their Fourier transform and then swap magnitudes and phases of their Fourier spectra.

Thus, we synthesize one spectrum from magnitude of the "Airplane" spectrum and phase of the "Lena" spectrum and another one from phase of the "Airplane" spectrum and magnitude of "Lena" spectrum. Let us now take the inverse Fourier transform from both synthesized spectra. The results are shown in Fig. 1.15c and Fig. 1.15d, respectively. It is very clearly visible that definitely those images were restored whose phases were used in the corresponding synthesized spectra. In Fig. 1.15c we see just the "Lena" image, while in Fig. 1.15d we see just the "Airplane" image. There is no single trace of those images whose magnitudes were used in the synthesized Fourier spectra from which images in Fig. 1.15c and Fig. 1.15d have been obtained.

Another interesting experiment is illustrated in Fig. 1.16. We took the Fourier spectra of the same original images "Lena" (Fig. 1.15a) and "Airplane" (Fig. 1.15b). Then magnitudes in both spectra were replaced by the constant 1, while phases were preserved. Thus, magnitudes became "unitary". Then we took the inverse Fourier transform from these modified spectra with "unitary" magnitude. The results are shown in Fig. 1.16a ("Lena") and Fig. 1.16b ("Airplane"). It is clearly seen that all edges, shapes, and even the smallest details from the original images are preserved. Since images in Fig. 1.16 were obtained just from phase (magnitude was eliminated by setting all its values to 1), this confirms that all information about the edges, shapes, objects and their orientation is contained only in phase.

 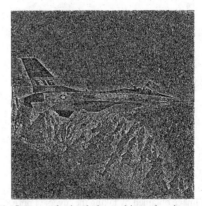

a) Image obtained by taking the inverse Fourier transform from the synthesized spectrum ("unitary" magnitude (constant 1) and phase of the "Lena" original spectrum)

b) Image obtained by taking the inverse Fourier transform from the synthesized spectrum ("unitary" magnitude (constant 1) and phase of the "Airplane" original spectrum)

**Fig. 1.16** The importance of phase

These wonderful properties of phase are determined by its physical nature. The Fourier transform express any signal in terms of the sum of its projections onto a set of basic functions that represent those electromagnetic waves, which form this

signal. Hence, the Fourier transform is the decomposition of a signal by these basic functions that are defined as

$$e^{i2\pi ut} = \cos(2\pi ut) + i\sin(2\pi ut), \qquad (1.41)$$

or in the discrete case

$$e^{i2\pi uk} = \cos\left(\frac{2\pi}{n}uk\right) + i\sin\left(\frac{2\pi}{n}uk\right); u,k = 0,1,...,n-1, \qquad (1.42)$$

where $u$ is the corresponding frequency. The Fourier spectrum of the continuous
signal $f(t)$ is

$$F(u) = \int f(t)e^{-i2\pi ut} = |F(u)|e^{i\varphi(u)}, \qquad (1.43)$$

where $|F(u)|$ is magnitude and $\varphi(u)$ is phase. For the discrete signal
$f(k), k = 0,1,...,n-1$, equation (1.43) is transformed as follows

$$F(u) = \sum_{k=0}^{n-1} f(k)e^{-i2\pi uk} = |F(\omega)|e^{i\varphi(u)}; u = 0,1,...,n-1, \qquad (1.44)$$

where each $|F(u)|e^{i\varphi(u)}, u = 0,1,...,n-1$ is referred to as a spectral coefficient
or a decomposition coefficient. $|F(u)|$ is the absolute value (magnitude) of the
$u$th spectral coefficient and $\varphi(u) = \arg F(u)$ is the argument (phase) of this
spectral coefficient. To reconstruct a signal from (1.43), we have to perform the
inverse Fourier transform

$$f(t) = \frac{1}{2\pi}\int F(\omega)e^{i2\pi ut}. \qquad (1.45)$$

To reconstruct a signal from (1.44) in the discrete case, we have to perform the inverse Fourier transform – to find a sum of basic functions (waves) (1.42) with the
coefficients (1.44):

$$f(k) = \frac{1}{n}\sum_{u=0}^{n-1} F(\omega)e^{i2\pi uk}; k = 0,1,...,n-1. \qquad (1.46)$$

In (1.41) and (1.42) that are the basic functions of the Fourier transform, the electromagnetic waves corresponding to all frequencies have a zero phase shift. Let us
set $2\pi u = \omega$ in (1.41). Then the corresponding basic function of the Fourier
transform is $e^{iut} = \cos(ut) + i\sin(ut)$. Respectively, (1.45) can be written as
follows

$$f(t) = \frac{1}{2\pi} \int F(\omega) e^{i\omega t} . \qquad (1.47)$$

Let us take a look at Fig. 1.17a. It shows a sinusoidal wave $\sin(2\pi ut)$ for $u = 1$. According to (1.46) and (1.47), after this sinusoidal wave is multiplied with the Fourier spectral coefficient $F(u)$, its absolute value (magnitude) is equal

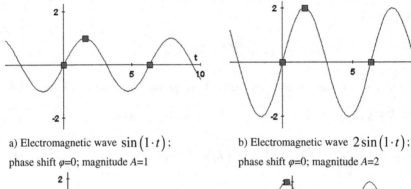

a) Electromagnetic wave $\sin(1 \cdot t)$;
phase shift $\varphi = 0$; magnitude $A=1$

b) Electromagnetic wave $2\sin(1 \cdot t)$;
phase shift $\varphi = 0$; magnitude $A=2$

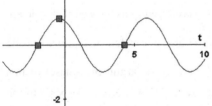
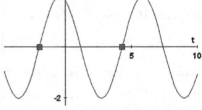

c) Electromagnetic wave $\sin((1 \cdot t) + 2)$;
phase shift $\varphi = 2$; magnitude $A=1$

d) Electromagnetic wave $2\sin((1 \cdot t) + 2)$;
phase shift $\varphi = 2$; magnitude $A=2$

**Fig. 1.17** A role of phase and magnitude in the Fourier transform. Phase in a Fourier transform coefficient shows the phase shift for the electromagnetic wave with the frequency corresponding to the given Fourier transform coefficient. The phase shift is a carrier of information about a signal concentrated in the wave with the corresponding frequency. Magnitude in a Fourier transform coefficient just shows the intensity (the "weight") of the wave corresponding to the given frequency in the formation of a signal[8]

---

[8] To create these pictures, we used a wonderful tool located at
http://www.ugrad.math.ubc.ca/coursedoc/math100/notes/trig/phase.html
(this is a site of University of British Columbia, Canada)

to $A|e^{i\alpha x}| = A|\cos(\omega t) + i\sin(\omega t)|$ where $A = |F(u)|$. Fig. 1.17b shows a

sinusoidal wave $A\sin(2\pi ut)$ for $u = 1$ and $A = |F(u)| = 2$. The phase shift of these both electromagnetic waves is equal to 0. Thus, if a sinusoidal wave has a basic form $A\sin(\omega t + \varphi)$, then waves in Fig. 1.17a and Fig. 1.17b have $\varphi = 0$.

It follows from (1.47) that $F(u)e^{i2\pi ut} = |F(u)|e^{i(\omega t + \arg(F(u)))}$ because the argument of the product of two complex numbers is equal to the sum of multipliers' arguments, while the magnitude of the product is equal to the product of magnitudes (take into account that $|e^{i2\pi ut}| = 1$). Fig. 1.17c and Fig. 1.17d show sinusoidal waves $A\sin(\omega t + \varphi)$ with the phase shift $\varphi = 2$. These waves could be obtained by multiplication of the "standard" wave $\sin(\omega t)$ by such $F(u)$ that $\arg F(u) = 2$. For the sinusoidal wave in Fig. 1.17c, $A = |F(u)| = 1$, while for the one in Fig. 1.17d, $A = |F(u)| = 2$. Hence, the sinusoidal waves in Fig. 1.17a, b have the same phase shifts and different magnitudes, and the sinusoidal waves in Fig. 1.17c, d have the same phase shifts and different magnitudes.

As we see, *the phase shift is nothing else than phase of the Fourier transform coefficient* corresponding to a wave with the certain frequency. This shift determines the contribution of this wave to the shape of a signal after its reconstruction from the Fourier transform while magnitude plays only subsidiary role.

Hopefully, it is clear now why all shapes and even all details of images in Fig. 1.16 were successfully reconstructed from the Fourier transform whose magnitude was completely eliminated and replaced by the constant 1. It is also clear now how images in Fig. 1.15 were reconstructed just from the original phases of their Fourier spectra, while their magnitudes were swapped.

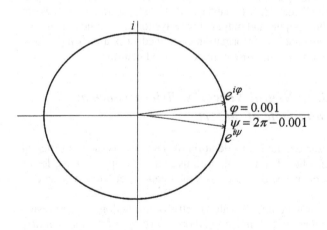

**Fig. 1.18** Importance of proper treatment of phase as an angle

This means that phase is a very important carrier of information about those objects that are presented by a signal. This information can be used, for example, for solving image recognition problems (we will consider this later, in Chapter 6). However, first of all it is absolutely important to treat properly phase and that information, which is concentrated in phase. We have to treat the phase $\varphi$ only as an angular value determining the complex number $e^{i\varphi}$ located on the unit circle. Any attempt to work with phases as with formal real numbers located either in interval $[0, 2\pi[$ or $[-\pi, \pi[$ without taking into account that they are angles that are in turn arguments of complex numbers, completely eliminates a physical nature of phase. If we do not treat phases properly (as arguments of complex numbers), then the information, which is contained in phase, is completely distorted.

For example, if we do not care of the nature of phase, we may treat numbers $\varphi = 0.001$ and $\psi = 2\pi - 0.001 = 6.282$ as such located in the opposite ends of the interval $[0, 2\pi[$. In this case, their formal difference is 6.282-0.001=6.281. But in fact, these numbers determine angles that are very close to each other, and the difference between them is just 0.002 radian. Respectively, these two phases determine two points on the unit circle $e^{i\varphi}$ and $e^{i\psi}$ that are located very close to each other (see Fig. 1.18).

Thus, to treat phases properly, they have to be considered only as arguments of complex numbers. To work only with that information concentrated in phase, it is enough to consider phases as arguments determining complex numbers located on the unit circle. In this case, we do not care of magnitude (like in the example presented in Fig. 1.16). We will see below (Chapter 2) that this is definitely the case of a multi-valued neuron whose inputs and output are always located on the unit circle.

Hence, to analyze phase and the information contained in phase, using neural networks, it is absolutely important to use complex-valued neurons.

### 1.4.4 Complex-Valued Neural Networks: Brief Historical Observation and State of the Art

Before we will move to the detailed consideration of multi-valued neurons, neural networks based on them, their learning algorithms and their applications, let us present a brief historical overview of complex-valued neural networks and state of the art in this area.

The first historically known complex-valued activation function was suggested in 1971 by Naum Aizenberg and his co-authors Yuriy Ivaskiv and Dmitriy Pospelov in [32]. Thus, complex-valued neural networks start their history form this seminal paper. A main idea behind this paper was to develop a model of multiple-valued threshold logic, to be able to learn and implement multiple-valued functions using a neural element similarly to learning and implementation of Boolean threshold functions using a neuron with the threshold activation function. Moreover, according to this new model, Boolean threshold logic should be just a particular case of multiple-valued threshold logic. We will consider this model in detail in Chapter 2. Now we just want to outline a basic approach.

As we have already seen, in neural networks the two-valued alphabet $E_2 = \{1, -1\}$ is usually used instead of the traditional Boolean alphabet $K_2 = \{0, 1\}$. This can easily be explained by two factors. First of all, unlike in $K_2$, in $E_2$ values of two-valued logic are normalized, their absolute value is equal to 1. Secondly, we have seen that, for example, in the error-correction learning rule (1.17), $x_i \in E_2$ is a very important multiplicative term participating in the adjustment of the weight $w_i, i = 1, ..., n$. If it was possible that $x_i = 0$, then the error-correction learning rule (1.17) could not be derived in that form, in which it exists.

In the classical multiple-valued ($k$-valued) logic, the truth values are traditionally encoded by integers from the alphabet $K = \{0, 1, ..., k-1\}$. They are not normalized. If we want to have them normalized, evidently, this problem can be solved neither within the set of integer numbers nor the set of real numbers for $k > 2$. However, Naum Aizenberg suggested a wonderful idea to jump to the field of complex numbers and to encode the values of $k$-valued logic by the $k$th roots of unity (see Fig. 1.19). Since there are exactly $k$ $k$th roots of unity, it is always possible and very easy to build a one-to-one correspondence between the set $K = \{0, 1, ..., k-1\}$ and the set $E_k = \{1, \varepsilon_k, \varepsilon_k^2, ..., \varepsilon_k^{k-1}\}$, where

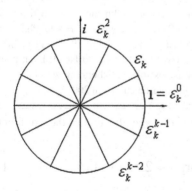

$\varepsilon_k = e^{i 2\pi / k}$ is the primitive $k^{\text{th}}$ root of unity ($i$ is an imaginary unity). We will consider later in detail, (Chapter 2, Section 2.1) a mathematical background behind this idea. Unlike in the set $K$, in the set $E_k$ the values of $k$-valued logic are normalized – their absolute values are equal to 1. Particularly, for two-valued logic, $E_2 = \{1, -1\}$, which corresponds to $K_2 = \{0, 1\}$, and we obtain a well known model of Boolean logic in the alphabet $E_2 = \{1, -1\}$.

**Fig. 1.19** Model of $k$-valued logic over the field of complex numbers. Values of $k$-valued logic are encoded by the $k$th roots of unity

Thus, in multiple-valued logic over the field of complex numbers, a multiple-valued ($k$-valued) function of $n$ variables becomes $f(x_1, ..., x_n): E_k^n \rightarrow E_k$. Naum Aizenberg and his co-authors suggested in [32] the following activation function, which they called

Naum Nisonovich Aizenberg (1928-2002)

Seminal ideas in the area of complex-valued neural networks and in multi-valued neurons were proposed and developed by Professor Naum N. Aizenberg. He was born in Kiev (Ukraine, that time USSR). From 1953 to 1998 he was with Uzhgorod National University (USSR until 1991 and then Ukraine) where he has started as a part time teaching assistant and then became a Professor. For a number of years he was a Chair of the Department of Cybernetics. His first love in research was Algebra, which formed a solid background for his further work in Computer Science and Engineering. His main result in Algebra is solution of the problem of computation of the wreath products of the finite groups. In early 1970s he developed a theory of multiple-valued threshold logic over the field of complex numbers, which became a background for complex-valued neural networks. He also developed an algebraic theory of signal processing in an arbitrary basis. His important accomplishment is also a theory of prime tests, which found many applications in Pattern Recognition. His 11 Ph.D. students got their Ph.D. degrees under his supervision. He retired in 1998 after he got a damaging heart attack. The same year he moved from Ukraine to Israel. Even being seriously ill, he continued his research as far as possible, collaborating with other colleagues. His last paper has been published right after he passed in 2002…

---

CSIGN[9] (keeping in mind that this is a specific generalization of the sgn function for the multiple-valued case)

$$\text{CSIGN}(z) = \varepsilon_k^j, \quad 2\pi j / k \leq \arg z < 2\pi (j+1) / k \qquad (1.48)$$

Function (1.48) divides complex plane into $k$ equal sectors (see Fig. 1.19). We will consider it and its properties in detail in Chapter 2. Now we can say that it follows form (1.48) that if the complex number $z$ is located in the sector $j$, then

---

[9] In the later work, where the multi-valued neuron was introduced, N. Aizenberg himself suggested to use another notation for the function CSIGN. Since in terms of logic this function is multiple-valued predicate, he suggested to use just a letter $P$ ("Predicate") for its notation considering that the initial CSIGN was not successful, because in fact a complex number does not have a sign. We will use the notation $P$ throughout the book except this section.

$\text{CSIGN}(z) = \varepsilon_k^j$. Then a notion of multiple-valued threshold function was introduced in [32]. A function $f(x_1, ..., x_n): E_k^n \to E_k$ is called a *multiple-valued threshold function* if there exist such $n+1$ complex numbers (weights) $w_0, w_1, ..., w_n$ that for all $(x_1, ..., x_n)$ from the domain of the function $f$

$$\text{CSIGN}(w_0 + w_1 x_1 + ... + w_n x_n) = f(x_1, ..., x_n). \qquad (1.49)$$

Paper [32] was then followed by two papers [33, 34] by N. Aizenberg and co-authors where a multi-valued threshold element was introduced as a processing element implementing (1.49) and, respectively, implementing a multiple-valued threshold function. A learning algorithm for this element was also introduced in [34]. By the way, papers [33, 34] originally published only in Russian (as well as [32]) are available now in English [35, 36] (the English version of the journal Cybernetics and Systems Analysis (previously Cybernetics) is published by Springer from late 1990s, and all the earlier journal issues are translated into English too and they are available online from the Springer website[10]). Papers [32-35] were followed in 1977 by the monograph [37] (also published only in Russian) by N. Aizenberg and Yu. Ivaskiv. In [37], all theoretical aspects of multiple-valued threshold logic over the field of complex numbers, multi-valued threshold elements, and their learning were comprehensively observed. It is important to mention that a word "neuron" was not used in those publications, but it is absolutely clear that a *multi-valued threshold element* is nothing else than the discrete multi-valued neuron formally named a neuron in 1992 [38] by N. Aizenberg and the author of this book.

It is difficult to overestimate the importance of the seminal publications [32-34, 37]. For the first time, a neural element introduced there, could learn multiple-valued input/output mappings $E_k^n \to E_k$ and $O^n \to E_k$ ($O$ is a set of points on the unit circle). This means that it was possible to use it for solving, for example, multi-class classification problems where the number of classes is greater than 2. Unfortunately, published only in Russian, these important results were unavailable to the international research community for many years. In 1988 (17 years later (!) after paper [32] was published) A. Noest even "re-invented" activation function (1.48) calling a neuron with this activation function a "phasor neuron" [39]. But in fact, this activation function was proposed in 1971 and we believe that A. Noest simply was not familiar with [32].

Since Chapters 2-6 of this book are completely devoted to multi-valued neurons and neural networks based on them, we will observe all publications devoted to MVN and MVN-based neural networks later as the corresponding topics will be deeply considered. However, we would like to observe briefly now other important works on complex-valued neural networks, not related to MVN.

---

[10] http://www.springer.com/mathematics/applications/journal/10559

Starting from early 1990s complex-valued neural networks became a very rapidly developing area. In 1991 and 1992, independently on each other, H. Leung and S. Haykin [40], and G. Georgiou and C. Koutsougeras [41], respectively, generalized the MLF backpropagation learning algorithms for the complex-valued case. They considered complex weights and complex-valued generalization of the sigmoid activation function and showed that complex backpropagation algorithm converges better than the real one.

Important contributions to CVNN are done by Akira Hirose He is the author of the fundamental monograph [42] with a detailed observation of the state of the art in the field, and the editor of the book [43] with a great collection of papers devoted to different aspects of complex-valued neural networks. He also was one of the first authors who considered a concept of fully-complex neural networks [44] and continuous complex-valued backpropagation [45].

Other interesting contributions to CVNN are done by Tohru Nitta. He has edited a recently published book on CVNN [46]. He also developed the original approach to complex backpropagation [47], and he is probably the first author who considered a quaternion neuron [48].

Very interesting results on application of complex-valued neural networks in nonlinear filtering are obtained by Danilo Mandic and under his supervision. Just a few of his and his co-authors important contributions are recently published fundamental monograph [49] and papers on different aspects of filtering [50] and prediction [51].

Important contributions to learning algorithms for complex-valued neural networks are done by Simone Fiori. We should mention here among others his generalization of Hebbian Learning for complex-valued neurons [52, 53] and original optimization method, which could be used for learning in complex-valued neural networks [54].

We should also mention recently published works by Md. F. Amin and his co-authors [55, 56] on solving classification problems using complex-valued neural networks.

It is also important to mention here interesting works by Sven Buchholz and his co-authors on neural computations in Clifford algebras where complex-valued and quaternion neurons are involved [57]. They also recently developed a concept of quaternionic feedforward neural networks [57, 58].

## 1.5  Concluding Remarks to Chapter 1

In this introductory Chapter, we have briefly considered a history of artificial neurons and neural networks. We have observed such turning-point classical solutions and concepts as the McCulloch-Pitts neuron, Hebbian learning, the Rosenblatt's perceptron, error-correction learning, a multilayer feedforward neural network, backpropagation learning, and linear separability/non-linear separability. We have paid a special attention to those specific limitations that characterize real-valued neural networks. This is first of all impossibility of a single real-valued neuron to learn non-linearly separable problems. This is also strict dependence of the backpropagation learning algorithm on the differentiability of an activation function.

This is also absence of some regular approach to representation of multiple-valued discrete input/output mappings.

We have shown that moving to the complex domain it is possible to overcome at least some of these disadvantages. For example, we have shown how a classical non-linearly separable problem XOR can be easily solved using a single complex-valued neuron without the extension of that 2-dimensional space where it is defined. We have also shown that complex-valued neurons can be extremely important for a proper treatment of phase, which in fact contains much more significant information about the objects presented by the corresponding signals.

We briefly presented the first historically known complex-valued activation function, which makes it possible to represent multiple-valued discrete input/output mappings.

We have also observed recent contributions in complex-valued neural networks. We have mentioned here just recent and perhaps the most cited works. Nevertheless, it follows from this observation that complex-valued neural networks have become increasingly popular. The reader may find many other papers devoted to different aspects of CVNN. Just, for example, take a look at [43, 46] where very good collections of papers are presented. There were also many interesting presentations in a number of special sessions on complex-valued neural networks organized just during last several years (IJCNN-2006, ICANN-2007, IJCNN-2008, IJCNN-2009, and IJCNN-2010). As the reader may see, there are different specific types of complex-valued neurons and complex-valued activation functions. Their common great advantage is that using complex-valued inputs/outputs, weights and activation functions, it is possible to improve the functionality of a single neuron and of a neural network, to improve their performance, and to reduce the training time (we will see later, for example, how simpler and more efficient is learning of MVN and MVN-based neural networks).

We hope that the reader is well prepared now to move to the main part of this book where we will present in detail the multi-valued neuron, its learning, and neural networks based on multi-valued neurons. We will also consider a number of examples and applications that will show great advantages of the multi-valued neuron with complex-valued weights and complex-valued activation function over its real-valued counterparts.

# Chapter 2
# The Multi-Valued Neuron

"The greatest challenge to any thinker is
stating the problem in a way that will allow a solution."

  Bertrand Russell

In this chapter, we introduce the multi-valued neuron. First of all, in Section 2.1 we consider the essentials of the theory of multiple-valued logic over the field of complex numbers. Then we define a threshold function of multiple-valued logic. In Section 2.2, we define the discrete-valued multi-valued neuron whose input/output mapping is always described by some threshold function of multiple-valued logic. Then we consider the continuous multi-valued neuron. In Section 2.3, we consider the edged separation of an $n$-dimensional space, which is determined by the activation function of the discrete multi-valued neuron, and which makes it possible to solve multi-class classification problems. In Section 2.4, we consider how the multi-valued neuron can be used for simulation of a biological neuron. Some concluding remarks are given in Section 2.5.

## 2.1 Multiple-Valued Threshold Logic over the Field of Complex Numbers

As we have seen in Section 1.1, the McCulloch-Pitts neuron, which is the first historically known artificial neuron, implements input/output mappings described by threshold Boolean functions. Hence, from the very beginning development of artificial neurons was closely connected with Boolean logic. In fact, the first neurons including the perceptron operated with Boolean input/output mappings. Later, when continuous inputs were introduced, a neuron still produced a binary output. A continuous output was introduced along with a sigmoid activation in 1980s. However, what about multiple-valued discrete input/output mappings? If two-valued input/output mappings can be presented in terms of Boolean logic, can we do the same with multiple-valued input/output mappings in terms of multiple-valued logic? Is so, can this representation be generalized for the continuous input/output mappings? If multi-valued input/output mappings can be represented in the way similar to Boolean input/output mappings, can such a representation be used for creation of a multi-valued neuron? Let us answer these questions.

I. Aizenberg: Complex-Valued Neural Networks with Multi-Valued Neurons, SCI 353, pp. 55–94.
springerlink.com                                    © Springer-Verlag Berlin Heidelberg 2011

## 2.1.1  Multiple-Valued Logic over the Field of Complex Numbers

From the time when a concept of multiple-valued logic was suggested by Jan Łu-kasiewicz in 1920 [59], the values of multiple-valued logic are traditionally encoded by integers. While in Boolean logic there are two truth values ("False" and "True" or 0 and 1), in multiple-value logic there are $k$ truth values. Thus, in Boolean logic, truth values are elements of the set $K_2 = \{0,1\}$. In $k$-valued logic, it was suggested to encode the truth values by elements of the set $K = \{0,1,...,k-1\}$. Particularly, it is important that for $k=2, K=K_2$. Thus, in classical multiple-valued ($k$-valued) logic, a function of $k$-valued logic is $f(x_1,...,x_n): K^n \to K$.

We have also seen that in many engineering applications and particularly in neural networks, it is very convenient to consider the Boolean alphabet $E_2 = \{1,-1\}$ instead of $K_2 = \{0,1\}$. For example, this is important for the ability to use both Hebbian learning rule (1.3) and error-correction learning rule (1.17), where a value of input is the essential multiplicative term participating in the weight adjustment. Since in $E_2$ values of two-valued logic are normalized (their absolute value is 1), it is possible to use them as multiplicative terms in the learning rule. In $K_2$, values of two-valued logic are not normalized. The alphabet $E_2 = \{1,-1\}$ is also more appropriate for the use if the sign function as an activation function of the threshold neuron. We have also seen (Section 1.1) that there is a simple linear connection between alphabets $K_2$ and $E_2$. (Just to recall: if $y \in K_2$ then $x = 1-2y \in E_2$, and if $x \in E_2$ then $y = -(x-1)/2 \in K_2$, respectively. Hence, $0 \leftrightarrow 1$, $1 \leftrightarrow -1$).

However, it should be very interesting to understand, does any deeper and more logical mathematical background behind the connectivity of alphabets $K_2$ and $E_2$ exist? If so, can the same mathematical background be used to create a normalized $k$-valued alphabet from $K = \{0,1,...,k-1\}$, which is not normalized?

A positive answer to this question could make it possible to generalize principles of Boolean logic and Boolean threshold logic for the multiple-valued case. This, in turn, could make it possible to consider a neuron whose input/output mapping is described by a function not of Boolean, but of multiple-valued logic, which could be very important, for example, for solving multi-class classification problems.

A very beautiful idea leaded to answers to these questions was suggested by Naum Aizenberg in early 1970s. It was presented in [32-37] and then summarized in [60].

Let $M$ be an arbitrary additive group[1] and its cardinality is not lower than $k$. Let $A_k = \{a_0, a_1, ..., a_{k-1}\}$, $A_k \subseteq M$ be a structural alphabet.

**Definition 2.3** [37, 60]. Let us call a function $f(x_1, ..., x_n) \mid f : A_k^n \to A_k$ of $n$ variables (where $A_k^n$ is the nth Cartesian power of $A_k$) a function of $k$-valued logic over group $M$.

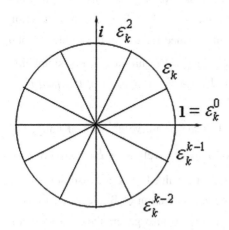

**Fig. 2.20** $k$th roots of unity $1 = \varepsilon_k^0, \varepsilon_k, \varepsilon_k^2, ..., \varepsilon_k^{k-1}$ are located on the unit circle. They form a structural alphabet of multiple-valued logic over the field of complex numbers

It is very easy to check, that a classical definition of a function of $k$-valued logic follows from Definition 2.3. Indeed, the set $K = \{0, 1, ..., k-1\}$ is an additive group with respect to mod $k$ addition. If $K$ is taken as a group $M$ and if the structural alphabet $A_k = K$, then any function $f(x_1, ..., x_n) : K^n \to K$ is a function of $k$-valued logic over the group $K$ according to Definition 2.3.

Let us now take the additive group of the field of complex numbers $\mathbb{C}$ as a group $M$. Evidently this group with respect to addition of complex numbers is infinite, and it contains all elements from $\mathbb{C}$ coinciding with it. As it is well known from algebra, there are exactly $k$ $k$th roots of unity. The root $\varepsilon_k = e^{i 2\pi / k}$ ($i$ is an imaginary unity) is called a *primitive* $k$th root of unity. The rest $k$-1 roots can be obtained from $\varepsilon_k$ by taking its 0th, 2nd, 3rd, ..., $k$-1st powers. Thus, we obtain the

---

[1] Let us recall just for those readers who are less familiar with abstract algebra that the set $A$ is called a *group* with respect to the operation $\circ$, if the following conditions hold: 1) this set is closed with respect to this operation; 2) this operation is associative ($\forall x, y, z \in A : (x \circ y) \circ z = x \circ (y \circ z)$); 3) there exist a neutral element $e \in A$ with respect to the operation $\circ$, which means that $\exists e \in A, \forall x \in A : x \circ e = x$; 4) each element from $A$ has an inverse element with respect to the operation $\circ$ ($\forall x \in A, \exists \tilde{x} \in A : x \circ \tilde{x} = e$). If additionally, $\forall x, y \in A : x \circ y = y \circ x$, $A$ is called a commutative (or Abelian) group. A group is called *additive* when it is a group with respect to addition, and it is called *multiplicative* when it is a group with respect to multiplication.

set $E_k = \left\{ 1 = \varepsilon_k^0, \varepsilon_k, \varepsilon_k^2, ..., \varepsilon_k^{k-1} \right\}$, of all $k$th roots of unity. Since the $j$th of $k$th

roots      of      unity      $\varepsilon_k^j = \left( e^{i2\pi/k} \right)^j = e^{i2\pi j/k}$; $j = 0, 1, ..., k-1$,      then

$\varepsilon_k^j = e^{i\varphi_j}$; $\varphi_j = 2\pi j / k$; $j = 0, 1, ..., k-1$ and therefore all $k$th roots of unity
are located on the unit circle (see Fig. 2.20, which is the same as Fig. 1.19, we just
put it here again for the reader's convenience).

Since the set $E_k$ contains exactly $k$ elements, we may use this set as a structural      alphabet      in      terms      of      Definition      2.3.      Thus,      any      function
$f\left( x_1, ..., x_n \right) : E_k^n \to E_k$ is a function of $k$-valued logic over the additive group
of the field of complex numbers $\mathbb{C}$ according to Definition 2.3. For simplicity,
we will call any function $f\left( x_1, ..., x_n \right) : E_k^n \to E_k$ a *function of k-valued logic*
*over the field of complex numbers* $\mathbb{C}$.

If $k=2$, then $E_2 = \left\{ 1, -1 \right\}$. Indeed, -1 is the primitive 2nd root of unity, and
$1 = -1^0$ is the second of two 2nd roots of unity. As well as in $E_2$ values of two-
valued logic are normalized, in $E_k$ values of $k$-valued logic are also normalized
for any $k$. Thus, Definition 2.3 generalizes consideration of Boolean functions in
both alphabets $K_2 = \left\{ 0, 1 \right\}$ and $E_2 = \left\{ 1, -1 \right\}$. It also generalizes consideration
of multiple-valued functions in both alphabets $K = \left\{ 0, 1, ..., k-1 \right\}$ and
$E_k = \left\{ \varepsilon_k^0, \varepsilon_k, \varepsilon_k^2, ..., \varepsilon_k^{k-1} \right\}$. Evidently, there is a one-to-one correspondence between sets $K$ and $E_k$, and any function $K^n \to K$ can be represented as a func-
tion $E_k^n \to E_k$ and vice versa.

As we have seen in Chapter 1, the use of the alphabet $E_2$ in the threshold
neuron is very important for definition of its activation function and for derivation
of its learning algorithms. We will show now how important is that approach,
which we have just presented, for multiple-valued threshold logic and multi-
valued neurons.

## 2.1.2  Multiple-Valued Threshold Functions over the Field of Complex Numbers

The threshold activation function is a function of the sign of its argument, the
function *sgn*. Depending on the sign of a weighted sum, the threshold activation
function equals 1 if the weighted sum is non-negative or to -1 if it is negative. As
we have already seen, a threshold function is also defined using the function *sgn*.
Let $f : T \to E_2$ (where $T = E_2^n$ or $T \subseteq \mathbb{R}^n$).

**Definition 2.4.** If there exist such real-valued weights $w_0, w_1, ..., w_n$ that for any $(x_1, ..., x_n) \in T$    $\operatorname{sgn}(w_0 + w_1 x_1 + ... + w_n x_n) = f(x_1, ..., x_n)$,    then the function $f(x_1, ..., x_n)$ is called a *threshold* function.

For a multiple-valued function $f(x_1, ..., x_n) : E_k^n \to E_k$, Definition 2.4 is not applicable. The function *sgn* is two-valued, while we consider now a $k$-valued function. Our function $f(x_1, ..., x_n) : E_k^n \to E_k$ is complex-valued. A complex number does not have a sign. However, a complex number has its argument. A codomain of our function is $E_k = \{\varepsilon_k^0, \varepsilon_k, \varepsilon_k^2, ..., \varepsilon_k^{k-1}\}$ where $\varepsilon_k = e^{i2\pi/k}$ and $\varepsilon_k^j = e^{i\varphi_j}; \varphi_j = 2\pi j / k; j = 0, 1, ..., k-1$. All the $\varepsilon_k^j; j = 0, 1, ..., k-1$ being the $k$th roots of unity are located on the unit circle (see Fig. 2.20) and $|\varepsilon_k^j| = 1; j = 0, 1, ..., k-1$. Thus, all $\varepsilon_k^j; j = 0, 1, ..., k-1$ have a unitary abso-

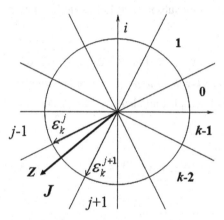

lute value. However, they all have different arguments, which determine their uniqueness! Indeed, $\varepsilon_k^j = e^{i\varphi_j}; j = 0, 1, ..., k-1$   has the argument $\varphi_j = 2\pi j / k$. Thus, while a value of a function of two-valued logic is determined by its sign, a value of a function of multiple-valued logic over the field of complex numbers is determined by its argument. This consideration leaded Naum Aizenberg and his co-authors [32] for the following definition of the "$k$-valued sign" (or as they initially called it "complex sign") function. If $z \in \mathbb{C}$, then

**Fig. 2.21** Definition of multiple-valued activation function (50). $P(z) = e^{i2\pi j/k}$

$$P(z) = \operatorname{CSIGN}(z) = \varepsilon_k^j, 2\pi j / k \leq \arg z < 2\pi (j+1)/k. \qquad (2.50)$$

Function (2.50) is illustrated in Fig. 2.21. The complex plane is divided into $k$ equal sectors by the lines passing through the origin and points on the unit circle corresponding to the $k$th roots of unity. Sectors are enumerated in the natural way: $0^{th}$, $1^{st}$, $2^{nd}$, ..., $k$-$1^{st}$. The $j$th sector is limited by the boarders originating in the origin and crossing the unit circle at the points corresponding to the $k$th roots of unity $\varepsilon_k^j$ and $\varepsilon_k^{j+1}$. If a complex number z is located in the jth sector, which means that $2\pi j / k \leq \arg z < 2\pi (j+1)/k$ , then $P(z) = e^{i2\pi j/k}$ .

In fact, function (2.50) is not a sign or "complex sign" function because a complex number does not have any sign. Considering that the "complex sign" name and, respectively, the CSIGN notation were not the most successful for this function, its inventor N. Aizenberg later suggested to call it a k-valued predicate and to use the notation $P(z)$ (P – predicate). This notation is used for function (2.50) from mid 1990s. Particularly, the function sgn is a two-valued predicate in terms of logic.

Let us consider a function $f(x_1,...,x_n):T \rightarrow E_k;T \subseteq E_k^n$ of k-valued logic. Thus, generally speaking $f(x_1,...,x_n)$ can be fully defined (if $T = E_k^n$) or partially defined (if $T \subset E_k^n$) function of n variables in k-valued logic. A fully defined function is defined on the whole nth Cartesian power of the set $E_k$, while a partially defined function is defined only on a subset of it. Let us define now a multiple-valued threshold function.

**Definition 2.5** [32-37, 60]. The function $f(x_1,...,x_n):T \rightarrow E_k;T \subseteq E_k^n$ of k-valued logic is called a *threshold function of k-valued logic* (or *multiple-valued (k-valued) threshold function*) if there exist n+1 complex numbers $w_0, w_1,..., w_n$ such that for any $(x_1,...,x_n) \in T$

$$f(x_1,...,x_n) = P(w_0 + w_1 x_1 +...+ w_n x_n),\qquad(2.51)$$

where $P(z)$ is function (2.50).

The vector $W = (w_0, w_1,..., w_n)$ is called a *weighting vector* of the function $f$. We also say that the weighting vector $W$ *implements* the function $f$.

It is important that on the one hand, Definition 2.5 "covers" Definition 2.4, but on the other hand, it drastically extends a set of functions that can be represented using n+1 weights by adding multiple-valued threshold functions to Boolean threshold functions. A Boolean threshold function is a particular case of multiple-valued threshold function.

Indeed, if $k = 2$ in (2.50), then this activation function is transformed to

$$P(z) = \begin{cases} 1; 0 \leq \arg z < \pi \\ -1; \pi \leq \arg z < 2\pi. \end{cases}\qquad(2.52)$$

It divides the complex plane into two sectors – the top half-plane ("1") and the bottom half-plane ("-1"). A Boolean function $f(x_1,...,x_n): E_2^n \rightarrow E_2$, which allows representation (2.51)-(2.52) was called [33, 35, 37, 60] a *Boolean complex-threshold function*. It was proven in [33, 35, 37, 60] that the set of Boolean complex-threshold functions coincides with the set of Boolean threshold functions (in terms of Definition 2.4). This means that for $k=2$, any $k$-valued threshold function is a Boolean threshold function and vice versa.

However, when $k>2$, Definition 2.5 and, respectively, representation (2.51)-(2.50) make it possible to represent multiply-valued functions using $n+1$ complex-valued weights in the same manner as Boolean threshold functions.

Let us consider the following example. The set $K = \{0,1,...,k-1\}$ is a linearly ordered set with respect to the "<" relation. In fact, $0 < 1 < 2 < ... < k-1$. Let us transfer this linear order onto the set $E_k$ in the following natural way: $\varepsilon_k^0 \prec \varepsilon_k \prec ... \prec \varepsilon_k^{k-1}$. Evidently, this is a linear order with respect to the "<" relation applied to the arguments of the corresponding $k$th roots of unity $\arg \varepsilon_k^0 < \arg \varepsilon_k < ... < \arg \varepsilon_k^{k-1}$. Let us consider well known Post functions $\max(y_1, y_2); y_i \in K; i=1,2$ and $\min(y_1, y_2); y_i \in K; i=1,2$. These functions become, respectively the following functions of $k$-valued logic over the field of complex numbers $f_{\max}(x_1, x_2) = \max(x_1, x_2); x_i \in E_k; i=1,2$ and $f_{\min}(x_1, x_2) = \min(x_1, x_2); x_i \in E_k; i=1,2$. Let $k=3$. It is easy to check that the weighting vector $W = (-2-4\varepsilon_3, 4+5\varepsilon_3, 4+5\varepsilon_3)$ implements the function $f_{\max}(x_1, x_2)$ (see Table 2.8 and Fig. 2.22a), and the weighting vector $W = (2+4\varepsilon_3, 5+4\varepsilon_3, 5+4\varepsilon_3)$ implements the function $f_{\min}(x_1, x_2)$ (see Table 2.9 and Fig. 2.22b), and therefore they are threshold functions of 3-valued logic.

These simple examples show that the extension of the set of threshold functions by multiple-valued functions opened absolutely new perspectives in threshold logic, in neural networks and in solving multi-class classification problems. This extension became possible after multiple-valued logic over the field of complex numbers was introduced. This bold jump to the complex domain was really historical. As we will see from the rest of this book it really resulted in many new efficient solutions and first of all in the creation of the multi-valued neuron, which will be considered in Section 2.2. In Section 2.3, we will consider in detail a linear separability of an $n$-dimensional space, which is determined by function (2.50) and a weighting vector implementing a corresponding multiple-valued threshold function.

**Table 2.8** Post function $f_{max}(x_1,x_2)$ is a 3-valued threshold function (a threshold function of 3-valued logic) with the weighting vector $W=(-2-4\varepsilon_3,4+5\varepsilon_3,4+5\varepsilon_3)$

| # | $x_1$ | $x_2$ | $z=w_0+w_1x_1+w_2x_2$ | $\arg(z)$ | $P(z)$ | $f_{max}(x_1,x_2)$ |
|---|---|---|---|---|---|---|
| 1 | $\varepsilon_3^0$ | $\varepsilon_3^0$ | $6+6\varepsilon_3$ | 1.0471 | $\varepsilon_3^0$ | $\varepsilon_3^0$ |
| 2 | $\varepsilon_3^0$ | $\varepsilon_3^1$ | $2+5\varepsilon_3+5\varepsilon_3^2$ | $\pi$ | $\varepsilon_3^1$ | $\varepsilon_3^1$ |
| 3 | $\varepsilon_3^0$ | $\varepsilon_3^2$ | $7+\varepsilon_3+4\varepsilon_3^2$ | 5.7596 | $\varepsilon_3^2$ | $\varepsilon_3^2$ |
| 4 | $\varepsilon_3^1$ | $\varepsilon_3^0$ | $2+5\varepsilon_3+5\varepsilon_3^2$ | $\pi$ | $\varepsilon_3^1$ | $\varepsilon_3^1$ |
| 5 | $\varepsilon_3^1$ | $\varepsilon_3^1$ | $2+4\varepsilon_3+10\varepsilon_3^2$ | 3.6652 | $\varepsilon_3^1$ | $\varepsilon_3^1$ |
| 6 | $\varepsilon_3^1$ | $\varepsilon_3^2$ | $-2+9\varepsilon_3^2$ | 4.5223 | $\varepsilon_3^2$ | $\varepsilon_3^2$ |
| 7 | $\varepsilon_3^2$ | $\varepsilon_3^0$ | $7+\varepsilon_3+4\varepsilon_3^2$ | 5.7596 | $\varepsilon_3^2$ | $\varepsilon_3^2$ |
| 8 | $\varepsilon_3^2$ | $\varepsilon_3^1$ | $3+9\varepsilon_3^2$ | 4.5223 | $\varepsilon_3^2$ | $\varepsilon_3^2$ |
| 9 | $\varepsilon_3^2$ | $\varepsilon_3^2$ | $8-4\varepsilon_3+8\varepsilon_3^2$ | 5.2359 | $\varepsilon_3^2$ | $\varepsilon_3^2$ |

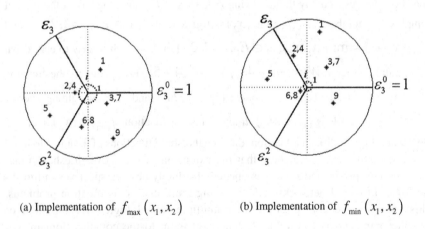

(a) Implementation of $f_{max}(x_1,x_2)$    (b) Implementation of $f_{min}(x_1,x_2)$

**Fig. 2.22** Implementation of the Post functions $\max(y_1,y_2)$ and $\min(y_1,y_2)$ as 3-valued threshold functions $f_{max}(x_1,x_2)$ and $f_{min}(x_1,x_2)$. Locations of the weighted sums in the complex plane are shown by *. Numbers nearby these locations are the numbers of the corresponding input samples (see Table 2.8 and Table 2.9).

Before we will move to the multi-valued neuron, it is important to mention that Definition 2.5 determines a discrete-valued threshold function of multiple-valued logic. Indeed, it is a function from $T$ to $E_k$, where $T \subseteq E_k^n$.

**Table 2.9** Post function $f_{\min}(x_1, x_2)$ is a 3-valued threshold function (a threshold function of 3-valued logic) with the weighting vector $W = (2+4\varepsilon_3, 4+5\varepsilon_3, 4+5\varepsilon_3)$

| # | $x_1$ | $x_2$ | $z = w_0 + w_1 x_1 + w_2 x_2$ | $\arg(z)$ | $P(z)$ | $f_{\min}(x_1, x_2)$ |
|---|---|---|---|---|---|---|
| 1 | $\varepsilon_3^0$ | $\varepsilon_3^0$ | $10 + 14\varepsilon$ | 1.3282 | $\varepsilon_3^0$ | $\varepsilon_3^0$ |
| 2 | $\varepsilon_3^0$ | $\varepsilon_3^1$ | $6 + 13\varepsilon + 5\varepsilon^2$ | 1.9794 | $\varepsilon_3^0$ | $\varepsilon_3^0$ |
| 3 | $\varepsilon_3^0$ | $\varepsilon_3^2$ | $11 + 9\varepsilon + 4\varepsilon^2$ | 0.7662 | $\varepsilon_3^0$ | $\varepsilon_3^0$ |
| 4 | $\varepsilon_3^1$ | $\varepsilon_3^0$ | $6 + 13\varepsilon + 5\varepsilon^2$ | 1.9794 | $\varepsilon_3^0$ | $\varepsilon_3^0$ |
| 5 | $\varepsilon_3^1$ | $\varepsilon_3^1$ | $2 + 12\varepsilon + 10\varepsilon^2$ | 2.9514 | $\varepsilon_3^1$ | $\varepsilon_3^1$ |
| 6 | $\varepsilon_3^1$ | $\varepsilon_3^2$ | $7 + 8\varepsilon + 9\varepsilon^2$ | 3.6652 | $\varepsilon_3^1$ | $\varepsilon_3^1$ |
| 7 | $\varepsilon_3^2$ | $\varepsilon_3^0$ | $11 + 9\varepsilon + 4\varepsilon^2$ | 0.7662 | $\varepsilon_3^0$ | $\varepsilon_3^0$ |
| 8 | $\varepsilon_3^2$ | $\varepsilon_3^1$ | $7 + 8\varepsilon + 9\varepsilon^2$ | 3.6652 | $\varepsilon_3^1$ | $\varepsilon_3^1$ |
| 9 | $\varepsilon_3^2$ | $\varepsilon_3^2$ | $12 + 4\varepsilon + 8\varepsilon^2$ | 5.7596 | $\varepsilon_3^2$ | $\varepsilon_3^2$ |

Since $E_k$ is a set of $k$th roots of unity, both domain and co-domain of a multiple-valued threshold function in terms of Definition 2.5 are discrete. However, in many classification problems features are continuous. In many prediction problems we need to have deal with time series of continuous data. Let us modify Definition 2.3 of a multiple-valued function over the field of complex numbers and Definition 2.5 of a multiple-valued threshold function, to be able to consider functions with a continuous domain and a continuous co-domain.

We have already seen that any function $f(x_1, ..., x_n) : K^n \to K$ of traditional multiple-valued logic can be easily represented as a function $f(x_1, ..., x_n) : E_k^n \to E_k$ of multiple-valued logic over the field of complex numbers, and there exist a one-to-one correspondence between functions of traditional multiple-valued logic and functions of multiple-valued logic over the field of complex numbers. A great advantage of the latter class of functions is that there are threshold functions among them.

Let $O$ be the continuous set of the points located on the unit circle. Let either $T \subseteq E_k$ or $T \subseteq O$.

**Definition 2.6** [61]. A function $f(x_1, ..., x_n):T^n \rightarrow E_k$ is called a function of $k$-valued logic over the field of complex numbers (or simply a $k$-valued function).

The co-domain of $f$ is discrete, while its domain can be either discrete or continuous. In general, its domain may be even hybrid. It should be mentioned that if some function $f(y_1, ..., y_n)$ is defined on the bounded subdomain $D^n, D \subset \mathbb{R}$, which means that this function is $f(y_1, ..., y_n):D^n \rightarrow K; y_j \in [a_j, b_j], a_j, b_j \in \mathbb{R}, j = 1, ..., n$, then it can be easily transformed to $f:O^n \rightarrow K$ by a linear transformation applied to each variable

$$y_j \in [a_j, b_j] \Rightarrow \varphi_j = \frac{y_j - a_j}{b_j - a_j} \alpha \in [0, 2\pi[,$$

$$j = 1, ..., n; 0 < \alpha < 2\pi, \tag{2.53}$$

and then $x_j = e^{i\varphi_j} \in O, j = 1, 2, ..., n$ is the complex number located on the unit circle[2]. Since, there exists a one-to-one correspondence between $K$ and $E_k$, then we obtain a function $f(x_1, ..., x_n):O^n \rightarrow E_k$. Now we can modify Definition 2.5 as follows.

**Definition 2.7.** A function $f(x_1, ..., x_n):T^n \rightarrow E_k$ is called a $k$-valued threshold function if there exists a complex-valued vector $(w_0, w_1, ..., w_n)$ such that for all $(x_1, ..., x_n)$ from the domain of the function $f(x_1, ..., x_n)$ equation (2.51) holds, so $f(x_1, ..., x_n) = P(w_0 + w_1 x_1 + ... + w_n x_n)$, where $P(z)$ is function (2.50).

In the next section, right after we will define the continuous multi-valued neuron, we will consider implementation of functions whose both domain and co-domain are continuous.

---

[2] The interval $[0, 2\pi[$ in (2.53) is open from the right side. This is important to avoid a collision, which follows from the fact that arguments 0 and $2\pi$ determine the same point on the unit circle. Since $0 < \alpha < 2\pi$ in (2.53), this guarantees that $0 \leq \varphi_j < 2\pi, j = 1, ..., n$, and a point on the unit circle corresponding to the maximal value of $y_j, j = 1, ..., n$ does not coincide with a point corresponding to its minimal value.

## 2.2 Multi-Valued Neuron (MVN)

### 2.2.1 Discrete MVN

The discrete multi-valued neuron was introduced in 1992 [38] by N. Aizenberg and the author of this book. To be more specific, we have to mention that the term "multi-valued neuron" was used in [38] for the first time. However, it was in fact introduced by N. Aizenberg and his co-authors 20 years earlier, in 1971 [33, 35], as an "element of multiple-valued threshold logic" or "multi-valued threshold element". Initially, the multi-valued neuron was considered as a neural element implementing only pure discrete input/output mapping $E_k^n \rightarrow E_k$. In 2007, the author of this book and Claudio Moraga introduced the continuous multi-valued neuron [62]. Taking into account considerations given in [60] and just extended in the previous section, the author of this book suggested in [61] the following adjusted definition of the discrete multi-valued neuron.

Let us consider a neuron (Fig. 1.3) with complex-valued inputs and output and with complex-valued weights. Inputs and output of this neuron are located on the unit circle. Moreover, its output is $k$-valued and it is one of the $k$th roots of unity $\varepsilon^j = e^{i2\pi j/k}$, $j \in \{0,1,...,k-1\}$, $i$ is an imaginary unity, while its inputs can be arbitrary complex numbers located on the unit circle. Weights can be arbitrary complex numbers.

**Definition 2.8.** The *discrete multi-valued neuron* (MVN) is a neuron with the activation function (2.50). It implements input/output mapping between $n$ inputs and a single output according to (2.51).

Hence, if $x_1,....,x_n$ are the MVN inputs, then the MVN output is

$$f(x_1,...,x_n) = P(w_0 + w_1 x_1 + ... + w_n x_n), \text{ where } P(z) \text{ is function (2.50) that}$$
is

$$P(z) = \varepsilon_k^j, 2\pi j / k \leq \arg z < 2\pi(j+1)/k.$$

It follows from Definition 2.7 that an input/output mapping of the discrete MVN is always described by some $k$-valued threshold function of $n$ variables $f(x_1,...,x_n)$.

Thus, if $z = w_0 + w_1 x_1 + ... + w_n x_n$ is the weighted sum of the MVN inputs, $P(z)$ is the MVN output. This means that any $k$-valued threshold function can be implemented by a single discrete MVN. Moreover, it follows from Definition 2.7 and Definition 2.8 that the discrete MVN cannot implement input/output mappings that are not described by some $k$-valued threshold function.

The functionality of the discrete MVN is higher than the functionality of other neurons. As we have seen, for $k=2$ it coincides with the functionality of the threshold neuron. But for $k > 2$ it is always higher because the discrete MVN can implement multiple-valued threshold functions, while a neuron with the threshold activation function cannot. Returning to the example with Post functions max and min considered in Section 2.1, we see that the discrete MVN really can easily implement multiple-valued input/output mappings. The MVN functionality is also higher than the one of a neuron with the sigmoid activation function. Additionally to the 2-D illustration of the activation function (2.50), let us consider its 3-D interpretation (see Fig. 2.23). There is an example of 3-D interpretation of the activation function (2.50) for $k = 16$. In Fig. 2.23, while the horizontal plane is the complex plane $\mathrm{Re}(z)\,\mathrm{Im}(z)$, the vertical axis is $\arg z$, the argument of the complex number $z$. The 3-D graph in Fig. 2.23 looks like infinite spiral stairs.

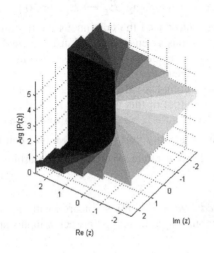

Each stair corresponds to one of the sectors in which the complex plane is divided by (2.50) (see also Fig. 2.21). Since, we consider $k$=16, there are exactly 16 stairs in Fig. 2.23. Each stair is limited by its corner and borders from two sides (the angle between these borders is $2\pi/16$ in the particular case of $k$=16 and $2\pi/k$ in general). However, each stair does not have any border from the third side, and it is infinite in the corresponding direction. This makes the multi-valued neuron much more flexible than other neurons because for each desired output

**Fig. 2.23** 3-D model of the discrete MVN activation function (50), $k$=16

$$\varepsilon_k^j, \quad j = 0, 1, ..., k-1, \quad \text{we}$$

have infinitely many opportunities to allocate a weighted sum in the $j$th stair (or sector in Fig. 2.21). This is a very important advantage of the multi-valued neuron over real-valued neurons.

For many years, there were no commonly used approaches to represent multiple-valued input/output mappings. For example, both sigmoid activation functions (logistic - (1.22) and hyperbolic tangent - (1.23) ) are continuous, but they are not suitable for representation of multiple-valued input/output mappings. If we need to have some certain desired output $d$, this means that there will be just a single acceptable value of the weighted sum, for which either function (1.22) or (1.23) takes the desired value $d$. Actually, this means that implementation of discrete input/output mappings that differ from either logistic or *tanh* function using a single neuron can be possible just occasionally, moreover very rarely. In 1995, J. Si and A. Michel proposed [63] to divide that part of a sigmoid curve, corresponding to the argument interval [0, 1] for the logistic function or, respectively, [-1, 1] for the *tanh* function, into $k$ subintervals, to be able to implement $k$-valued discrete input/output mappings. In this case, there is some limited level of flexibility, because to ensure that a neuron produces a desired output value, it is possible to fit a weighted sum into some small interval (of the length $1/k$ for the logistic function or $(1-(-1))/k = 2/k$ for the *tanh* function). However, these intervals are in fact very small and therefore the flexibility of a single neuron to adapt to some highly nonlinear input/output mapping is very limited. This means that to learn those input/output mappings that are different from a sigmoid activation function, a network will be needed because the functionality and the flexibility of a single neuron are limited.

The multi-valued neuron is incompatibly more flexible because for any desired output we have infinite amount of opportunities to fit a weighted sum in an infinite sector corresponding to the desired output. This is possible due to the fact that the MVN operates in the complex domain. This advantage of the discrete MVN is illustrated in Fig. 2.24.

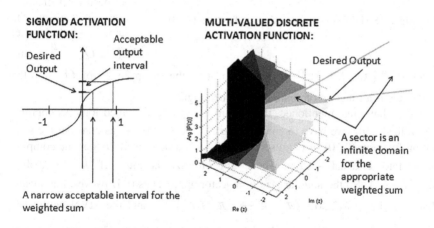

**Fig. 2.24** Multi-valued activation function vs. sigmoid activation function. The multi-valued neuron is much more flexible than a sigmoidal neuron.

## 2.2.2  Continuous MVN

The ability to implement input/output mappings that take continuous values is very important for solving prediction problems because, for example, time series that we may need to predict are typically continuous-valued.

The continuous MVN was introduced in 2007 by the author of this book and Claudio Moraga in [62]. The discrete MVN implements an input/output mapping described by a threshold function of $k$-valued logic. If $k \rightarrow \infty$, then $k$-valued logic becomes continuous-valued. Since we consider multiple-valued logic over the field of complex numbers, let us encode values of continuous-valued logic by numbers located on the unit circle like we do this for finite-valued logic. If in the latter case, we use $k$th roots of unity to encode values of $k$-valued logic, in the continuous-valued logic over the field of complex numbers its values can be arbitrary complex numbers located on the unit circle. Let $O$ be the set of numbers located on the unit circle. Let $T \subset O^n$. Then any

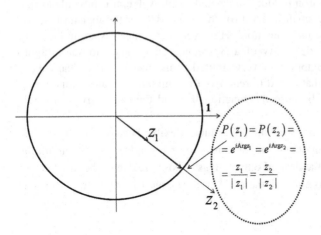

**Fig. 2.25** MVN continuous activation function (2.54)

function $f(x_1,...,x_n):T \to O$, which maps the subspace $T$ of $O^n$ into the unit circle is a function of continuous-valued logic.

As we have just mentioned, if we want to consider $k$-valued logic as continuous-valued logic, we have to consider the case $k \to \infty$. Let us consider the activation function (2.50) of the discrete multi-valued neuron. It divides the complex plane into $k$ equal sectors whose angular size is $2\pi/k$. If $k \to \infty$, then $2\pi/k \to 0$. So the angular size of a sector approaches 0. However, this means that according to (2.50) $2\pi j/k \underset{k \to \infty}{\to} 2\pi(j+1)/k$, and the MVN output becomes a projection of the weighted sum onto the unit circle. This leads us to the following definition of the continuous MVN activation function

$$P(z) = e^{iArg\ z} = z / |z|, \tag{2.54}$$

where Arg $z$ is the main value of the argument of the complex number $z$ (the main value of the argument is located in the interval $[0, 2\pi[$) and $|z|$ is its absolute value. The activation function (2.54) is illustrated in Fig. 2.25. This activation function determines the *continuous-valued MVN*, which is a neuron with complex-valued weights (that can be arbitrary complex numbers) and inputs and output that are complex numbers located on the unit circle.

Suppose $D \subset \mathbb{R}$ and $D^n$ is the domain of some real-valued function $f(y_1,\ ...,\ y_n)$ bounded on its whole domain. This means that the function's co-domain is $Y = [a,b]; a,b \in \mathbb{R}$. Let the function's domain is also bounded, that is each variable is located within some bounded interval. Hence $f(y_1, ..., y_n) : D^n \rightarrow Y; y_j \in [a_j, b_j], a_j, b_j \in \mathbb{R}, j = 1, ..., n$. To implement this function using the continuous MVN, we need to transform it to a function $f : O^n \rightarrow O$. This can be easily done by a linear transformation (2.53) applied to each variable and by the similar transformation applied to the function values:

$$y_j \in [a_j, b_j] \Rightarrow \varphi_j = \frac{y_j - a_j}{b_j - a_j} \alpha \in [0, 2\pi[;$$

$$j = 1, ..., n; 0 < \alpha < 2\pi;$$

$$f(y_1, ..., y_n) \in [a,b] \Rightarrow \varphi = \frac{f(y_1, ..., y_n) - a}{b - a} \beta \in [0, 2\pi[;$$

$$0 < \beta < 2\pi,$$

(2.55)

and then $x_j = e^{i\varphi_j} \in O, j = 1, 2, ..., n$ and $e^{i\varphi} \in O$ are the complex numbers located on the unit circle. It is important to mention that the interval $[0, 2\pi[$ in (2.55) is open from the right side as it was in (2.53). This is important to avoid a collision, which follows from the fact that arguments 0 and $2\pi$ determine the same point on the unit circle.

It follows from the last considerations that any bounded real-valued function of real variables defined on the bounded domain can be transformed into the complex-valued function $f : O^n \rightarrow O$. Then we can speak about the implementation of such a function either using the continuous MVN or MVN-based neural network. It is important that it is always possible to invert the linear transformation (2.55) and to return back from the complex-valued function $f : O^n \rightarrow O$ to the real-valued function $f : D^n \rightarrow Y$ if all $a_j, b_j, j = 1, ..., n$ and $a, b, \alpha, \beta$ are known. For example, if $f(x_1, ..., x_n) = e^{i\varphi} \in O$ then

$$y = \frac{\varphi(b - a)}{\beta} + a \in Y = [a, b]$$

(2.56)

is the value of the initial real-valued function. Transformation (2.56) is very useful, for example, when we need to transform predicted data to their initial real-valued scale. We will use this transformation in Chapters 5-6.

A 3-D model of the continuous MVN activation function (2.54) is shown in Fig. 2.26. Comparing it to the model of the discrete MVN activation function (Fig. 2.23), we see that there are no more spiral stairs (since there are no more sectors on the complex plane), and there is a continuous and infinite helical surface. Likewise in Fig. 2.23, the horizontal plane in Fig. 2.26 is the complex plane $\mathrm{Re}(z)\,\mathrm{Im}(z)$, and the vertical axis is $\arg z$

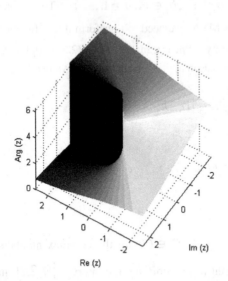

**Fig. 2.26** 3-D model of the continuous MVN activation function (2.54)

The functionality of a single continuous MVN is higher than the one of a single sigmoidal neuron. The continuous MVN is also more flexible than the sigmoidal neuron. Suppose we need to implement some continuous-valued function using a single sigmoidal neuron. To ensure that the neuron produces the desired output $d$, we have to ensure that the weighted sum equals to a single acceptable value, which is determined by the inverse function of either (1.22) ($\varphi^{-1}$) or (1.23) (*arctanh*) depending on which one of them is used. It follows from this consideration that it is quite difficult to implement using a single sigmoidal neuron any continuous-valued function except its own activation function. If we need to solve the same task using the continuous MVN, we have the unlimited choice of potential acceptable values of weighted sums. If our desired output is $e^{i\varphi} \in O$, then the ray $L_\varphi$ generating the angle $\varphi$ (counting in the counterclockwise direction) between the real axis and itself, is the set of acceptable weighted sums. Indeed, $\forall z \in L_\varphi \ \arg z = \varphi$, and according to (2.54) the output of the continuous MVN is $e^{i\varphi}$. For example, in Fig. 2.25, the weighted sums $z_1$ and $z_2$ produce the same output $e^{i\varphi}$, where $\varphi = \arg z_1 = \arg z_2$.

It is important to mention that the implementation of some input/output mapping using a neuron means that this mapping should be learned by the neuron. In Chapter 3, we will consider learning algorithms for MVN.

But first we would like to consider the separation of an $n$-dimensional space, which is established by the MVN and its discrete and continuous activation functions.

## 2.3  Edged Separation of *n*-Dimensional Space

We have introduced the discrete multi-valued neuron as a neural element whose input/output mapping is described by a multiple-valued threshold function. We told that MVN *implements* a multiple-valued threshold function. What does it mean from the geometrical point of view? How the discrete MVN separates an *n*-dimensional space where a multiple-valued threshold function representing its input/output mapping is defined? How the continuous MVN separates an *n*-dimensional space where its input/output mapping is defined? Answers to these questions are very important for understanding of how MVN works and for solving classification problems using MVN and MVN-based neural networks.

We have to mention that this section naturally contains necessarily mathematical considerations that are deeper than the ones in other sections of this book. Those readers who do not want to go to the mathematical details may skip over detailed proofs. Those who interested just in application of MVN and MVN-based neural networks may even skip over this whole section. However, that mathematics, which we use here, does not go beyond linear algebra and basic analytical geometry and we believe that its reading and understanding should not be difficult. But its understanding should be very important for understanding of those specific advantages that characterize MVN and MVN-based neural networks.

Let us first recall that when a neuron with the threshold activation function implements some mapping from *n* inputs to a single output, from the geometrical point of view this means (see Section 1.1, Fig. 1.4a) that there exists a hyperplane, which separates the "1" outputs from the "-1" outputs. The neuron's $n+1$ weights determine the coefficients of the hyperplane equation. This hyperplane separates the *n*-dimensional space $\mathbb{R}^n$ over the field of real numbers into the two subspaces. One of them contains all the points $x_1^{(1)}, x_2^{(1)}, ..., x_n^{(1)}$ marked by 1s (corresponding to the neuron output 1), while another one contains all the points $x_1^{(-1)}, x_2^{(-1)}, ..., x_n^{(-1)}$ marked by -1s (corresponding to the neuron output -1). If our threshold neuron solves some two-class classification problem, this means that a hyperplane determined by the neuron's weights separates two classes. Objects belonging to the first class are located on the one side from the hyperplane, while objects belonging to the second class are located on the opposite side from the hyperplane. Another popular machine learning tool, the support vector machine, works similarly. What about MVN?

### 2.3.1  *Important Background*

Let us return first to the definition of the activation function $P(z)$ (see (2.50) ). It is not defined in the $z=(0,0)$. Let us agree that we always can define $P(0)$ assigning it equal to $\varepsilon_k' \in E_k = \{\varepsilon_k^0, \varepsilon_k, ..., \varepsilon_k^{k-1}\}$. In this case, if $(0,...,0)$ is a weighting vector of some function $f(x_1,...,x_n)$, then there exists $w_0 \in \mathbb{C}, w_0 \neq 0$ such that $(w_0, 0, ..., 0)$ is also a weighting vector of the same function $f(x_1,...,x_n)$. Indeed, if $P(0) = \varepsilon_k'$, then it is always possible to find $w_0 \in \mathbb{C}, w_0 \neq 0$ such that $P(w_0) = \varepsilon_k'$.

We need to consider one important extension of the definition of a $k$-valued threshold function. Let us extend the Definition 2.7 in the following way.

**Definition   2.9.**   We   will   call   any   complex-valued   function $f(x_1,...,x_n):T \to \mathbb{C}; T \subset \mathbb{C}^n$, a complex-valued threshold function, if it is possible to find a complex-valued weighting vector $W = (w_0, w_1, ..., w_n)$, and to define $P(0)$ in such a way that the equation

$$P(f(x_1,...x_n)) = P(w_0 + w_1 x_1 + ... + w_n x_n) \qquad (2.57)$$

(where $P$ is the function (2.50)) is true for all $(x_1,...,x_n) \in T$ (for all $(x_1,...,x_n)$ from the domain of the function $f$).

Evidently, a $k$-valued threshold function (a threshold function of $k$-valued logic, see Definition 2.7) is a particular case of complex-valued threshold function. It is also evident that the set of all $k$-valued threshold functions is a subset of the set of all complex-valued threshold functions for any fixed value of $k$.

It is clear that if $(0, 0, ..., 0)$ is a weighting vector of a function $f$, then such a $w_0 \neq 0$ exists that $(w_0, 0, ..., 0)$ is also a weighting vector of the function $f$. Really, if $P(0) = \varepsilon^t$, then $w_0 \neq 0$ should be chosen in such a way that $P(w_0) = \varepsilon^t$. Moreover, the following statement is true.

**Theorem 2.1.** If $(w_0, 0, ..., 0)$ is a weighting vector of function $f(x_1,...,x_n)$, which is a complex-valued threshold function, and if the domain of $f(x_1,...,x_n)$ is bounded, then a complex number $w_0'$ and a real number $\delta > 0$ exist such that for all the $w_j$, $j = 1,2,...,n$, for which $|w_j| < \delta$, the vector $(w_0', w_1, ..., w_n)$ also is a weighting vector of the function $f$.

**Proof.** Let $P(w_0) = \varepsilon^t$ and let without loss of generality $w_0 = \varepsilon^t$ (see Fig. 2.27). Then $\forall X = (x_1,...,x_n) \in T$:

$P\big(f(X)\big) = P(w_0 + 0 \cdot x_1 + ... + 0 \cdot x_n) = \varepsilon^t$, therefore $P\big(f(X)\big) = const.$

Let $w_0' = \varepsilon^{t+\frac{1}{2}}$, which corresponds to the rotation of $w_0$ by the angle $\dfrac{\pi}{k}$ (see Fig. 2.27). Since the domain of the function $f(x_1, ..., x_n)$ is bounded, $N > 0$ exists such that $|x_j| < N$, $j = 1, 2, ,..., n$. Let $\delta = \dfrac{\sin(\pi / k)}{Nn}$, and taking all the $w_j$, $j = 1, 2, ..., n$ in such a way that $|w_j| < \delta$, we obtain the following:

$$\big|w_1 x_1 + ... + w_n x_n\big| \leq |w_1| \, \| \, x_1 \, | + ... + | \, w_n \, \| \, x_n \, |< N \left(| \, w_1 \, | + ... + | \, w_n \, |\right) < Nn\delta =$$
$$= \sin\left(\pi / k\right),$$

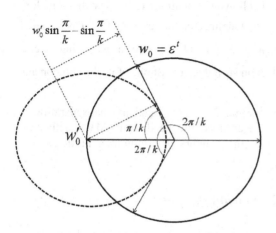

where $\sin\left(\pi / k\right)$ is the distance from the point $w_0'$ to the bound of a sector which it belongs to (see Fig. 2.27). Therefore the sum of vectors (on the complex plane) corresponding to the complex numbers $w_0'$ and $w_1 x_1 + ... + w_n x_n$ for any $\left(x_1, ..., x_n\right) \in T$ is always located within the same semi-open sector that the vector corresponding to $w_0'$. This means that the following correspondence also holds for the

**Fig. 2.27** Illustration to Theorem 2.1

complex numbers $w_0'$ and $w_0 + w_1 x_1 + ... + w_n x_n$: $P(w_0') = P(w_0 + w_1 x_1 + ... + w_n x_n)$. Taking now into account that $P(w_0') = P(w_0)$, it is easy to conclude that if $(w_0, 0, ..., 0)$ is a weighting vector then the vector $(w_0', w_1, ..., w_n)$ also is a weighting vector of the function $f$. Theorem is proven.

According to Theorem 2.1, it is possible to find such a weighting vector $\left(w_0, w_1, ..., w_n\right)$ for a threshold function with a bounded domain that at least one of the components $w_1, ..., w_n$ is not equal to zero. It is also evident that it is possible to find a weighting vector with the same property not only for a $k$-valued threshold function with a bounded domain, but for a function with an arbitrary domain upon a condition that $P(f(x_1, ... x_n)) \neq const$. Really, the last condition involves the following: it is possible to find such vectors $\alpha, \beta \in T$, $\alpha = (\alpha_1, ..., \alpha_n) \neq (\beta_1, ..., \beta_n) = \beta$ that

$$P(f(\alpha)) = P(W(\alpha)) \neq P(W(\beta)) = P(f(\beta)).$$

From the last expression it is evident that $W(\alpha) \neq W(\beta)$, and $W(\alpha) - W(\beta) = w_1(\alpha_1 - \beta_1) + ... + w_n(\alpha_n - \beta_n)$. Taking into account that at least one of the differences $\alpha_j - \beta_j \neq 0; j = 1, ..., n$, we have to conclude that at least one complex number from $w_1, ..., w_n$ is not equal to zero. The following theorem has been proven by the latest considerations.

**Theorem 2.2.** If at least one of the following conditions is true for the complex-valued threshold function $f(x_1, ..., x_n)$ defined on the set $T \subset \mathbb{C}^n$ : $T$ is bounded or $P(f(x_1, ... x_n)) \neq const$, then it is possible to find a weighting vector $W = (w_0, w_1, ..., w_n)$ for this function such that at least one of its the components $w_1, ..., w_n$ is not equal to zero.

Now, using Theorem 2.1 and Theorem 2.2 we can investigate that separation of an $n$-dimensional space, which is determined by the discrete MVN activation function (2.50) in conjunction with the weights determining a particular input/output mapping.

## 2.3.2 Separation of an n-Dimensional Space

So let one of the numbers $w_1, ..., w_n$ be nonzero. Let us consider a linear function $W(X) = w_0 + w_1 x_1 + ... + w_n x_n$, which is defined on $\mathbb{C}^n$ without connection with some $k$-valued threshold function. Let us investigate where those points $(x_1, ..., x_n) \in \mathbb{C}^n$ are located, for which $P(W(X)) = e^{i\varphi}$. Thus, our goal is to clarify a geometrical interpretation of the set

$$\Pi_\varphi = \{\alpha \in C^n \mid P(W(\alpha)) = e^{i\varphi}\} \qquad (2.58)$$

where $\alpha = (\alpha_1, ..., \alpha_n) \in \mathbb{C}^n$, $W(\alpha) = w_0 + w_1\alpha_1, ..., w_n\alpha_n$, $\varphi \in [0, 2\pi[$, and $w_j \neq 0$ at least for one of $j=1, 2, ..., n$. Since $P(W(\alpha)) = e^{i\varphi}$, then for the continuous MVN this means that $\arg(W(X)) = \varphi$, and for the discrete MVN this means that $\varphi \leq \arg(W(X)) < \varphi + 2\pi / k$. Let us consider for simplicity, but without loss of generality the case $\arg(W(X)) = \varphi$. This equality can be expressed as the equivalent system of equalities

$$\text{Re}(W(X)) = t\cos\varphi$$
$$\text{Im}(W(X)) = t\sin\varphi \qquad (2.59)$$

where $t \in \mathbb{R}$.

Let $x_j = a_j + ib_j$, $j = 1,2,\ldots,n$ and $w_j = u_j + iv_j$, $j = 0,1,\ldots,n$, where $i$ is an imaginary unity. Then system (2.59) may be transformed to

$$u_0 + u_1 a_1 + \ldots + u_n a_n - v_1 b_1 - \ldots - v_n b_n = t\cos(\varphi)$$
$$v_0 + v_1 a_1 + \ldots + v_n a_n + u_1 b_1 + \ldots + u_n b_n = t\sin(\varphi).$$

After an elimination of the parameter $t$ from the last system of equations, we obtain the following linear equation:

$$(u_1 \sin\varphi - v_1 \cos\varphi)a_1 + \ldots + (u_n \sin\varphi - v_n \cos\varphi)a_n -$$
$$-(u_1 \cos\varphi + v_1 \sin\varphi)b_1 - \ldots - \qquad (2.60)$$
$$-(u_n \cos\varphi + v_n \sin\varphi)a_n + u_0 \sin\varphi - v_0 \cos\varphi = 0.$$

Taking into account that $w_j \neq 0$ for at least one of $j=1, 2, \ldots, n$, we can conclude that at least one of the coefficients under the variables $a_1,\ldots,a_n,b_1,\ldots,b_n$ is not equal to zero as well. It means that the rank of (2.60) as of a system of linear algebraic equations is equal to 1.

Really, let us suppose that the opposite is true, so that all the coefficients under the variables $a_1,\ldots,a_n,b_1,\ldots,b_n$ are equal to zero. Then we obtain a system of $2n$ homogenous equations with $2n$ unknowns $u_1,\ldots,u_n,v_1,\ldots,v_n$ which can be broken up into $n$ pairs of the equations as follows:

$$u_j \sin\varphi - v_j \cos\varphi = 0,$$
$$u_j \cos\varphi + v_j \sin\varphi = 0, \quad j = 1, 2,\ldots, n.$$

The determinant of the last system is equal to $\sin^2(\varphi) + \cos^2(\varphi) = 1$. Therefore $u_j = 0$, $v_j = 0$; $j = 1,2,\ldots,n$. But this means that $w_1 = w_2 = \ldots = w_n = 0$, which contradicts to choice of the $w_j$ in (2.58).

Thus, we just proved that equation (2.60) defines the hyperplane $T_\varphi$ within the space $\mathbb{R}^{2n}$. But system (2.59) is not equivalent to equation (2.60) for $t > 0$, and $\Pi_\varphi$ does not coincide with the hyperplane $T_\varphi$.

**Theorem 2.3.** There is the plane $T_0 = \{X \in C^n \mid W(X) = 0\}$ of a dimension $2n-2$ in $\mathbb{R}^{2n}$, which separates the hyperplane $T_\varphi$ into two half-hyperplanes. One of them is that "half" $T_\varphi^{t>0}$ of the hyperplane $T_\varphi$, for which $t > 0$ in (2.59) and another one is that "half" $T_\varphi^{t<0}$ of $T_\varphi$, where $t < 0$ in (2.59). The plane $T_0$ is defined by the equations

$$\begin{aligned} \operatorname{Re}(W(X)) &= 0, \\ \operatorname{Im}(W(X)) &= 0, \end{aligned}$$

(2.61)

or (which is equivalent):

$$\begin{aligned} u_0 + u_1 a_1 + \ldots + u_n a_n - v_1 b_1 - \ldots - v_n b_n &= 0, \\ v_0 + v_1 a_1 + \ldots + v_n a_n + u_1 b_1 + \ldots + u_n b_n &= 0. \end{aligned}$$

(2.62)

**Proof.** First of all we have to show that $T_0$ is a plane within $\mathbb{R}^{2n}$. This follows from the fact that the rank of the system of equations (2.62) is equal to 2. Really, since at least for some $j = 1, \ldots, n$, $w_j \neq 0$, then the determinant

$$\begin{vmatrix} u_j & -v_j \\ v_j & u_j \end{vmatrix} = u_j^2 + v_j^2 = |w_j|^2 > 0$$

for at least some value of $j = 1, \ldots, n$. It is also evident that $T_0$ does not depend on a value of $\varphi$ (see (2.62)) and additionally $\forall \varphi \in [0, 2\pi[\ T_0 \cap T_\varphi \neq \varnothing$. This means that $T_0 = \bigcap_{0 \leq \varphi < 2\pi} T_\varphi$. Theorem is proven.

It should be noted that $T_0$ is an analytical plane [64] of a dimension $n-1$ within the space $C^n$ because $T_0$ is defined by the linear equation $W(X) = 0$ in $C^n$, in other words its equation depends on the variables $x_1, \ldots, x_n$, and does not depend on the conjugate variables $\bar{x}_1, \ldots, \bar{x}_n$.

Let $T_0^1$ be a linear subspace of a dimension $2n-2$ of the space $\mathbb{R}^{2n}$ corresponding to the plane $T_0$. Therefore, $T_0^1$ is the space of solutions of the system, which consists of the following two homogenous equations

$$\begin{aligned} u_1 a_1 + \ldots + u_n a_n - v_1 b_1 - \ldots - v_n b_n &= 0, \\ v_1 a_1 + \ldots + v_n a_n + u_1 b_1 + \ldots + u_n b_n &= 0. \end{aligned}$$

Let also $S_\varphi^0$ be the orthogonal complement[3] [65] of some subspace of the space $T_0^1$. Then $S^0 = \bigcup\limits_{0 \le \varphi < 2\pi} S_\varphi^0$ is the orthogonal complement to the whole space $T_0^1$.

Each of $S_\varphi^0$ is defined by the following system of 2n-2 algebraic equations

$$s_1^\varphi(a_1,...,a_n, \ b_1,...,b_n) = 0,$$

$$.............................. \qquad (2.63)$$

$$s_{2n-2}^\varphi(a_1,...,a_n, \ b_1,...,b_n) = 0.$$

Evidently, all of $S_\varphi^0$ are two-dimensional subspaces of $\mathbb{R}^{2n}$, and therefore they are two-dimensional planes in $\mathbb{R}^{2n}$.

**Theorem 2.4.** The planes $S_\varphi^0, 0 \le \varphi < 2\pi$ and $T_0$ have a single common point $M^0$ (see Fig. 2.28).

**Proof.** Intersection of the planes $S_\varphi^0, 0 \le \varphi < 2\pi$ and $T_0$ is defined by the following system of linear equations:

$$u_0 + u_1 a_1 + ... + u_n a_n - v_1 b_1 - ... - v_n b_n = 0,$$
$$v_0 + v_1 a_1 + ... + v_n a_n + u_1 b_1 + ... + u_n b_n = 0,$$
$$s_1^{\varphi_1}(a_1,...,a_n, \ b_1,...,b_n) = 0,$$

$$................................$$

$$s_{2n-2}^{\varphi_1}(a_1,...,a_n, \ b_1,...,b_n) = 0,$$

$$................................ \qquad (2.64)$$

$$s_1^{\varphi_j}(a_1,...,a_n, \ b_1,...,b_n) = 0,$$

$$................................$$

$$s_{2n-2}^{\varphi_j}(a_1,...,a_n, \ b_1,...,b_n) = 0,$$

$$................................$$

which is a result of the union of the systems of linear algebraic equations (2.62) and (2.63), which define $T_0$ and all the $S_\varphi^0, 0 \le \varphi < 2\pi$, respectively.

---

[3] The orthogonal complement $V'$ of the subspace $V$ of the space $W$ is the set of all vectors in $W$ that are orthogonal to every vector in $V$ [65].

On the one hand, it is evident that the rank of the matrix of the system (2.64) does not depend on the number of hyperplanes $S_\varphi^0$, and it is equal to $2n$ (the sum of dimensions of the subspace $T_0^1$ and its orthogonal supplements $S_\varphi^0$). On the other hand, this is a system with respect to exactly $2n$ unknowns $a_1^0, ..., a_n^0, b_1^0, ..., b_n^0$. This means that it has a single solution

$$M^0 = \left( a_1^0, ..., a_n^0, b_1^0, ..., b_n^0 \right)$$

that is the point of intersection of all the orthogonal complements $S_\varphi^0, 0 \le \varphi < 2\pi$ and the plane $T_0$. Theorem is proven.

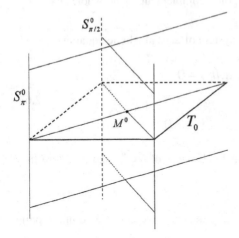

**Fig. 2.28** Illustration to Theorem 2.4. Planes $T_0$, $S_{\pi/2}^0$, and $S_\pi^0$ have a single common point $M^0$.

Theorem 2.4 is illustrated in Fig. 2.28. The planes $T_0$, $S_{\pi/2}^0$, and $S_\pi^0$ have a single common point $M^0$ where they intersect. Evidently, all the planes $S_\varphi^0, 0 \le \varphi < 2\pi$ have the same common point $M^0$, which is also the point of their intersection with the plane $T_0$.

**Theorem 2.5.** The plane $S_\varphi^0$ of a dimension 2 intersects with the half-hyperplane $T_\varphi^{t>0}$ of the hyperplane $T_\varphi$ by the ray $L_\varphi$ originating at the point $M^0$ (see Fig. 2.29).

**Proof.** Intersection of the plane $S_\varphi^0$ with the hyperplane $T_\varphi$ is a plane within $\mathbb{R}^{2n}$, which is defined by the following system of linear algebraic equations:

$$s_1^\varphi(a_1, ..., a_n, b_1, ..., b_n) = 0,$$

$$................................................$$

$$s_{2n-2}^\varphi(a_1, ..., a_n, b_1, ..., b_n) = 0,$$

$$(u_1 \sin \varphi - v_1 \cos \varphi)a_1 + ... + (u_n \sin \varphi - v_n \cos \varphi)a_n -$$
$$-(u_1 \cos \varphi + v_1 \sin \varphi)b_1 - ... - (u_n \cos \varphi + v_n \sin \varphi)a_n +$$
$$+u_0 \sin \varphi - v_0 \cos \varphi = 0$$

$$(2.65)$$

that is the result of the union of the systems of linear algebraic equations (2.63) and (2.60), which define the plane $S_\varphi^0$ and the hyperplane $T_\varphi$, respectively. The rank of the system (2.65) is equal to $2n$-1 for any $\varphi \in [0, 2\pi[$. This means that all the equations of this system are linearly independent. Really, suppose that the opposite is true. Let the equations in (2.65) are linearly dependent. Since the equation (2.60) is a linear combination of the equations (2.62), the linear dependence of the equations of (2.65) follows to the conclusion that the equations of the system (2.64) also are linearly dependent. The last conclusion contradicts to the fact, which has been shown in the proof of Theorem 2.4. This means that the rank of the system (2.65) is equal to $2n$-1, and therefore this system defines a plane of a dimension 1, that is a line. Finally, $T_\varphi^{t>0}$ is separated from the rest part of the hyperplane $T_\varphi$ by the plane $T^0$, which intersects with the plane $S_\varphi^0$ in the single point $M^0$ (see Fig. 2.29). Therefore, a part of the line defined by (65) forms the

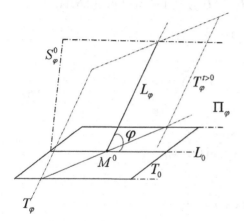

ray    $$L_\varphi = S_\varphi^0 \cap T_\varphi = S_\varphi^0 \cap \Pi_\varphi$$

originating at the point $M^0$. Theorem is proven.

Let us choose a polar coordinate system with the center $M^0$ and the polar axis $L_0$ $(\varphi = 0)$, which is the intersection of the planes $S_\varphi^0$ and $T^0$.

We will consider the smaller angle from the two angles between the rays $L_\varphi$ and $L_\psi$ as angle between these rays. Since $k \geq 2$, a value of the angle between $L_\varphi$ and $L_\psi$ is

**Fig. 2.29** Illustration to Theorem 2.5

always less than or equal to $\pi$.

**Theorem 2.6.** A value of the angle between the polar axis $L_0$ and the ray $L_\varphi$ is equal to $\varphi$ (see Fig. 2.29).

*Remark.* Evidently, the angle between the half-hyperplanes $T_\varphi^{t>0}$ and $T_\psi^{t>0}$ is measured by the linear angle between $L_\varphi$ and $L_\psi$. So, this theorem is equivalent to the following one: the angle between the half-hyperplanes $\Pi_\varphi \cap T_0$ and $T_\varphi^{t>0}$ is equal to $\varphi$ (see Fig. 2.29).

**Proof.** The space $\mathbb{R}^{2n}$ should be considered as $2n$-dimensional Euclidian space with the obvious dot product $(a,b) = \sum_{j=1}^{2n} a_j b_j$, where $a = (a_1,...,a_{2n})$, $b = (b_1,...,b_{2n})$. Therefore, cosine of the angle $\gamma$ between the vectors $a, b \in R^{2n}$ may be evaluated as follows:

$$\cos \gamma = \frac{(a,b)}{|a\|b|} \tag{2.66}$$

where $|a| = \sqrt{(a,a)}$ is the Euclidian norm of the vector $a$. The angle between the half-hyperplanes $\Pi_\varphi \cap T_0$ and $T_\varphi^{t>0}$ is equal to the angle between the normal vectors $\vec{n}_0$ and $\vec{n}_\varphi$.

Let $\vec{n}_0 = (a_1,...,a_{2n})$ and $\vec{n}_\varphi = (b_1,...,b_{2n})$. It follows from (2.60) that

$$a_1 = u_1 \sin 0 - v_1 \cos 0 = -v_1,$$

$$\dots\dots\dots\dots\dots\dots\dots$$

$$a_n = u_n \sin 0 - v_n \cos 0 = -v_n,$$
$$a_{n+1} = -u_1 \cos 0 - v_1 \sin 0 = -u_1, \tag{2.67}$$

$$\dots\dots\dots\dots\dots\dots\dots$$

$$a_{2n} = -u_n \cos 0 - v_n \sin 0 = -u_n$$

and

$$b_1 = u_1 \sin \phi - v_1 \cos \phi = -v_1,$$

$$\dots\dots\dots\dots\dots\dots\dots$$

$$b_n = u_n \sin \phi - v_n \cos \phi = -v_n,$$
$$b_{n+1} = -u_1 \cos \phi - v_1 \sin \phi = -u_1, \tag{2.68}$$

$$\dots\dots\dots\dots\dots\dots\dots$$

$$b_{2n} = -u_n \cos \phi - v_n \sin \phi = -u_n.$$

Then taking into account (2.66), we obtain $\cos \gamma = \dfrac{(\vec{n}_0, \vec{n}_\varphi)}{|\vec{n}_0\| \vec{n}_\varphi|}$. Let us substitute the coordinates of the vectors $\vec{n}_0$ and $\vec{n}_\varphi$ from the equations (2.67)-(2.68) in the last equation. Then we obtain the following:

$$\cos\gamma = \frac{\sum_{j=1}^{n}\left[(-v_j)(u_j\sin\varphi - v_j\cos\varphi) + (-u_j)(-u_j\cos\varphi - v_j\sin\varphi)\right]}{\sqrt{\sum_{j=1}^{n}(v_j^2 + u_j^2)}\sqrt{\left[(u_j\sin\varphi - v_j\cos\varphi)^2 + (-u_j\cos\varphi - v_j\sin\varphi)^2\right]}},$$

and after simplification:

$$\cos\gamma = \frac{\cos\varphi\sum_{j=1}^{n}(v_j^2 + u_j^2)}{\sqrt{\sum_{j=1}^{n}(v_j^2 + u_j^2)}\sqrt{\sum_{j=1}^{n}\left[u_j^2(\sin^2\varphi + \cos^2\varphi) + v_j^2(\cos^2\varphi + \sin^2\varphi)\right]}} =$$

$$= \cos\varphi$$

This means that the angle between $\Pi_\varphi \cap T_0$ and $T_\varphi^{t>0}$ is equal to $\varphi$, and the proof is completed.

The following statement, which generalizes Theorem 2.6, is important.

**Theorem 2.7.** The angle between the half-hyperplanes $T_\varphi^{t>0}$ and $T_\psi^{t>0}$ is equal to $\psi - \varphi$.

A proof of this theorem is similar to the proof of the previous one. It is just necessary to apply (2.66) to the vectors $\vec{n}_\varphi$ and $\vec{n}_\psi$. It also clearly follows from Theorem 2.3 - Theorem 2.7 that all hyperplanes $T_{\frac{2\pi}{k}j}$; $j = 0,1,...,\lceil k/2\rceil$

(where $\lceil a \rceil$ is the ceiling function returning the smallest integer, which is larger than $a$) have a single common point $M^0$ where they intersect.

Let now

$$Q_j = \bigcup_{j \le \frac{k\varphi}{2\pi} < j+1} \Pi_\varphi, \quad j = 0,1,...,k-1. \tag{2.69}$$

Taking into account that $S_\varphi^0$ is a two-dimensional plane, we can conclude that $S_\varphi^0 \cap Q_j$, $j = 0,1,...,k-1$ is a semi-open sector with the center $M^0$, and bounded by the rays $L_{\frac{2\pi}{k}j}$ and $L_{\frac{2\pi}{k}(j+1)}$ within the plane $S_\varphi^0$. The first of these rays belongs to the mentioned sector (without the point $M^0$), and the second one does not belong to this sector. According to Theorem 2.7, the angle between these sectors is equal to $2\pi/k$.

### 2.3.3 k-Edge

Thus, we have realized that for any fixed $k$, an $n$-dimensional space is separated by the hyperplanes $T_{\frac{2\pi}{k}j}; j = 0,1,...,\lceil k/2 \rceil$ into the subspaces

$Q_j$, $j = 0,1,...,k-1$, which are determined by (2.69).

**Definition 2.10.** An ordered collection of the sets $Q = \{Q_0, Q_1, ..., Q_{k-1}\}$ is called a *k-edge* corresponding to the given linear function $W(X) = w_0 + w_1 x_1 + ... + w_n x_n$. The set $Q_j$ is the $j$th edge of the $k$-edge $Q$. The set $T_0$ (a plane of a dimension $2n$-2) is called a *sharp point* of the $k$-edge $Q$.

Let us for simplicity use the notation $S^0$ for the plane $S^0_{2\pi/k}$. It should be noted that the edges $Q_0, Q_1, ..., Q_{k-1}$ of the $k$-edge $Q$ are enumerated by such a way that their intersections $Q_j \cap S^0$, $j = 0,1,...,k-1$ with the plane $S^0$ induce a sequence

$$(Q_0 \cap S^0, Q_1 \cap S^0, ..., Q_{k-1} \cap S^0) \tag{2.70}$$

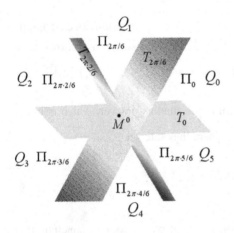

$Q_1$

$\Pi_{2\pi/6}$

$T_{2\pi\cdot2/6}$

$T_{2\pi/6}$

$Q_2$ $\Pi_{2\pi\cdot2/6}$

$\Pi_0$ $Q_0$

$M^0$

$T_0$

$Q_3$ $\Pi_{2\pi\cdot3/6}$

$\Pi_{2\pi\cdot5/6}$ $Q_5$

$\Pi_{2\pi\cdot4/6}$

$Q_4$

**Fig. 2.30** $k$-edge for $k=6$, $n=2$

of the sectors of the same angular size $2\pi/k$ within the two-dimensional plane $S^0$, and following one by one in a positive direction. We assume that the positive direction is the direction of the rotation around the center $M^0$ within the plane $S^0$ from the ray $L_0$ to the ray $L_{2\pi/k}$ (counterclockwise). Thus, our considerations (Theorem 2.1-Theorem 2.7) finally lead us to Definition 2.10 of the $k$-edge. *We can conclude now that if the discrete MVN implements some* input/output mapping, its weights along with its activation function determine the separation of that $n$-dimensional space where this mapping is defined by the $k$-edge $Q$. This $k$-edge is formed by the family of separating hyperplanes $T_{\frac{2\pi}{k}j}; j = 0,1,...,\lceil k/2 \rceil$ that have a single

common point $M^0$. Thus, the $k$-edge separates the $n$-dimensional space into $k$

subspaces $\Pi_{\frac{2\pi}{k}j}$ ; $j=0,1,...,k-1$ corresponding to the edges $Q_0,Q_1,...,Q_{k-1}$

of the $k$-edge. In Fig. 2.30, we see the example of the $k$-edge for $k$=6, $n$=2. The

space $\mathbb{C}^n$ is divided by the hyperplanes $T_0, T_{\frac{2\pi}{6}}$ and $T_{\frac{2\pi}{6} \cdot 2}$ into six subspaces

$\Pi_{\frac{2\pi}{k}j}$ ; $j=0,1,...,5$ corresponding to the edges $Q_0,Q_1,...,Q_5$ of the 6-edge

created by the hyperplanes $T_{\frac{2\pi}{k}j}$ ; $j=0,1,2$. In Fig. 2.31, the same 6-edge is

shown together with the plane $S_\varphi^0$, which passes through the $k$-edge creating the

rays

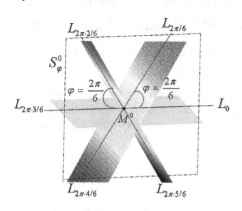

$L_{\frac{2\pi}{k}j}$ ; $j=0,1,...,k-1$ as a result

of its intersection with the edges
$Q_0,Q_1,...,Q_5$ of the 6-edge. The
corresponding       six       rays
$L_0, L_{\frac{2\pi}{k}},..., L_{\frac{2\pi}{k} \cdot 5}$ are also shown

in Fig. 2.31. As we have shown earlier (Theorem 2.6 and Theorem 2.7)
the smaller angle between the hy-
perplanes $T_{\frac{2\pi}{k}j}$ and $T_{\frac{2\pi}{k}(j+1)}$ , as

well as the angle between all adja-
cent rays $L_{\frac{2\pi}{k}j}$ and $L_{\frac{2\pi}{k}(j+1)}$ is

equal to $\varphi$.

**Fig. 2.31** $k$-edge for $k$=6 and its projection into
the plane $S_\varphi^0$. This projection "generates" the
activation function $P(z)$

Particularly,  in  Fig.  2.31,  for  $k$=3,  $\varphi = 2\pi / 6$.  The  rays

$L_{\frac{2\pi}{k}j}$ ; $j=0,1,...,k-1$ belong to the plane $S_\varphi^0$. This means that the $k$-edge

corresponding  to  the  given  linear  function  $W(X)=w_0+w_1x_1+...+w_nx_n$

"projects" all points $(x_1,...,x_n)$ from the corresponding $n$-dimensional space on

the plane $S_\varphi^0$. Since $W(X)\in \mathbb{C}$, we may associate the plane $S_\varphi^0$ with the com-

plex plane. The point $M^0$ corresponds to the complex number (0, 0) (the origin).

The ray $L_0$ in this case can be treated as the positive direction of the real axis. As

a result, the rays $L_{\frac{2\pi}{k}j}$ ; $j=0,1,...,k-1$ divide the plane $S_\varphi^0$ (and the complex plane, respectively) into $k$ equal sectors. This absolutely corresponds to the discrete MVN activation function (2.50), which divides the complex plane into $k$ equal sectors.

Thus, answering the question about a geometrical interpretation of the set $\Pi_\varphi$ determined by (2.58), we can say that for the discrete MVN and discrete-valued activation function (2.50) this is a bounded subspace of the space $\mathbb{C}^n$ such that for any $(x_1,...,x_n)\in\Pi_\varphi$

$$\varphi\le\arg\left(W(w_0+w_1x_1+...+w_nx_n)\right)<\varphi+2\pi/k.$$

For the continuous MVN and the continuous activation function (2.54), $k\to\infty$ and $\varphi\to 0$. Therefore, the $k$-edge consists of the infinite amount of hyperplanes still intersecting in a single common point $M^0$. In this case, $\Pi_\varphi$, which is determined by (2.58), is a bounded subspace of the space $\mathbb{C}^n$ such that for any $(x_1,...,x_n)\in\Pi_\varphi$ $\arg\left(W(w_0+w_1x_1+...+w_nx_n)\right)=\varphi$.

Let us consider now some important properties of the $k$-edge and how these properties can be employed for solving multi-class classification problems using the discrete MVN and approximation of continuous-valued functions using the continuous MVN.

### 2.3.4  Properties of the k-Edge

**Definition 2.11.** An *ordered decomposition* of the non-empty set $M$ is a sequence $M_1,...,M_s$ of non-empty subsets of the set $M$, which are mutually disjoint, and union of which is equal to the set $M$.

Let us denote an ordered decomposition via $[M_1,...,M_s]$ and the fact that it is an exact decomposition of the set $M$ should be written as $M=[M_1,...,M_s]$. $W(\alpha)=0$ for $\alpha\in T_0$ according to the equations (2.61). As it was mentioned above, we have to set some value from the set $E_k$ as a value of the function $P$ on the zero ($P(0)$). It will be natural to connect the sharp point $T_0$ of the $k$-edge with the fixed edge $Q_t$. We can use the following notation in such a case $Q_t\cap S^0=\bar{Q}_t$. An ordered collection of the sets $\{Q_0, Q_1, ...,Q_t, ..., Q_{k-1}\}$ may be denoted as $\bar{Q}_t$, or $\bar{Q}$, if a number $t$ is unknown, or, if its exact value is not important. In such a case $\bar{Q}$ and $Q$ form the same $k$-edge corresponding to a linear function $W(X)$. The notations

$Q = \{Q_0,\ Q_1,\ ...,Q_t,\ ...,\ Q_{k-1}\}$   and   $\bar{Q} = \{Q_0,\ Q_1,\ ...,\bar{Q}_t,\ ...,\ Q_{k-1}\}$   are equivalent in such a case. In other words, we will denote sometimes the component $Q_t \cap T_0$ of the $k$-edge $\bar{Q}_t$ as $Q_t$. Therefore one of the $k$-edges $Q$ or $\bar{Q}$ shows, is $Q_t$ connected with one of them. We can conclude now that $\bar{Q}$ forms an ordered decomposition of the space $\mathbb{C}^n$

$$\mathbb{C}^n = \left[ Q_0,\ Q_1,\ ...,\bar{Q}_t,\ ...,\ Q_{k-1} \right].$$

**Theorem 2.8.** If the vector $\alpha = (\alpha_1,...,a_n)$ from the space $\mathbb{C}^n$ belongs to the $j$th edge of the $k$-edge $\bar{Q}$, then $P(W(\alpha)) = \varepsilon^j$.

**Proof.** $\forall \alpha \in \Pi_\varphi\ P(W(\alpha)) = e^{i\varphi}$ according to (2.58). It follows from (2.69) that $2\pi j / k \le \varphi < 2\pi(j+1)/k$ for $\Pi_\varphi \in Q_j$.

Therefore we obtain $2\pi j / k \le \arg(W(\alpha)) < 2\pi(j+1)/k$ for $\alpha \in Q_j$. If $\alpha \in T_0$ (it also means that $\alpha \in \bar{Q}_t$) then $W(\alpha) = 0$, and therefore $P(W(\alpha)) = \varepsilon^t$, which completes the proof.

**Theorem 2.9.** If the condition $P(W(\alpha)) = \varepsilon^j$ is true for some $\alpha = (\alpha_1,...,a_n) \in \mathbb{C}^n$, then $\alpha \in Q_j$.

**Proof.** Let $W(\alpha) \ne 0$, and $P(W(\alpha)) = \varepsilon^j$. Then $2\pi j / k \le \arg(W(\alpha)) < 2\pi(j+1)/k$. The edges $Q_0$, $Q_1$, ..., $Q_{k-1}$ form an ordered decomposition of the space $\mathbb{C}^n$, and therefore $\alpha$ belongs to one of the edges, and to one of the half-hyperplanes $\Pi_\varphi$, exactly to such that $2\pi j / k \le \varphi < 2\pi(j+1)/k$. It means that $\alpha \in Q_j$ according to (2.58) and (2.69). Let now $P(W(\alpha)) = \varepsilon^t$, and $W(\alpha) = 0$. Then $\alpha \in T_0$ according to (2.61), and therefore also $\alpha \in \bar{Q}_t$.

We are ready now to consider a very important correspondence between a complex-valued threshold function (or a $k$-valued threshold function) and the $k$-edge corresponding to this function. This relationship is important for understanding how multi-class classification problems can be solved using MVN or MVN-based neural networks.

**Theorem 2.10.** Let $f(x_1,...,x_n)$ be a complex-valued threshold function, which is defined on the set $T \subset \mathbb{C}^n$, and at least one of the following conditions holds:

1) The set $T$ is bounded; 2) $P\big(f\left(x_1,...,x_n\right)\big)\neq const$. Then such a $k$-edge

$\bar{Q}=\{Q_0,\ Q_1,\ ...,\ Q_{k-1}\}$ exists that

$$\forall\ \alpha=(\alpha_1,...,\alpha_n)\in T\cap Q_j\ \ P(f(\alpha))=\varepsilon^j.$$

**Proof.** Let $W=(w_0,\ w_1,\ ...,\ w_n)$ be such a weighting vector of the given function $f$ that at least one of its components $w_1,\ ...,\ w_n$ is not equal to zero. According to the conditions of the theorem and to the above considerations it is always possible to find such a weighting vector for the function $f$.

Let us consider a linear function $W(X)=w_0+w_1x_1+...+w_nx_n$. Let $\bar{Q}=\{Q_0,\ Q_1,\ ...,\ Q_{k-1}\}$ be a $k$-edge corresponding to this linear function. Since $f$ is a threshold function, then $P(f(X))=P(W(X))$. Therefore, if $X=\left(x_1,...,x_n\right)\in T\cap Q_j$, then according to Theorem 2.8 $P(f(\alpha))=\varepsilon^j$. Theorem is proven.

The inverse theorem is also true. Moreover, it is not necessary to put the restrictions on the domain of a function $f$.

**Theorem 2.11.** Let $f\left(x_1,...,x_n\right)$ be a complex-valued function, which is defined on the set $T\subset\mathbb{C}^n$, and assume that a $k$-edge $\bar{Q}=\{Q_0,\ Q_1,\ ...,\ Q_{k-1}\}$ exists such that $\forall\ \alpha=(\alpha_1,...,\alpha_n)\in T\cap Q_j\ \ P(f(\alpha))=\varepsilon^j$. Then $f\left(x_1,...,x_n\right)$ is a complex-valued threshold function.

**Proof.** Let $W(X)=w_0+w_1x_1+...+w_nx_n$ be a linear function, to which $k$-edge $\bar{Q}$ corresponds. If $\alpha\in T$, and $P(f(\alpha))=\varepsilon^j$, then according to Theorem 2.9 $\alpha\in Q_j$. Therefore, $\forall\alpha\in T\ P(f(\alpha))=P(W(\alpha))$, and this means that the function $f\left(x_1,...,x_n\right)$ is a threshold function.

**Definition 2.12.** A $k$-edge $\bar{Q}$, which satisfies the conditions of Theorem 2.11, is called a *k-edge of the function* $f\left(x_1,...,x_n\right)$.

Let $f\left(x_1,...,x_n\right)$ is an arbitrary complex-valued function of the complex-valued variables which is defined on the set $T\subset\mathbb{C}^n$. Let us introduce the following sets of $n$-dimensional vectors (or a set of points in an $n$-dimensional space)

$$A_j=\{\alpha=(\alpha_1,...,\alpha_n)\in T\mid P(f(\alpha))=\varepsilon^j\};\ \ j=0,\ 1,\ ...,\ k-1. \quad (2.71)$$

In general, some of the sets $A_j$, $j = 1, ..., k-1$ can be empty.

**Definition 2.13.** An ordered collection of the sets $A_0$, $A_1$, ..., $A_{k-1}$ (see (2.71) ) is called an *edged decomposition* of the domain $T$ of the function $f(x_1, ..., x_n)$.

We will use the notation $[A_0, A_1, ..., A_{k-1}]$ for the edged decomposition. Similarly to the case of the ordered decomposition of the arbitrary set $M$ (see the Definition 2.11 above) we will write that $T = [A_0, A_1, ..., A_{k-1}]$. Evidently, the elements $A_0$, $A_1$, ..., $A_{k-1}$ of the edged decomposition are the complete prototypes of the values of the function $P(f(X))$, which is defined on the set $T$.

The following theorem is very important for understanding the geometrical meaning of a complex-valued threshold function and of a $k$-valued threshold function, in particular.

**Theorem 2.12.** The complex-valued function $f(x_1, ..., x_n)$, which is defined on the set $T \subset \mathbb{C}^n$, is a complex-valued threshold function if and only if a $k$-edge $\bar{Q} = \{Q_0, Q_1, ..., Q_{k-1}\}$ exists such that its elements (edges) are related to the sets $A_0$, $A_1$, ..., $A_{k-1}$ of the edged decomposition $T = [A_0, A_1, ..., A_{k-1}]$ as follows $A_j \subseteq Q_j$; $j = 0, 1, ..., k-1$.

**Proof.** Necessity. Let the function $f(x_1, ..., x_n)$ be threshold, and $(w_0, w_1, ..., w_n)$ is such a weighting vector of the function $f$ that $w_t \neq 0$ at least for some $t \in \{1, 2, ..., n\}$. Let also $\bar{Q} = \{Q_0, Q_1, ..., Q_{k-1}\}$ be a $k$-edge corresponding to the linear function $W(X) = w_0 + w_1 x_1 + ... + w_n x_n$. Then according to Theorem 2.9 and Definition 2.12 (see also (2.71) ) we obtain $A_t \subset Q_t$.

Sufficiency. Let $A_t \subset Q_t$ for all $t \in \{0, 1, ..., k-1\}$, and $W(X) = w_0 + w_1 x_1 + ... + w_n x_n$ be a linear function. Let $\bar{Q} = \{Q_0, Q_1, ..., Q_{k-1}\}$ be a $k$-edge corresponding to this linear function. Then according to Theorem 2.8 the condition $P(W(\alpha)) = \varepsilon^t$ is true for $\alpha \in A_t \subset Q_t$. However, it also follows from (2.71) that $P(f(\alpha)) = \varepsilon^t$. So $\forall \alpha \in Z$ $P(f(\alpha)) = P(W(\alpha))$ and this means that $f(x_1, ..., x_n)$ is a $k$-valued threshold function with the weighting vector $W$.

**Definition 2.14.** The sets $A_0$, $A_1$, ..., $A_{k-1}$ establish an *edge-like sequence*, if such a $k$-edge $\bar{Q} = \{Q_0, Q_1, ..., Q_{k-1}\}$ exists, that $A_j \subset Q_j$ for all $j \in \{0, 1, ..., k-1\}$. In such a case the edged decomposition $[A_0, A_1, ..., A_{k-1}]$ of some set $A = A_0 \cup A_1 \cup ... \cup A_{k-1} \subset \mathbb{C}^n$ is an *edge-like decomposition*.

The following theorem follows directly from Theorem 2.12, Definition 2.13, and Definition 2.14.

**Theorem 2.13.** A complex-valued function $f(x_1, ..., x_n)$, which is defined on the set $T \subset \mathbb{C}^n$, is a complex-valued threshold function if and only if the edged decomposition of the set $T$ corresponding to the given $k$ is an edge-like decomposition.

Let us clarify a geometrical meaning of the $k$-edge corresponding to the linear function $w_0 + w_1 x$ $(w_1 \neq 0)$ of a single complex variable.

In such a case ($n=1$) the sequence $(Q_0 \cap S^0, Q_1 \cap S^0, ..., Q_{k-1} \cap S^0)$ (see (2.70) ) coincides with the $k$-edge $Q = \{Q_0, Q_1, ..., Q_{k-1}\}$ (see Definition 2.10), and the set $T_0$, which is a sharp point of the $k$-edge, is a plane of the dimension $2n - 2 = 2 \cdot 1 - 2 = 0$. Therefore, $T_0$ is a point, and it coincides with the point $M^0$. Thus, the $k$-edge $Q$ for $n=1$ is a sequence of the sectors on the obvious complex plane $C = S^0$ (see Fig. 2.20), which are created by $k$ rays, of the same angular size $2\pi / k$ and the same center $M^0$. These rays follow each other in the positive direction (counterclockwise), and each sector contains only the first of two rays, which are its boundaries.

After the connection of the sharp-point $M^0$ with the fixed edge $Q_t$ we obtain the $k$-edge $\bar{Q} = \{Q_0, Q_1, ..., \bar{Q}_t, ..., Q_{k-1}\}$ corresponding to the linear function $w_0 + w_1 x$ on the plane $C = S^0$. According to (71) it is possible to say that the edges of the $k$-edge establish the ordered decomposition of all the points of the complex plane $C = [Q_0, Q_1, ..., Q_t, ..., Q_{k-1}]$.

To complete the geometrical interpretation of $k$-valued threshold functions, complex-valued threshold functions and the edged separation of an $n$-dimensional space established by the multi-valued neuron, let us consider one more generalization of a $k$-valued threshold function.

**Definition 2.15.** A complex-valued function $f(x_1,...,x_n):T \to \mathbb{C}$, where $T \subseteq \mathbb{C}^n$, is called a *complex-valued threshold function*, if it is possible to find such a complex-valued weighting vector $W = (w_0, w_1,..., w_n)$, and to define $P(0)$ in such a way that (2.57)

$$P\big(f(x_1,...x_n)\big) = P(w_0 + w_1 x_1 +...+ w_n x_n)$$

(where $P$ is the function (2.50) ) holds for all $(x_1,...,x_n)$ from the domain of the function $f$.

Definition 2.9 and Definition 2.15 of the complex-valued threshold function differ from each other in the following aspect. Definition 2.9 employs the function $P$, which is not defined in (0,0), while Definition 2.15 employs the function $P$ which is additionally defined in (0,0). These both definitions are equivalent because determine the same class of complex-valued threshold functions.

**Theorem 2.14.** Equivalence of Definition 9 and Definition 15 of the complex-valued threshold function $f(x_1,...,x_n)$, which is defined on the set $T \subset \mathbb{C}^n$, does not depend on the definition of the function $P$ in (0,0) ( $P(0)$ ) if and only if a weighting vector $W = (w_0, w_1, ..., w_n)$ exists for the threshold function $f$ (according to Definition 9) such that the set $\{\alpha \in \mathbb{C}^n \,|\, W(\alpha) = 0\}$ does not intersect with the domain $T$ of the function $f$:

$$T \cap \{\alpha \in \mathbb{C}^n \,|\, W(\alpha) = 0\} = \varnothing \qquad (2.72)$$

**Proof.** The sufficiency of the condition (2.72) is evident because according to Definition 9 $\forall \alpha \in T$ $P(f(\alpha)) = P(W(\alpha))$, and therefore, taking into account (72), we obtain that $\forall \alpha \in T$ $W(\alpha) \neq 0$.

Necessity. Let both definitions of the complex-valued threshold function are equivalent with no matter of the definition of $P(0)$. Let us assume that a weighting vector $W = (w_0, w_1, ..., w_n)$ exists for the threshold function $f$ (according to Definition 15) such that $T \cap \{\alpha \in \mathbb{C}^n \,|\, W(\alpha) = 0\} = A \neq \varnothing$. This means that $\forall \alpha \in A, P(W(\alpha))$ takes some specific value, such that $P(W(\alpha)) = P(0)$. However, this contradicts to arbitrariness of the definition of the value, which the function $P(z)$ takes in (0,0).

It follows from Theorem 14 that it is possible either to keep $P(z)$ undefined in (0,0) or it is possible to define it arbitrarily if (2.72) holds.

## 2.3.5 Edged Separation and MVN: Summary

Let us summarize briefly how the edged separation of an $n$-dimensional space is closely related to the implementation of some input/output mapping by MVN.

1) If some input/output mapping, which we need to learn, is described by the $k$-valued function $f(x_1,...,x_n): T \rightarrow E_m; T \subseteq E_k^n \vee T \subseteq O^n$, then this input/output mapping can be implemented using a single MVN if the function $f(x_1,...,x_n)$ is a $k$-valued threshold function.

2) $f(x_1,...,x_n)$ is a $k$-valued threshold function if its domain $T$ allows the edged decomposition $T = [A_0, A_1, ..., A_{k-1}]$ into disjoint subsets $A_j, j = 1,...,k-1$ such that (2.71) holds. In other words, $A_j = \{\alpha = (\alpha_1,...,\alpha_n) \in T \mid P(f(\alpha)) = \varepsilon^j\}$; $j = 0,1,...,k-1$. The sets $A_0, A_1, ..., A_{k-1}$ that form the edged decomposition are the complete prototypes of the values of the function $P(f(X))$.

3) The existence of the edged decomposition $T = [A_0, A_1, ..., A_{k-1}]$ means the existence of the $k$-edge $\bar{Q} = \{Q_0, Q_1, ..., Q_{k-1}\}$ such that its edges include the sets $A_0, A_1, ..., A_{k-1}$: $A_j \subseteq Q_j$; $j = 0,1,...,k-1$. In this case, the $k$-edge $\bar{Q}$, is a $k$-edge of the function $f(x_1,...,x_n)$ and the sets $A_0, A_1, ..., A_{k-1}$ establish an edge-like sequence.

Thus, a problem of the implementation of the input/output mapping described by the $k$-valued function $f(x_1,...,x_n)$ is a problem of finding the $k$-edge $\bar{Q}$ such that its edges include the sets $A_0, A_1, ..., A_{k-1}$ that in turn establish an edge-like sequence. According to Definition 2.10 of the $k$-edge, to find the $k$-edge $\bar{Q}$, we have to find coefficients of the linear function $W(X) = w_0 + w_1 x_1 + ... + w_n x_n$ that is the weights $w_0, w_1, ..., w_n$. To find them, we should use a learning algorithm. In Chapter 3, we will consider a learning algorithm for MVN and three learning rules.

It is also important to mention that the same considerations work for the case of $k \rightarrow \infty$. Thus for continuous MVN and its input/output mapping, which is described by the function $f(x_1,...,x_n): T \rightarrow O; T \subseteq E_k^n \vee T \subseteq O^n$.

As it follows from Definition 2.9-Definition 2.15 and Theorem 2.1-Theorem 2.14, all these conclusions are applicable to input/output mappings described not only by

$k$-valued functions $f(x_1,...,x_n):T \to E_m; T \subseteq E_k^n \vee T \subseteq O^n$, but by arbitrary complex-valued functions $f(x_1,...,x_n):T \to \mathbb{C}; T \subseteq \mathbb{C}^n$. However, learning algorithms for the latter case have not been developed yet.

It is very important to mention that if the domain of some function $f(x_1,...,x_n):T \to E_m; T \subseteq E_k^n \vee T \subseteq O^n$ does not allow the edged decomposition $T = [A_0, A_1, ..., A_{k-1}]$, it is possible that for some $m > k$, the edged decomposition $\tilde{T} = \left[ \tilde{A}_0, \tilde{A}_1, ..., \tilde{A}_{k-1}, \tilde{A}_k, \tilde{A}_{k+1},..., \tilde{A}_{m-1} \right]$ exists such that

$$A_j = \bigcup_{i=1}^{t_j} \tilde{A}_i^j, j = 0,...,k-1; 1 \le t_j < m; \tilde{A}_t^j \cap \tilde{A}_s^j = \varnothing, t \neq s$$

for the function $\tilde{f}(x_1,...,x_n):T \to E_m; T \subseteq E_k^n \vee T \subseteq O^n$ whose domain coincides with the domain of the initial $k$-valued function $f$. In this case $\tilde{f}(x_1,...,x_n)$ is a partially defined $m$-valued threshold function (partially defined in terms of $m$-valued logic). We will consider this case in detail in Chapter 5 where MVN with a periodic activation function will be presented.

Finally, we should outline one generalization of MVN and its activation function, which also follows from the edged separation. It should be mentioned that we can also consider the following generalization of the $k$-valued threshold function. Let $\mathbb{C} \setminus \{0\} = [s_0, s_1,..., s_{k-1}]$ be an ordered decomposition of the set of points of the complex plane excluding the origin $(0, 0)$. This is the decomposition of the set $\mathbb{C} \setminus \{0\}$ into a set of mutually disjoint sectors $s_0, s_1,..., s_{k-1}$, which are bounded by the rays $l_0, l_1,..., l_{k-1}$ originating at the origin. Only the first of those two rays, which form a sector, is included to this sector. The direction of the ray $l_0$ coincides with the direction of the positive real semi-axis. The angular sizes $\varphi_0, \varphi_1, ..., \varphi_{k-1}$ of the sectors $s_0, s_1,..., s_{k-1}$ are arbitrary and should only satisfy the condition $\varphi_0 + \varphi_1 + ... + \varphi_{k-1} = 2\pi$. Let us define the following function

$$\tilde{P}(z) = \varepsilon^j \text{ if } z \in s_j; \ j = 0, 1,..., k-1, \ \varepsilon = 2\pi i / k.$$

We may use this function $\tilde{P}(z)$ as the MVN activation function instead of the function $P(z)$ defined by (2.50). These two functions differ from each other by the angular size of the sectors, to which they divide the complex plane. The function $P(z)$ separates the complex plane into $k$ equal sectors of the angular size $2\pi / k$, while the function $\tilde{P}(z)$ separates the complex plane into $k$ sectors, which may have different angular sizes. MVN with the activation function $\tilde{P}(z)$ was not considered so far, but it should be attractive to consider it in the further work.

## 2.4  MVN and a Biological Neuron

We have already introduced MVN and explained why this neuron and complex-valued neurons in general are important. MVN is more functional and more flexible than its real-valued counterparts. MVN makes it possible to implement multiple-valued input/output mappings. It can also implement continuous input/output mappings. It can treat properly phase and the information contained in phase.

Before we will move to the MVN learning algorithm and learning rules and before we will consider MVN-based neural networks and their applications, we would like to outline another important feature of MVN. Let us consider how the MVN phenomenon can be used for simulation of a biological neuron.

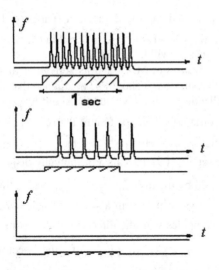

For many years, artificial neurons and artificial neural networks have been developing separately from the study of their biological counterparts. However, it was always very attractive to simulate a biological neuron and its behavior, as well as to understand how the most sophisticated neural network (the brain) works. A biological neuron is not the threshold neuron. The signal transmission in the nervous system does not have a digital nature in which a neuron is assumed to be either fully active or inactive [66]. Thus, a popular assumption that a biological neuron can "fire" only when its excitation exceeds some threshold is nothing more than simplification. The level of the biological neuron "excitation" and, respectively, the information, which the neuron transmits to

**Fig. 2.32** A biological neuron generates a sequence of spikes (a spike train). Spikes' magnitude is a constant, while their frequency can be high ("excitation", top graph), moderate (middle graph) and zero ("inhibition", no spikes, bottom graph)

other neurons, is coded in the frequency of pulses (spikes) [67], which form the output signal of the neuron (also referred to as a spike train). Hence, a better conclusion would be to interpret the biological neural systems as using a form of pulse frequency modulation to transmit the information. The nerve pulses passing along the axon of a particular neuron are of approximately constant magnitude (see Fig. 2.32), but the number of generated pulses (spikes) and their time spacing (that is the frequency of generated pulses) contains that information, which is transmitted to other neurons [66-69]. This is illustrated in Fig. 2.32.

The understanding of these facts initiated a number of works, e.g., [70-73], where a problem of simulation of a biological neuron is studied.

The most popular model, which is used now, is a spiking neuron model [70-73]. Spiking neural model incorporates the concept of time in order to simulate a real behavior of a biological neuron. However, in our view this approach is rather biological than mathematical. It simulates those electro-chemical processes that take place in a biological neuron. Of course, this is very important for understanding of these processes. But it could be very attractive and important to be able to simulate also the informational processes in biological neurons. This simulation is perhaps the most interesting for understanding the mechanisms of thinking and functioning of the human brain.

As it was mentioned above, the information transmitted by a biological neuron to other neurons is contained only in the frequency of generated impulses, while their magnitude is a constant. To adapt this model for simulation of the different cortical cells whose typical firing frequency is different, the different versions of a spiking neuron are suggested to be used [73], because there is no universal model. The most significant drawback of a spiking neuron model in terms of information processing is that the frequency of spikes is rather treated there as some "abstract number" than a specific physical feature. In fact, since definitely frequency is meaningful in the neural information processing, it is necessary to consider such a model of a biological neuron, which operates with frequency not as with an abstract real or integer number, but takes into account its physical nature.

Let us consider how the continuous MVN can be employed to simulate a biological neuron. As we already know, the continuous MVN inputs and output are located on the unit circle. Their unitary magnitude is a constant (as well as the magnitude of the spikes generated by a biological neuron), and they are determined by their arguments (phases). Thus, they are the numbers $e^{i\varphi}, 0 \le \varphi < 2\pi$.

Let us return to a biological neuron. As we have mentioned, the information transmitted by a biological neuron to other neurons and accepted by the latter ones is completely contained in the frequency of the generated spikes, while their magnitude is a constant. Let $f$ be the frequency. As it is commonly known from the oscillations theory, if $t$ is the time, and $\varphi$ is the phase, then

$$\varphi = \theta_0 + 2\pi \int f dt = \theta_0 + \theta(t).$$

If the frequency $f$ is fixed for some time interval $\Delta t$, then the last equation may be transformed as follows

$$\varphi = \theta_0 + 2\pi f \Delta t$$

Thus, if the frequency $f$ of spikes generated by a biological neuron is known, it is very easy to transform it to the phase $\varphi$. But then the phase $\varphi$ can be easily transformed to the complex number $e^{i\varphi}$ located on the unit circle!

As we already know, exactly such numbers $e^{i\varphi}, 0 \le \varphi < 2\pi$, which are located on the unit circle, are the continuous MVN inputs. The opposite is also true: having any complex number located on the unit circle, which is the MVN output, it is possible to transform it to the frequency. This means that all signals generated by the biological neurons may be unambiguously transformed into the form acceptable by the MVN, and vice versa, preserving a physical nature of the signals.

Thus, the continuous MVN can be used to simulate a biological neuron. This will be not biologically, but mathematically inspired simulation, which might be very useful for understanding of informational processes in a biological neuron. Respectively, an MVN-based network can be used for simulation of informational processes in a biological neural network. This interesting idea has not been developed yet, but it looks very attractive and natural. Its development will be a good subject for the further work.

## 2.5  Concluding Remarks to Chapter 2

In this Chapter, we have first observed a theoretical background behind MVN – multiple-valued threshold logic over the field of complex numbers. In this model of $k$-valued logic, its values are encoded by the $k$th roots of unity. We introduced a $k$-valued threshold function as a function, which can be represented using the specific complex-valued weighting parameters and the universal $k$-valued predicate $P$.

Then we have introduced the discrete MVN as a neuron whose input/output mapping is presented by some $k$-valued threshold function. The $k$-valued predicate $P$ is the discrete MVN activation function. We have also introduced the continuous MVN. The MVN inputs and output are located on the unit circle. The MVN activation function depends only on the argument of the weighted sum and does not depend on its magnitude.

We have deeply considered the edged separation of an $n$-dimensional space, which is implemented by MVN. We showed that the MVN weights determine the $k$-edge, which is resulted from the division of an $n$-dimensional space by a set of hyperplanes into $k$ subspaces.

We have also outlined how the continuous MVN can be used for simulation of a biological neuron.

So we are ready now to consider how MVN learns.

# Chapter 3
# MVN Learning

"Since we cannot know all that there is to be known about anything,
we ought to know a little about everything."

Blaise Pascal

In this Chapter, we consider all aspects of the MVN learning. We start in Section 3.1 from the specific theoretical aspects of MVN learning and from the representation of the MVN learning algorithm. Then we describe the MVN learning rules. In Section 3.2, we consider the first learning rule, which is based on the adjustment of the weights depending on the difference (in terms of the angular distance) between the arguments of the current weighted sum and the desired output. In Section 3.3, we present the error-correction learning rule for MVN. For both learning rules presented in Sections 3.2 and 3.3, we prove theorems about the convergence of the learning algorithm based on these rules. In Section 3.4, we discuss the Hebbian learning rule for MVN. Section 3.5 contains some concluding remarks.

## 3.1 MVN Learning Algorithm

In Chapter 2, we have introduced discrete and continuous MVN. We also have considered the edged separation of an $n$-dimensional space, which is implemented by MVN using the $k$-edge generated by the weights that implement a corresponding input/output mapping.

As any other neuron, MVN creates the weights implementing its input/output mapping during the learning process. The ability to learn from its environment is a fundamental property of a neuron. We have already observed fundamentals of learning in Section 1.2. According to Definition 1.2, the MVN learning as well as any other neuron learning is the iterative process of the adjustments of the weights using a learning rule. In other words, it is reduced to the adaptation of the neuron to its input/output mapping thorough the adjustment of the weights using a learning rule every time, when for some learning sample the neuron's actual output does not coincide with the desired output.

In this Chapter, we present the MVN learning algorithm, which can be based on the two learning rules. One of the rules is based on the estimation of the closeness of the actual output to the desired one in terms of angular distance. Another one is the error-correction learning rule. We will also consider the Hebbian learning for MVN.

I. Aizenberg: Complex-Valued Neural Networks with Multi-Valued Neurons, SCI 353, pp. 95–132.
springerlink.com                                                    © Springer-Verlag Berlin Heidelberg 2011

### 3.1.1  Mechanism of MVN Learning

Let us now consider what MVN has to do when it learns, what is behind the learning process. It is also important to mention that those fundamentals of MVN learning, which we consider here are important not only for a single MVN, but for MLMVN (multilayer neural network based on multi-valued neurons), whose learning algorithm with the error backpropagation we will consider in Chapter 4.

Let $A$ be a learning set with the cardinality $|A| = N$ , thus the learning set contains $N$ learning samples. They are such samples $\left( x_1^i, ..., x_n^i \right) \to d_i, i = 1, ..., N$ for which the exact desired output $f \left( x_1^i, ..., x_n^i \right) = d_i, i = 1, ..., N$ is known.

If we return to the threshold neuron, its learning process can be presented in the following way. Its learning set can always be presented as $A = A_1 \bigcup A_{-1}$ , where $A_1$ is a subset of the learning samples where the neuron's output has to be equal to 1, and $A_{-1}$ is a subset of the learning samples where the neuron's output has to be equal to -1. As we have seen, learning in this case is reduced to the search for a hyperplane, which separates the subsets $A_1$ and $A_{-1}$ of the learning set in that $n$-dimensional space where a problem to be learned is defined. The coefficients of a hyperplane equation are the weights implementing a corresponding input/output mapping.

Let us consider now how the discrete MVN learns. Taking into account that the discrete MVN implements a $k$-valued input/output mapping, it is easy to conclude that a learning set should consist of $k$ classes. Let $k > 2$ be some integer. Let us consider $(n+1)$ - dimensional vectors $X = \left( 1, x_1, ..., x_n \right)$ , $\left( x_1, ..., x_n \right) \in T \subseteq O^n$ , where $O$ is the set of points located on the unit circle. The $0^{\text{th}}$ coordinate (constant 1) can be considered as a pseudo input corresponding to the weight $w_0$ . We introduce it just to be able to consider a weighted sum $w_0 + w_1 x_1 + ... + w_n x_n$ as a dot product of two $(n+1)$ - dimensional vectors $X = \left( 1, x_1, ..., x_n \right)$ and $W = \left( w_0, w_1, ..., w_n \right)$ .

Let $A_j$ be a learning subset $\left\{ X_1^{(j)}, ..., X_{N_j}^{(j)} \right\}$ of the input neuron states corresponding to the desired output $\varepsilon^j, j = 0, ..., k-1$ . In such a case we can present the entire learning set $A$ as a union of the learning subsets $A_j, j = 0, 1, ..., k-1$ as follows $A = \bigcup_{0 \le j \le k-1} A_j$ . In general, some of the sets $A_j, j = 0, 1, ..., k-1$ may be empty if for some $j = 0, ..., k-1$ there is no learning sample whose desired output is $\varepsilon^j$. It is also clear that $A_i \cap A_j = \varnothing$ for any $i \neq j$. A practical content behind this mathematical representation is, for example, a $k$-class classification problem, which is presented by the learning set $A$ containing learning subsets $A_j, j = 0,1,...,k-1$ such that each of them contains learning samples belonging only to one of $k$ classes labeled by the corresponding class membership label $\varepsilon^j, j = 0,...,k-1$.

**Definition 3.16.** The sets $A_0, A_1,..., A_{k-1}$ are called $k$-*separable*, if it is possible to find a permutation $R = (\alpha_0, \alpha_1,..., \alpha_{k-1})$ of the elements of the set $K = \{0, 1, ..., k-1\}$, and a weighting vector $W = (w_0, w_1,..., w_n)$ such that

$$P(X, \overline{W}) = \varepsilon^{\alpha_j} \qquad (3.73)$$

for each $A_j, j = 0,1,...,k-1$. Here $\overline{W}$ is a vector with the components complex-conjugated to the ones of the vector $W$, $(X, \overline{W})$ is a dot product of the $(n+1)$-dimensional vectors within the $(n+1)$-dimensional unitary space, $P$ is the MVN activation function (2.50). Without loss of generality we may always supply (2.50) by $P(0,0) = \varepsilon^0 = 1$. This means that the function $P$ is now determined on the entire set $\mathbb{C}$ of complex numbers.

Let $A$ be a learning set and $A = \bigcup_{0 \leq j \leq k-1} A_j$ is a union of the $k$-separable disjoint subsets $A_0, A_1,..., A_{k-1}$. On the one hand, this means that (3.73) holds for any $X \in A$. On the other hand, the MVN input/output mapping presented by such a learning set is described by the $k$-valued function $f(x_1, ..., x_n): A \rightarrow E_k$. It follows from the fact that the learning subsets $A_0, A_1,..., A_{k-1}$ are $k$-separable and (3.73) holds for any $X \in A$ that (2.51) also holds for the function $f(x_1, ..., x_n)$ on the its entire domain $A$, which means that according to Definition 2.7 this function is a $k$-valued threshold function. Hence, we proved the following theorem.

**Theorem 15.** If the domain of a $k$-valued function $f(x_1, ..., x_n): A \rightarrow E_k$ can be represented as a union of $k$-separable disjoint subsets $A_0, A_1,..., A_{k-1}$ (some of them can be empty in general), then the function $f(x_1, ..., x_n)$ is a $k$-valued threshold function.

It directly follows from Definition 3.16 and (3.73) that the problem of MVN learning should be reduced to the problem of the $k$-separation of learning subsets. In other words, the learning problem for the given learning subsets $A_0, A_1, ..., A_{k-1}$ can be formulated as a problem, how to find a permutation $(\alpha_0, \alpha_1, ..., \alpha_k)$ and a weighting vector $W = (w_0, w_1, ..., w_n)$ such that (3.73) holds for the entire learning set $A$.

On the other hand, we see that the notion of $k$-separation of the learning subsets is closely related to the notion of edge-like sequence (see Definition 2.14). Evidently, if the sets $A_0, A_1, ..., A_{k-1}$ are $k$-separable, and the permutation $R = (\alpha_0, \alpha_1, ..., \alpha_{k-1})$ is applied to them, then the edge-like sequence results from such a permutation. From the geometrical point of view, the $k$-separation means that elements from only one learning subset (one class) $A_j = \{X \mid f(X) = \varepsilon^j\}$ belong to each edge of the $k$-edge. Moreover, the elements belonging to the same class cannot belong to the different edges.

We will say that any infinite sequence $S_u$ of objects $u_0, u_1, ..., u_{k-1}$ such that $u_j \in S_u \Rightarrow u_j \in A$, $u \in A \Rightarrow u = u_j$ for some $j$ taking its values from the infinite set, forms a learning sequence from the set $A$.

The MVN learning process should be defined as a process of finding such a permutation $R = (\alpha_0, \alpha_1, ..., \alpha_{k-1})$ of the elements of the set $K = \{0, 1, ..., k-1\}$ and such a weighting vector $W = (w_0, w_1, ..., w_n)$ that (3.73) holds for the entire learning set $A$.

Let us suppose that the permutation $R = (\alpha_0, \alpha_1, ..., \alpha_{k-1})$ is already known. Then the learning process is reduced to obtaining the sequence $S_w$ of the weighting vectors $W_0, W_1, ...$ such that starting from the some $m_0$ $W_{m_0} = W_{m_0+1} = W_{m_0+2} = ...$ and (3.73) holds. Each weighting vector in the sequence $S_w$ corresponds to the next learning sample. The process of finding the sequence $S_w$ of the weighting vectors is iterative. One iteration of the learning process consists of the consecutive checking for all the learning samples whether (3.73) holds for the current learning sample. If so, the next learning sample should be checked. If not, the weights should be adjusted according to a learning rule (we did not consider learning rules yet). One *learning iteration* (*learning epoch*) is a complete pass over all the learning samples $(x_1^i, ..., x_n^i) \rightarrow d_i, i = 1, ..., N$.

Stabilization of the sequence $S_w$ in conjunction with the fact that (3.73) holds means that the learning process converges with the zero error.

As well as for any other neuron, there can be situations when MVN learning with the zero error is not reasonable. It depends on the particular problem, which is necessary to learn. If errors for some learning samples are acceptable (which means that (3.73) may not hold for some learning samples), the mean square error (MSE) (1.20) or the root mean square error (RMSE) (1.21) criteria should be used to stop the learning process. It is important to understand that for the discrete MVN both MSE and RMSE should be applied to the errors in terms of numbers of sectors (see Fig. 2.21), thus not to the elements of the set $E_k = \left\{ \varepsilon_k^0, \varepsilon_k, ..., \varepsilon_k^{k-1} \right\}$, but to the elements of the set $K = \{0,1,...,k-1\}$ or to their arguments $\left\{ \arg \varepsilon_k^0, \arg \varepsilon_k, ..., \arg \varepsilon_k^{k-1} \right\}$. Hence, in this case the error for the $s$th learning sample $\gamma_s, s = 1, ..., N$ is either of

$$\gamma_s = \left( \alpha_{j_s} - \alpha_s \right) \bmod k; s = 1, ..., N ,$$

(3.74)

$$\gamma_s = \left( \arg \varepsilon^{\alpha_{j_s}} - \arg \varepsilon^{\alpha_s} \right) \bmod 2\pi; s = 1, ..., N ,$$

(3.75)

where $\varepsilon^{\alpha_{j_s}}$ is the desired neuron's output for the $s$th learning sample, and $\varepsilon^{\alpha_s}$ is the actual MVN output.

The learning process continues until either of MSE or RMSE drops below some pre-determined acceptable minimal value $\lambda$. For the reader's convenience, let us adapt here the expressions (1.20) and (1.21) for MSE and RMSE over all $N$ learning samples for the local errors (3.74) and (3.75). Equations (1.20) and (1.21) are transformed to, respectively

$$MSE = \frac{1}{N} \sum_{s=1}^{N} \gamma_s^2 < \lambda,$$

(3.76)

$$RMSE = \sqrt{MSE} = \sqrt{\frac{1}{N} \sum_{s=1}^{N} \gamma_s^2} < \lambda.$$

(3.77)

If either of MSE (3.76) or RMSE (3.77) criteria is used to stop the learning process, we have to take into account that the sequence of weighting vectors $S_w$ may have left not stabilized when $\lambda > 0$.

The learning process for the continuous MVN does not differ from the one for the discrete MVN. Anyway, the learning set $A$ even for the continuous-valued input/output mapping $f(x_1,...,x_n):T \to O, T \subseteq O^n$ is finite, which means that it can be represented as a union $A = \bigcup_{0 \le j \le k-1} A_j$ of the learning subsets $A_j, j = 0,1,...,k-1$, where $k$ is the number of different values of the function $f(x_1,...,x_n)$ corresponding to the elements of the learning set. In other words, $k$ is the cardinality of the range of the function $f(x_1,...,x_n)$ with respect to the learning set $A$. If the continuous MVN should learn not with the zero error, then either of MSE (3.76) or RMSE (3.77) criteria with respect to the local errors (3.75) should be applied.

## 3.1.2  Learning Strategy

Let us have the learning set containing $N$ learning samples $(x_1^i,...,x_n^i) \to d_i, i = 1,...,N$. As we told, one iteration of the learning process consists of the consecutive checking for all the learning samples whether (3.73) holds for the current learning sample. If it does not hold, the weights should be adjusted using a learning rule. This process continues either until the zero error is reached or one of (3.76) or (3.77) holds. It should be mentioned that in the latter case criterion (3.73) can be replaced by either of

$$\gamma_i = (\alpha_{j_i} - \alpha) \bmod k < \beta; i = 1,...,N , \tag{3.78}$$

$$\gamma_i = \left( \arg \varepsilon^{\alpha_{j_i}} - \arg \varepsilon^{\alpha} \right) \bmod 2\pi < \beta; i = 1,...,N , \tag{3.79}$$

where $\beta$ is some acceptable error level for a single learning sample in terms of sectors numbers (3.78) or angular distance (3.79).

If (3.73) does not hold or one of (3.78) or (3.79) does not hold (depending on which error criterion is used), then the MVN weights must be adjusted using a learning rule. Two learning rules will be considered below (Sections 3.2 and 3.3). Geometrically, adjustment of the weights means movement of the weighted sum from the incorrect sector $s$ (discrete MVN, see Fig. 3.33a) to the correct sector $q$ or from the incorrect ray $OY$ (continuous MVN, see Fig. 3.33b) to the correct ray $OD$.

Thus, the following learning algorithm should be used for MVN learning.

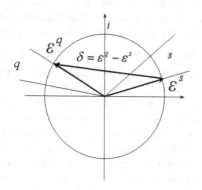 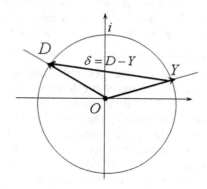

(a) Discrete MVN. $\varepsilon^q$ is the desired output, $\varepsilon^s$ is the actual output

(b) Continuous MVN. $D$ is the desired output, $Y$ is the actual output.

**Fig. 3.33** Geometrical interpretation of the weights adjustment. The weighted sum has to be moved from the incorrect domain to the correct one.

Let $X_j^s$ be the $s$th element of the learning set $A$ belonging to the learning subset $A_j$. Let $N$ be the cardinality of the set $A$, $|A| = N$.

Let *Learning* be a flag, which is 'True' if the weights adjustment is required and "False", if it not required, and $r$ be the number of the weighting vector in the sequence $S_w$.

Step 1. The starting weighting vector $W_0$ is chosen arbitrarily (e.g., real and imaginary parts of its components can be random numbers); $r=0$; $t=1$; *Learning* = 'False';

Step 2. Check (3.73) or one of (3.78) or (3.79) (depending on the error criterion, which is used) for $X_j^s$ :

    *if* (3.73) or one of (3.78) or (3.79) holds
    *then go to* the step 4
    *else begin Learning* = 'True'; *go to* Step 3 *end*;

Step 3. Obtain the vector $W_{r+1}$ from the vector $W_r$ by the learning rule (to be considered);

Step 4. $t = t+1$;     *if* $t \leq N$
        *then go to* Step 2
        *else if Learning* = 'False'
            *then* the learning process is finished successfully
            *else begin* $t=1$; *Learning* = 'False'; *go to* Step 2; *end*.

A learning rule, which should be applied on Step 3, is a key point of the learning algorithm. It determines the correction of the weights. It should ensure that after the weights adjustment a weighting vector resulted from this adjustment approaches us closer to the stabilization of the sequence $S_w$ of the weighting vectors. In other words, a learning rule should approach the convergence of the learning algorithm and ensure decreasing of the error after each learning step.

We will consider here two learning rules. Their wonderful property is that they are *derivative-free*. The MVN learning based on this rules should not be considered as the optimization problem. Yes, we have to minimize the neuron error or even to reduce it to zero. But we will see that both learning rules, which we will consider here, generalize the Novikoff's approach to the threshold neuron error-correction learning [12] where the distance from the current weighting vector to the desired weighting vector is decreasing during the learning process without involvement of any optimization technique.

Let us consider the MVN learning rules in detail.

## 3.2 MVN Learning Rule Based on the Closeness to the Desired Output in Terms of Angular Distance

### 3.2.1 Basic Fundamentals

Both learning rules (that we consider in this section and in the following section) are based on the compensation of the MVN error by adding the adjusting term to each component of the current weighting vector.

While the second learning rule, which we will consider below in Section 3.3, can be considered as a direct generalization of the Rosenblatt error-correction leaning rule for the threshold neuron, the first learning rule, which we are going to consider now, is based on the compensation of the error between the desired and actual neuron outputs in terms of angular distance between their arguments. This learning rule was initially proposed by N. Aizenberg and his co-authors in [34, 36], then the convergence of the learning algorithm based on this rule was proven in [37], some more adjustments were made in [38] and [60]. Here we will present the most comprehensive description of this algorithm with its deeper analysis (compared to earlier publications).

Let the discrete MVN has to learn some input/output mapping $f(x_1,...,x_n):T \to E_k;T \subseteq O^n$. We have already mentioned that in the case of the continuous MVN and continuous-valued input/output mapping $f(x_1,...,x_n):T \to O;T \subseteq O^n$, all considerations can be reduced to the discrete case because since a learning set with the cardinality $N$ contains exactly $N$ learning samples and the neuron may have at most $k \leq N$ different outputs. So we may consider the learning algorithm just for the discrete MVN.

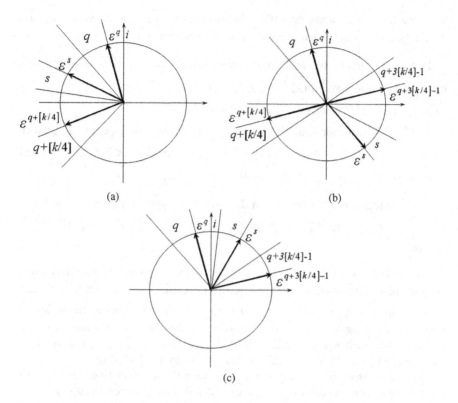

**Fig. 3.34** Thee cases of mutual location of the desired ($\varepsilon^q$) and actual ($\varepsilon^s$) MVN outputs

Let us define the following partial order relation on the set $E_k = \left\{\varepsilon_k^0, \varepsilon_k, ..., \varepsilon_k^{k-1}\right\}$. Let us for simplicity use the following notation $\varepsilon = \varepsilon_k$. We will say that $\varepsilon^\alpha$ *precedes* to $\varepsilon^\beta$ ($\varepsilon^\alpha \prec \varepsilon^\beta$) if and only if the following condition holds

$$\left(\alpha \bmod k \leq \beta \bmod k\right) \wedge \left(0 \leq \arg \varepsilon^\beta - \arg \varepsilon^\alpha < \pi\right) \text{ or}$$

$$\left(\alpha \bmod k \geq \beta \bmod k\right) \wedge \left(-\pi \leq \arg \varepsilon^\beta - \arg \varepsilon^\alpha < 0\right),$$

where $\alpha, \beta \in K = \left\{0, 1, ..., k-1\right\}$. In other words $\varepsilon^\alpha \prec \varepsilon^\beta$ if and only if $\varepsilon^\alpha$ is located "lower" than $\varepsilon^\beta$ in the clockwise direction from $\varepsilon^\beta$ in the "right" half-plane from the line crossing the origin and the point corresponding to $\varepsilon^\alpha$ on the unit circle.

Let $\varepsilon^q$ be the desired MVN output and $\varepsilon^s$ be the actual MVN output (see Fig. 3.33a). Let us discover how far they can be located from each other in terms of the angular distance accurate within the angle $\pi/2$. Let us consider the case $k \geq 4$ (see Fig. 3.34). There are three possible situations.

1) $\varepsilon^s$ is located to the "left" (in the counterclockwise direction) from $\varepsilon^q$ such that $\left(\arg \varepsilon^s - \arg \varepsilon^q\right) \bmod 2\pi \leq \pi/2$ and $\varepsilon^{(q+1)\bmod k} \prec \varepsilon^s \prec \varepsilon^{(q+[k/4])\bmod k}$ (see Fig. 3.34a, [k/4] is an integer part of k/4).

2) $\varepsilon^s$ is located approximately across the unit circle with respect to $\varepsilon^q$, which means that $\pi/2 < \left(\arg \varepsilon^s - \arg \varepsilon^q\right)\bmod 2\pi < 3\pi/2$ and $\varepsilon^{(q+[k/4]+1)\bmod k} \prec \varepsilon^s \prec \varepsilon^{(q+3[k/4]-1)\bmod k}$ (see Fig. 3.34b).

3) $\varepsilon^s$ is located to the "right" (in the clockwise direction) from $\varepsilon^q$ such that $\left(\arg \varepsilon^s - \arg \varepsilon^q\right) \bmod 2\pi \leq \pi/2$ and $\varepsilon^{(q+3[k/4])\bmod k} \prec \varepsilon^s \prec \varepsilon^{(q+k-1)\bmod k}$ (see Fig. 3.34c).

The goal of a learning rule is to correct the error, which is approximately equal in terms of angular distance for the three just considered cases $-\pi/2$, $\pi$, and $\pi/2$, respectively. Thus, to correct the error, we need to "rotate" the weighted sum, compensating this error. This "rotation" must be done by the adjustment of the weights in such a way, that the adjusted weighting vector moves the weighted sum either exactly where we need or at least closer to the desired sector.

This means that our learning rule has to contain some "rotating" term, which should vary depending on which of the considered above error cases takes place.

The following learning rule was proposed in [34, 36, 37] to correct the weighting vector $W = \left(w_0, w_1, ..., w_n\right)$

$$\tilde{w}_i = w_i + \frac{1}{n+1}C_r\omega_r\varepsilon^q\bar{x}_i; i = 0,1,...,n, \tag{3.80}$$

where $n$ is the number of the neuron inputs, $w_i$ is the $i$th component of the weighting vector before correction, $\tilde{w}_i$ is the same component after correction, $\bar{x}_i$ is the $i$th neuron input complex-conjugated ($x_0 \equiv 1$ as we agreed above is a pseudo-input corresponding to the bias $w_0$), $C_r$ is the learning rate, $\varepsilon^q$ is the desired output, and $\omega_r$ is a rotating coefficient, which has to compensate the angular error, which we have just considered. To choose $\omega_r$, we have to consider the following cases corresponding to the error cases considered above. Let $i$ be an imaginary unity, and $\varepsilon = \varepsilon_k = e^{i2\pi/k}$.

Case 1. $\omega_m = -i\varepsilon$, if $\varepsilon^s = \varepsilon^{q+1}$ for $k$=2 and $k$=3, or

$\varepsilon^{(q+1)\bmod k} \prec \varepsilon^s \prec \varepsilon^{(q+[k/4])\bmod k}$ for $k\geq4$.

Case 2. $\omega_m = 1$, if $\varepsilon^{(q+[k/4]+1)\bmod k} \prec \varepsilon^s \prec \varepsilon^{(q+3[k/4]-1)\bmod k}$ for $k\geq4$ (for $k<4$ such a case is impossible).

Case 3. $\omega_m = i$, if $\varepsilon^s \prec \varepsilon^{q+2}$ for $k$=3, or

$\varepsilon^{(q+3[k/4])\bmod k} \prec \varepsilon^s \prec \varepsilon^{(q+k-1)\bmod k}$ for $k\geq4$ (for $k$=2 such a case is impossible).

Thus, adjusting the weights according to (3.80), we obtain the following equation for the $r+1^{st}$ weighting vector belonging to the sequence $S_w$ from the $r$th weighting vector belonging to the same sequence

$$W_{r+1} = W_r + \frac{1}{n+1}C_r\omega_r\varepsilon^q\bar{X} , \qquad (3.81)$$

where $r$ is the number of the current weighting vector in the sequence $S_w$, addition is component-wise, and $\bar{X} = (1,\bar{x}_1,...,\bar{x}_n)$ is the vector of neuron inputs with the complex-conjugated components.

Before justification of the choice of $\omega_r$ let us first clarify a very important role of the multiplier $\frac{1}{n+1}$ in the learning rule (3.80) and let us see how the weighted sum changes after the weights are corrected. Let us find the updated weighted sum after the weights are corrected according to (3.80). The current weighted sum is $z = w_0 + w_1 x_1 +...+ w_n x_n$. Suppose for simplicity, but without loss of generality that $C_m = 1$. For the updated weighted sum, taking into account that $x_i \bar{x}_i = 1; i = 0,1,...,n$ (since all $x_i$ are located on the unit circle) we obtain

$$\tilde{z} = \tilde{w}_0 + \tilde{w}_1 x_1 +...+ \tilde{w}_n x_n = \left(w_0 + \frac{1}{n+1}\omega_r\varepsilon^q\right) + \left(w_1 + \frac{1}{n+1}\omega_r\varepsilon^q\bar{x}_1\right)x_1 +$$

$$+...+\left(w_n + \frac{1}{n+1}\omega_r\varepsilon^q\bar{x}_n\right)x_n =$$

$$= \underbrace{w_0 + w_1 x_1 +...+ w_n x_n}_{z} + \underbrace{\frac{1}{n+1}\omega_r\varepsilon^q +...+ \frac{1}{n+1}\omega_r\varepsilon^q}_{n+1\ \text{times}} = z + \omega_r\varepsilon^q.$$

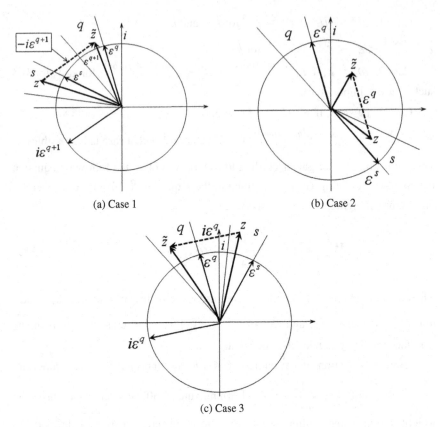

(a) Case 1                                              (b) Case 2

(c) Case 3

**Fig. 3.35** Movement of the weighted sum $z$ after the correction of the weights according to (3.80)

This means that after the weights are corrected, the weighted sum is changed by $\omega_r \varepsilon^q$. The multiplier $\dfrac{1}{n+1}$, which can be considered as a constant learning rate is important to avoid change of the weighted sum by $(n+1)\omega_r \varepsilon^q$, which may lead to the jump over a desired output. Thus, using the multiplier $\dfrac{1}{n+1}$, we share the adjusting term among all the weights. In fact, we do not know which of the weights contributes more to the error. Hence, if we assume that each of them contributes uniformly, this assumption is natural.

We can clarify now the choice of $\omega_r$ in (3.80) (see Fig. 3.35).

In the Case 1 (see Fig. 3.35a), the current weighted sum should be rotated clockwise, that is to the right side from its current location, because the actual output $\varepsilon^s$ is located to the left from the desired output $\varepsilon^q$ and the difference between the arguments of the actual and desired outputs does not exceed $\pi/2$. Choosing $\omega_r = -i\varepsilon$, we ensure that after the correction of the weights the updated weighted sum $z + \omega_r \varepsilon^q = z - i\varepsilon\varepsilon^q = z - i\varepsilon^{q+1}$ moves closer or exactly to the desired sector $q$ and the MVN output moves closer or exactly becomes equal to $\varepsilon^q$, respectively.

In the Case 3 (see Fig. 3.35c), situation is similar, the difference between the arguments of the actual and desired outputs does not exceed $\pi/2$, but we need to rotate the current weighted sum counterclockwise, that is to the left side from its current location, because the actual output $\varepsilon^s$ is located to the right from the desired output $\varepsilon^q$. Choosing $\omega_r = i$, we ensure that after the correction of the weights the updated weighted sum $z + \omega_r \varepsilon^q = z + i\varepsilon^q$ moves closer or exactly to the desired sector $q$ and the MVN output moves closer or exactly becomes equal to $\varepsilon^q$, respectively.

In the Case 2 (see Fig. 3.35b), the current weighted sum should be flipped because the actual output $\varepsilon^s$ is about the opposite to the desired output $\varepsilon^q$ and the difference between the arguments of the actual and desired outputs exceeds $\pi/2$. Since in this case we cannot have any preference where to rotate the weighted sum (clockwise or counterclockwise), we should simply rotate it such that it will be moved to the desired output. Choosing $\omega_r = 1$, we ensure that after the correction of the weights the updated weighted sum $z + \omega_r \varepsilon^q = z + \varepsilon^q$ moves closer or even exactly to the desired sector $q$ and the MVN output moves closer or may exactly becomes equal to $\varepsilon^q$, respectively.

### 3.2.2  Convergence Theorem

Now we are ready to formulate and prove the theorem about the convergence of the MVN learning algorithm with the learning rule (3.80). We will provide the reader with a new proof of this theorem compared to [37] and [60]. This new proof is shorter and more elegant.

Suppose that the permutation $R = (\alpha_0, \alpha_1, ..., \alpha_{k-1})$ such that (3.73) holds for the entire learning set $A$ is known.

**Theorem 16** (About the convergence of the learning algorithm with the learning rule (3.80)). If the learning subsets $A_0, A_1, ..., A_{k-1}$ of the learning set $A$ ($A = \bigcup\limits_{0 \le j \le k-1} A_j$) are $k$-separable for the given value of $k$ according to Definition 3.16 (which means that the corresponding MVN input/output mapping is a $k$-valued threshold function), then the MVN learning algorithm with the rule (3.80) converges after a finite number of steps.

**Proof.** Suppose that the conclusion of the theorem is false. This means that the sequence $S_w$ of the weighting vectors is infinite. Therefore the weights correction using the rule (3.80) gives the infinite amount of the new weighting vectors, which do not satisfy condition (3.73) at least for one element from some learning subset. According to our assumption the learning subsets $A_0, A_1, ..., A_{k-1}$ are $k$-separable. Therefore the weighting vector $W$ exist such that (3.73) holds for the entire learning set $A$.

For simplicity and without loss of generality, let us start the learning process from the zero vector $W_1 = ((0,0),(0,0),...,(0,0))$, where $(a,b)$ is a complex number $a+bi$, ($i$ is an imaginary unity). Let $S_X = (X_1, X_2,..., X_N)$ be a learning sequence of input vectors $X_j = (1, x_1^j,..., x_n^j), j = 1,..., N$, and $S_W = (W_1, W_2,..., W_r,...)$ be a sequence of weighting vectors, which appear during the learning process. We have to prove that this sequence cannot be infinite. Let us remove from the learning sequence those vectors for which $W_{r+1} = W_r$, in other words, those input vectors, for which (3.73) hold without any learning. Let $S_{\tilde{W}}$ be the reduced sequence of the weighting vectors. The Theorem will be proven if we will show that the sequence $S_{\tilde{W}}$ is finite. Let us suppose that the opposite is true: the sequence $S_{\tilde{W}}$ is infinite. So from the assumption that $S_{\tilde{W}}$ is infinite, we have to get the contradiction with the conditions of the theorem.

Without loss of generality we can take $C_r = 1$ in (3.80). This leads to the following transformation of (3.81):

$$\tilde{W}_{r+1} = \tilde{W}_r + \frac{1}{n+1} \omega_r \varepsilon^q \bar{X}_r, \tag{3.82}$$

Thus, our sequence of the weighting vectors $S_{\tilde{W}}$ is obtained according to (3.82). The theorem will be proven, if we can prove that the sequence $S_{\tilde{W}}$ is finite.

Suppose the desired MVN output for the first learning sample does not coincide with the actual MVN output. Thus, we have to adjust the weights according to (3.82):

$$\tilde{W}_2 = \frac{1}{n+1} \omega_{m_1} \bar{\bar{X}}_1, \text{ and then for the next correction we obtain}$$

$$\tilde{W}_3 = \tilde{W}_2 + \frac{1}{n+1}\omega_{m_2}\bar{\tilde{X}}_2 = \frac{1}{n+1}\left[\omega_{m_1}\bar{\tilde{X}}_1 + \omega_{m_2}\bar{\tilde{X}}_2\right],....$$

Applying (3.82) to obtain the $r+1^{st}$ vector from the learning sequence, we have the following

$$\tilde{W}_{r+1} = \frac{1}{n+1}\left[\omega_{m_1}\bar{\tilde{X}}_1 + ... + \omega_{m_r}\bar{\tilde{X}}_r\right]. \tag{3.83}$$

where $m_j \in \{1,2,3\}$; $j = 1,...,r$, and every time $m_j$ is chosen depending on which of three cases (see above) for the angular error takes place.

Let us find a dot product of both parts of (3.83) with the weighting vector $W$, which exists according to the condition of the theorem (subsets $A_0, A_1, ..., A_{k-1}$ are $k$-separable):

$$\left(\tilde{W}_{r+1}, \bar{W}\right) = \frac{1}{n+1}\left[\left(\omega_{m_1}\bar{\tilde{X}}_1, \bar{W}\right) + ... + \left(\omega_{m_r}\bar{\tilde{X}}_r, \bar{W}\right)\right]. \tag{3.84}$$

Let us now estimate the absolute value $\left|\left(\tilde{W}_{r+1}, \bar{W}\right)\right|$ of the dot product $\left(\tilde{W}_{r+1}, \bar{W}\right)$:

$$\left|\left(\tilde{W}_{r+1}, \bar{W}\right)\right| = \frac{1}{n+1}\left|\left[\left(\omega_{m_1}\bar{\tilde{X}}_1, \bar{W}\right) + ... + \left(\omega_{m_r}\bar{\tilde{X}}_r, \bar{W}\right)\right]\right|. \tag{3.85}$$

Since for any complex number $\beta$ $|\beta| \geq |\mathrm{Re}\,\beta|$ and $|\beta| \geq |\mathrm{Im}\,\beta|$, then the absolute value of the sum in the right-hand side of (3.85) is always greater than or equal to the absolute values of the real and imaginary parts of this sum. Let $a = \min\limits_{j=1,...,r}\left|\mathrm{Re}\left(\omega_{m_j}\bar{\tilde{X}}_j, \bar{W}\right)\right|$. Then it follows from (3.85) that

$$\left|\left(\tilde{W}_{r+1}, \bar{W}\right)\right| \geq \frac{ra}{n+1}. \tag{3.86}$$

According to the fundamental Schwarz inequality [74] the squared dot product of the two vectors does not exceed the product of the squared norms of these vectors or in other words, the norm of the dot product of the two vectors does not exceed the product of the norms of these vectors $\left\|\left(V_1, V_2\right)\right\| \leq \|V_1\| \cdot \|V_2\|$. Thus, according to the Schwartz inequality

$$\left|\left(\tilde{W}_{r+1}, \bar{W}\right)\right| \leq \left\|\tilde{W}_{r+1}\right\| \cdot \|W\|. \tag{3.87}$$

Taking into account (3.86), we obtain from (3.87) the following

$$\frac{ra}{n+1} \le \left| \left( \tilde{W}_{r+1}, \bar{W} \right) \right| \le \left\| \tilde{W}_{r+1} \right\| \cdot \left\| W \right\|.$$

Then it follows from the last inequality that

$$\left\| \tilde{W}_{r+1} \right\| \ge \frac{ra}{\left\| W \right\| (n+1)}. \qquad (3.88)$$

Let for simplicity $\dfrac{a}{n+1} = \tilde{a}$ . Then (3.88) is transformed as follows:

$$\left\| \tilde{W}_{r+1} \right\| \ge \frac{r\tilde{a}}{\left\| W \right\|}. \qquad (3.89)$$

As we told, $W$ is a weighting vector, which exist according to the condition of the Theorem. According to our assumption, the sequence $S_{\tilde{W}}$ of the weighting vectors is infinite. Since $r$ is the number of the weighting vector in the sequence $S_{\tilde{W}}$ , let us consider (3.89) when $r \to \infty$ . $\left\| \tilde{W}_{r+1} \right\|$ is a norm of the vector and therefore it is a non-negative finite real number, $\left\| W \right\|$ is a norm of the vector and it is a finite positive real number ( $\left\| W \right\| \ne 0$ because vector $W$ is a weighting vector satisfying (3.73) and therefore at least one of its components is not equal to 0), and $\tilde{a}$ is a finite positive real number. It follows from this analysis that $\dfrac{r\tilde{a}}{\left\| W \right\|} \xrightarrow[r \to \infty]{} \infty$ .

However, this means that from (3.89) we obtain

$$\left\| \tilde{W}_{r+1} \right\| \ge \frac{r\tilde{a}}{\left\| W \right\|} \to \infty. \qquad (3.90)$$

Inequality (3.90) is contradictory. Indeed, the norm of a vector, which is in the left-hand side, is a finite non-negative real number. However, it has to be greater than or equal to the infinity in the right-hand side of (3.90), which is impossible. This means that (3.90) is contradictory. This means in turn that either it is impossible that $r \to \infty$ or the vector $W$ does not exist. The latter contradicts to the condition of the Theorem. Hence, $r \not\to \infty$ and it is always a finite integer number. Thus, our assumption that the sequence $S_{\tilde{W}}$ of the weighting vectors is infinite, is false, which means that it is always finite. Theorem is proven.

So the MVN learning algorithm with the learning rule (3.80) converges after a finite number of learning iterations. As we see, this learning algorithm is derivative-free. It is not considered as the optimization problem of the minimization of the error functional. It is important that a famous local minima problem, which is typical for those learning rules that are based on the optimization technique and which we have considered in Section 1.3 (see Fig. 11), does not exist for the MVN learning algorithm based on the learning rule (3.80). The error in this MVN learning algorithm decreases because each following weighting vector in the sequence $S_{\tilde{W}}$ should be closer to the "ideal" weighting vector $W$, which exists if the MVN input/output mapping is described by some $k$-valued threshold function. According to (3.86) the absolute value of the dot product of the vector $W$ and the weighting vector $W_{r+1}$ in the sequence $S_{\tilde{W}}$ must be greater than or equal to the finite number proportional to $r$, which is the number of the correction. On the one hand, since $r$ increases, this means that $\left|\left(\tilde{W}_{r+1}, \overline{W}\right)\right|$ at least does not decrease. On the other hand, as we have proven, $\left|\left(\tilde{W}_{r+1}, \overline{W}\right)\right|$ cannot increase to infinity. This means that the learning algorithm converges when $\left|\left(\tilde{W}_{r+1}, \overline{W}\right)\right|$ reaches its maximum. This means that vectors $W_{r+1}$ and $W$ are as close to each other as it is possible. Ideally, they are collinear or close to collinearity. It follows from (3.85) and (3.86) that $\left|\left(\tilde{W}_{r+1}, \overline{W}\right)\right|$ cannot decrease during the learning process. It may only increase or remain the same. If it increases, this means that the error decreases. This means that geometrically, the MVN learning algorithm based on the learning rule (3.80) "rotates" the initial weighting vector such that $\left|\left(\tilde{W}_{r+1}, \overline{W}\right)\right|$ should be maximized. It follows from this that the worst starting condition for the learning process is when the vectors $W_1$ (the starting weighting vector) and $W$ are orthogonal to each other and $\left|\left(\tilde{W}_{r+1}, \overline{W}\right)\right|=0$, while the best starting condition is when the same vectors are about collinear. The closer they are to the collinearity, the smaller is the error and the shorter way is required for the convergence of the learning process. This kind of "non-optimization" learning is based on the same idea, which was developed in [12] by A. Novikoff for the threshold neuron and its error-correction learning.

Let us make one more remark. If the permutation $R=\left(\alpha_0, \alpha_1, ..., \alpha_{k-1}\right)$ such that (3.73) holds for the entire learning set $A$ is not known, it is possible to find such a permutation by $k!$ means. This follows from the fact that there are exactly $k!$ different permutations from the elements of the set $A$.

## 3.3  MVN Error-Correction Learning Rule

### 3.3.1  Basic Fundamentals

We have considered in Section 1.2 the error-correction learning rule (1.17) for the threshold neuron. For the reader's convenience we repeat it here

$$\tilde{w}_0 = w_0 + \alpha \delta;$$
$$\tilde{w}_i = w_i + \alpha \delta x_i, i = 1, ..., n,$$

where $\delta = d - y$ is the error, which is according to (1.5) the difference between the desired neuron output and its actual output, and $\alpha$ is the learning rate. Is it possible to apply the same idea for MVN? Yes, the MVN error-correction learning rule was justified in [60] where the convergence of the MVN learning algorithm with the error-correction learning rule was also proven.

The MVN error-correction learning rule was proposed in 1995 by the author of this book, Naum Aizenberg, and Georgy Krivosheev in [110] (then the convergence of the learning algorithm based on it was proven by the author of this book, N. Aizenberg, and J. Vandewalle in [60]), as follows

$$\tilde{w}_i = w_i + \frac{C_r}{(n+1)}\left(\varepsilon^q - \varepsilon^s\right)\bar{x}_i; i = 0, 1, ..., n, \qquad (3.91)$$

where $n$ is the number of the neuron inputs, $w_i$ is the $i$th component of the weighting vector before correction, $\tilde{w}_i$ is the same component after correction, $\bar{x}_i$ is the $i$th neuron input complex-conjugated ($x_0 \equiv 1$ as we agreed above is a pseudo-input corresponding to the bias $w_0$), $C_r$ is the learning rate, $\varepsilon^q$ is the desired output, $\varepsilon^s$ is the actual output. $\delta = \varepsilon^q - \varepsilon^s$ is the error. It should be mentioned that throughout this Section we still use the notation $\varepsilon = \varepsilon_k$ for simplicity. As well, as the MVN learning algorithm based on the angular error compensation learning rule, the MVN learning algorithm based on the error-correction learning rule is also reduced to the straightening of the sequence $S_w$ of the weighting vectors. Thus, adjusting the weights according to (3.91), we obtain the following equation for the $r+1^{st}$ weighting vector belonging to the sequence $S_w$ from the $r$th weighting vector belonging to the same sequence

$$W_{r+1} = W_r + \frac{C_r}{(n+1)}\left(\varepsilon^q - \varepsilon^s\right)\bar{X}, \qquad (3.92)$$

where $r$ is the number of the current weighting vector in the sequence $S_w$, addition is component-wise, and $\overline{X} = (1, \overline{x}_1, ..., \overline{x}_n)$ is the vector of inputs with the complex-conjugated components. As we have done this earlier, we introduce the pseudo-input $x_0 \equiv 1$ corresponding to the weight $w_0$.

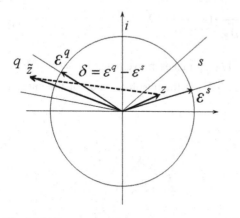

Let us find the updated weighted sum after the weights are corrected according to (3.91). The current weighted sum is $z = w_0 + w_1 x_1 + ... + w_n x_n$. Suppose for simplicity, but without loss of generality that $C_r = 1$.

If $\delta = \varepsilon^q - \varepsilon^s$ is the error, then for the updated weighted sum, taking into account that $x_i \overline{x}_i = 1; i = 0, 1, ..., n$ (since all $x_i$ are located on the unit circle) we obtain

**Fig. 3.36** Movement of the weighted sum $z$ after the correction of the weights according to (3.91)

$$\tilde{z} = \tilde{w}_0 + \tilde{w}_1 x_1 + ... + \tilde{w}_n x_n =$$

$$\left( w_0 + \frac{1}{n+1}\delta \right) + \left( w_1 + \frac{1}{n+1}\delta\overline{x}_1 \right) x_1 + ... + \left( w_n + \frac{1}{n+1}\delta\overline{x}_n \right) x_n =$$

$$= \underbrace{w_0 + w_1 x_1 + ... + w_n x_n}_{z} + \underbrace{\frac{1}{n+1}\delta + ... + \frac{1}{n+1}\delta}_{n+1 \text{ times}} = z + \frac{n+1}{n+1}\delta = \quad (3.93)$$

$$= z + \delta.$$

This means that after the weights are corrected, the weighted sum is changed exactly by $\delta$ that is by the error. This is illustrated in Fig. 3.36. The current weighted sum located in the sector $s$ has to be moved to the sector $q$. The direction of this movement is determined by the error $\delta = \varepsilon^q - \varepsilon^s$, which is equal to the difference between the desired output $\varepsilon^q$ and actual output $\varepsilon^s$. Correcting the weights according to (3.91) (or (3.92), which is the same), we ensure that after the correction of the weights the updated weighted sum $z + \delta$ moves closer or exactly to the desired sector $q$ and the MVN output moves closer or exactly becomes equal to $\varepsilon^q$, respectively.

In [62], it was suggested to modify the learning rule (3.92). A variable learning rate, which is equal to $1/|z_r|$, the inverse absolute value of the current weighted sum, was introduce there. This modification should be reasonable for those input/output mappings that are described by highly nonlinear functions with many irregular jumps. With this modification, the learning rule (3.92) becomes

$$W_{r+1} = W_r + \frac{C_r}{(n+1)|z_r|}\left(\varepsilon^q - \varepsilon^s\right)\bar{X}. \tag{3.94}$$

The use of the variable learning rate $1/|z_r|$ makes movements of the weighted sum "softer". This is illustrated in Fig. 3.37. If the absolute value of the current

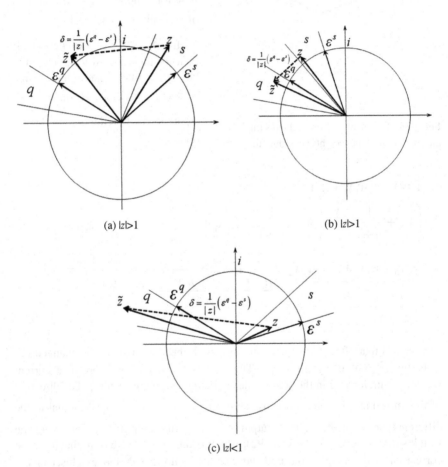

(a) |z|>1                                                    (b) |z|>1

(c) |z|<1

**Fig. 3.37** Movement of the weighted sum $z$ after the correction of the weights according to (3.94)

weighted sum $z$ is greater than 1 (it is located outside of the unit circle, see Fig. 3.37a), then $\dfrac{1}{|z|} < 1$, and the weighted sum moves shorter than by $\varepsilon^q - \varepsilon^s$.

It cannot reach in this way its target, the sector $q$, but it also could not reach it moving by $\varepsilon^q - \varepsilon^s$. However, it moved closer to the sector $q$, and on the next learning step it moves exactly there (see Fig. 3.37b). Moving by $\varepsilon^q - \varepsilon^s$, the adjusted weighted sum would need even one more step to move to the desired sector, while moving by $\dfrac{1}{|z|}\left(\varepsilon^q - \varepsilon^s\right)$, it does not need it. In Fig. 3.37c, the absolute value of the current weighted sum $z$ is less than 1 (it is located inside the unit circle). Therefore $\dfrac{1}{|z|} > 1$ and the weighted sum moves further than by $\varepsilon^q - \varepsilon^s$.

It not only reaches its target – the desired sector $q$, but $\tilde{z}$ is located even more distant from the sector borders than $z$. The use of the learning rule (3.94) is especially reasonable for highly nonlinear input/output mappings, which may have multiple jumps, peaks, etc. Correcting the weights carefully, this learning rule makes the learning process more adaptive, which may lead to faster convergence of the learning algorithm. However, we will see below that the convergence of the learning algorithm with the error-correction learning rule does not depend on the particular form of this rule - (3.92) or (3.94). Moreover, in [61] another modification of the learning rules (3.92) or (3.94) was proposed by the author of this book.

To learn highly nonlinear input/output mappings, it might be reasonable to calculate the error not as a difference between the desired and actual outputs, but as a difference between the desired output and the projection of the current weighted sum on the unit circle $\delta = \varepsilon^q - \dfrac{z}{|z|}$.

This leads to the following modification of the learning rules (3.92) and (3.94), respectively:

$$W_{r+1} = W_r + \frac{C_r}{(n+1)}\left(\varepsilon^q - \frac{z_r}{|z_r|}\right)\bar{X}\,, \tag{3.95}$$

$$W_{r+1} = W_r + \frac{C_r}{(n+1)|z_r|}\left(\varepsilon^q - \frac{z_r}{|z_r|}\right)\bar{X}\,. \tag{3.96}$$

For the continuous MVN, the error-correction learning rule is derived from the same considerations. Just the error $\delta$ is not a difference of the $k$th roots of unity, but it is a difference of the desired output $D$ and actual output $Y$, which can be arbitrary numbers located on the unit circle. Taking into account the $\delta = D - Y$ and that

$Y = \dfrac{z_r}{|z_r|}$ , we obtain from (3.95) and (3.96) the following error-correction learn-

ing rules for the continuous MVN, respectively

$$W_{r+1} = W_r + \frac{C_r}{(n+1)}(D-Y)\bar{X} = W_r + \frac{C_r}{(n+1)}\left(D - \frac{z_r}{|z_r|}\right)\bar{X} , \qquad (3.97)$$

$$W_{r+1} = W_r + \frac{C_r}{(n+1)|z_r|}(D-Y)\bar{X} = W_r + \frac{C_r}{(n+1)|z_r|}\left(D - \frac{z_r}{|z_r|}\right)\bar{X}. \qquad (3.98)$$

### 3.3.2  Convergence Theorem

Now we can formulate and prove the theorem about the convergence of the learning algorithm for the discrete MVN with the learning rules (3.92), and
(3.94)-(3.98). For the discrete MVN learning rules (3.92), and (3.94)-(3.96), the
proof, which will be given here, was done by the author of this book in [61]. For
the continuous MVN learning rules (3.97) and (3.98), the proof is based on the
same approach and it is very similar. It will be given here for the first time. It is
important to mention that for the discrete MVN learning algorithm with the learning rule (3.92) the convergence theorem was proven in [60]. The proof, which was
done in [61] and will be presented here, is shorter and more elegant.

So let us have the learning set $A$. Suppose that the permutation
$R = (\alpha_0, \alpha_1, ..., \alpha_{k-1})$ such that (3.73) holds for the entire learning set $A$ is
known. We use the learning algorithm presented in Section 3.1. Let us just make
one important remark about the continuous MVN case. We have already mentioned in Section 3.1 that any learning set $A$ for the continuous-valued input/output
mapping $f(x_1, ..., x_n) : T \rightarrow O, T \subseteq O^n$ is finite, which means that it can be
represented    as    a    union    $A = \bigcup\limits_{0 \leq j \leq k-1} A_j$    of    the    learning    subsets
$A_j, j = 0, 1, ..., k-1$, where $k$ is the number of different values of the function
$f(x_1, ..., x_n)$ representing a continuous input/output mapping.

**Theorem 3.17** (About the convergence of the learning algorithm with the error-
correction    learning    rules    (3.92),    (3.94)-(3.98)).    If    the    learning    subsets
$A_0, A_1, ..., A_{k-1}$ of the learning set $A$ ( $A = \bigcup\limits_{0 \leq j \leq k-1} A_j$ ) are $k$-separable for the

given value of $k$ according to Definition 3.16 (which means that the corresponding
MVN input/output mapping is a $k$-valued threshold function), then the MVN learning

algorithm with either of the learning rules (3.92), (3.94)-(3.98) converges after a finite number of steps.

**Proof.** This proof is based on the same idea as the one of Theorem 3.16 and it is in major similar. Let us first proof the Theorem for the rule (3.92). Then we will prove it for the learning rules (3.94)-(3.98).

Since we are given a condition that our learning subsets $A_0, A_1, ..., A_{k-1}$ are $k$-separable, this means that there exists a weighting vector $W = (w_0, w_1, ..., w_n)$ such that (3.73) holds for any $X = (x_1, ..., x_n)$ from the domain of $f$ and at least one of the weights is non-zero.

Let us now look for a weighting vector applying the learning rule (3.92) according to our learning algorithm. We may set $C_r = 1$ in (3.92) for any $r$. For simplicity and without loss of generality, let us start learning process from the zero vector $W_1 = ((0,0), (0,0), ..., (0,0))$, where $(a,b)$ is a complex number $a + bi$, where $i$ is an imaginary unity. Let $S_X = (X_1, X_2, ..., X_N)$ be a learning sequence of input vectors $X_j = (x_1^j, ..., x_n^j)$, $j = 1, ..., N$, and $S_W = (W_1, W_2, ..., W_r, ...)$ be a sequence of weighting vectors, which appear during the learning process. We have to prove that this sequence cannot be infinite. Let us remove from the learning sequence those vectors for which $W_{r+1} = W_r$, in other words, those input vectors, for which (3.73) hold without any learning. Let $S_{\tilde{W}}$ be the reduced sequence of the weighting vectors. The Theorem will be proven if we will show that the sequence $S_{\tilde{W}}$ is finite. Let us suppose that the opposite is true: the sequence $S_{\tilde{W}}$ is infinite. Let $\varepsilon^{s_1}$ be the actual output for the input vector $X_1$ and the weighting vector $W_1$ and $\varepsilon^{q_1}$ be the desired MVN output for the same input vector $X_1$. Since the desired and actual outputs do not coincide with each other, we have to apply the learning rule (3.92) to adjust the weights. According to (3.92) we obtain $\tilde{W}_2 = \frac{1}{n+1} (\varepsilon^{q_1} - \varepsilon^{s_1}) \bar{\tilde{X}}_1$,

$$\tilde{W}_3 = \tilde{W}_2 + \frac{1}{n+1} (\varepsilon^{q_2} - \varepsilon^{s_2}) \bar{\tilde{X}}_2 =$$

$$\frac{1}{n+1} \left[ (\varepsilon^{q_1} - \varepsilon^{s_1}) \bar{\tilde{X}}_1 + (\varepsilon^{q_2} - \varepsilon^{s_2}) \bar{\tilde{X}}_2 \right], ....$$

$$\tilde{W}_{r+1} = \frac{1}{n+1}\left[\left(\varepsilon^{q_1} - \varepsilon^{s_1}\right)\bar{\tilde{X}}_1 + ... + \left(\varepsilon^{q_r} - \varepsilon^{s_r}\right)\bar{\tilde{X}}_r\right]. \tag{3.99}$$

Let us find a dot product of both parts of (3.99) with $W$:

$$\left(\tilde{W}_{r+1}, \bar{W}\right) = \frac{1}{n+1}\left[\left(\left(\varepsilon^{q_1} - \varepsilon^{s_1}\right)\bar{\tilde{X}}_1, \bar{W}\right) + ... + \left(\left(\varepsilon^{q_r} - \varepsilon^{s_r}\right)\bar{\tilde{X}}_r, \bar{W}\right)\right].$$

Let $\varepsilon^{q_j} - \varepsilon^{s_j} = \omega_j, j = 1, ..., r$.

Then the last equation may be rewritten as follows:

$$\left(\tilde{W}_{r+1}, \bar{W}\right) = \frac{1}{n+1}\left[\left(\omega_1\bar{\tilde{X}}_1, \bar{W}\right) + ... + \left(\omega_r\bar{\tilde{X}}_r, \bar{W}\right)\right]. \tag{3.100}$$

Let us estimate the absolute value $\left|\left(\tilde{W}_{r+1}, \bar{W}\right)\right|$:

$$\left|\left(\tilde{W}_{r+1}, \bar{W}\right)\right| = \frac{1}{n+1}\left|\left[\left(\omega_1\bar{\tilde{X}}_1, \bar{W}\right) + ... + \left(\omega_r\bar{\tilde{X}}_r, \bar{W}\right)\right]\right|. \tag{3.101}$$

Since for any complex number $\beta$ $|\beta| \geq |\mathrm{Re}\,\beta|$ and $|\beta| \geq |\mathrm{Im}\,\beta|$, then the absolute value of the sum in the right-hand side of (3.101) is always greater than or equal to the absolute values of the real and imaginary parts of this sum. Let $a = \min\limits_{j=1,...,r}\left|\mathrm{Re}\left(\omega_j\tilde{X}_j, \bar{W}\right)\right|$. Then it follows from (3.101) that

$$\left|\left(\tilde{W}_{r+1}, \bar{W}\right)\right| \geq \frac{ra}{n+1}. \tag{3.102}$$

According to the fundamental Schwarz inequality [74] the squared dot product of the two vectors does not exceed the product of the squared norms of these vectors or in other words, the norm of the dot product of the two vectors does not exceed the product of the norms of these vectors $\left\|\left(V_1, V_2\right)\right\| \leq \|V_1\| \cdot \|V_2\|$. Thus, according to the Schwartz inequality

$$\left|\left(\tilde{W}_{r+1}, \bar{W}\right)\right| \leq \|\tilde{W}_{r+1}\| \cdot \|W\|. \tag{3.103}$$

Taking into account (3.102), we obtain from (3.103) the following

$$\frac{ra}{n+1} \leq \left|\left(\tilde{W}_{r+1}, \bar{W}\right)\right| \leq \|\tilde{W}_{r+1}\| \cdot \|W\|.$$

Then it follows from the last inequality that

$$\left\|\tilde{W}_{r+1}\right\| \geq \frac{ra}{\|W\|(n+1)}.$$

(3.104)

Let for simplicity $\dfrac{a}{n+1} = \tilde{a}$ . Then (3.104) is transformed as follows:

$$\left\|\tilde{W}_{r+1}\right\| \geq r\tilde{a} / \|W\|.$$

(3.105)

As we told, $W$ is some weighting vector, which exists for our input/output mapping. This vector exists according to the condition of the Theorem because the learning subsets $A_0, A_1, \ldots, A_{k-1}$ are $k$-separable. According to our assumption, the sequence $S_{\tilde{W}}$ of the weighting vectors is infinite. Since $r$ is the number of the learning step, let us consider (3.105) when $r \rightarrow \infty$. $\left\|\tilde{W}_{r+1}\right\|$ is a non-negative finite real number, $\|W\|$ is a finite positive real number ($\|W\| \neq 0$ because vector $W$ is a weighting vector for our input/output mapping, and this means that at least one of the weights is not equal to 0), and $\tilde{a}$ is a finite positive real number. It follows from this analysis that

$$\frac{r\tilde{a}}{\|W\|} \underset{r \to \infty}{\rightarrow} \infty.$$

However, this means that from (3.105) we obtain

$$\left\|\tilde{W}_{r+1}\right\| \geq \frac{r\tilde{a}}{\|W\|} \rightarrow \infty.$$

(3.106)

Inequality (3.106) is contradictory. Indeed, the norm of a vector, which is in the left-hand side, is a finite non-negative real number. However, it has to be greater than or equal to the infinity in the right-hand side of (3.106), which is impossible. This means that (3.106) is contradictory. This means in turn that either it is impossible that $r \rightarrow \infty$ or the vector $W$ does not exist. The latter means that the learning subsets $A_0, A_1, \ldots, A_{k-1}$ are not $k$-separable. However, this contradicts to the condition of the Theorem. Hence, $r \not\rightarrow \infty$ and it is always a finite integer number. Thus, our assumption that the sequence $S_{\tilde{W}}$ of the weighting vectors is infinite, is false, which means that it is always finite. Hence, the learning algorithm with the learning rule (3.92) converges after a finite number of steps.

Let us now prove that the learning algorithm also converges when either of the learning rules (3.94)-(3.98) is used. For these three learning rules the proof of the convergence of the learning algorithm is almost identical to the proof we have just presented, accurate within specifics of some equations. Let us demonstrate this.

If we apply the learning rule (3.94), we obtain the following equation instead of (3.99)

$$\tilde{W}_{r+1} = \frac{1}{n+1}\left[\frac{1}{|z_1|}\left(\varepsilon^{q_1} - \varepsilon^{s_1}\right)\bar{\tilde{X}}_1 + ... + \frac{1}{|z_r|}\left(\varepsilon^{q_r} - \varepsilon^{s_r}\right)\bar{\tilde{X}}_r\right].$$

Then putting $\frac{1}{|z_j|}\left(\varepsilon^{q_j} - \varepsilon^{s_j}\right) = \omega_j$, $j = 1, ..., r$, we obtain (3.100) and from that

moment the proof continues with no changes.

If we apply the learning rule (3.95), then (3.99) is substituted by the following expression

$$\tilde{W}_{r+1} = \frac{1}{n+1}\left[\left(\varepsilon^{q_1} - \frac{z_1}{|z_1|}\right)\bar{\tilde{X}}_1 + ... + \left(\varepsilon^{q_r} - \frac{z_r}{|z_r|}\right)\bar{\tilde{X}}_r\right].$$

Then putting $\varepsilon^{q_j} - \frac{z_j}{|z_j|} = \omega_j$, $j = 1, ..., r$, we obtain (3.100) and from that mo-

ment the proof continues again with no changes.

If we apply the learning rule (3.96), then we again have to substitute (3.99), this time as follows

$$\tilde{W}_{r+1} = \frac{1}{n+1}\left[\frac{1}{|z_1|}\left(\varepsilon^{q_1} - \frac{z_1}{|z_1|}\right)\bar{\tilde{X}}_1 + ... + \frac{1}{|z_r|}\left(\varepsilon^{q_r} - \frac{z_r}{|z_r|}\right)\bar{\tilde{X}}_r\right].$$

Then putting $\frac{1}{|z_j|}\left(\varepsilon^{q_j} - \frac{z_j}{|z_j|}\right) = \omega_j$, $j = 1, ..., r$, we obtain (3.100) and from

that moment the proof continues again with no changes.

If we apply the learning rule (3.97), then we again have to substitute (3.99), this time as follows

$$\tilde{W}_{r+1} = \frac{1}{n+1}\left[\left(D - \frac{z_1}{|z_1|}\right)\bar{\tilde{X}}_1 + ... + \left(D - \frac{z_r}{|z_r|}\right)\bar{\tilde{X}}_r\right].$$

Then putting $\left(D - \frac{z_j}{|z_j|}\right) = \omega_j$, $j = 1, ..., r$, we obtain (3.100) and from that

moment the proof continues again with no changes.

If we apply the learning rule (3.98), then (3.99) should be substituted as follows

$$\tilde{W}_{r+1} = \frac{1}{n+1}\left[\frac{1}{|z_1|}\left(D - \frac{z_1}{|z_1|}\right)\bar{\bar{X}}_1 + ... + \frac{1}{|z_r|}\left(D - \frac{z_r}{|z_r|}\right)\bar{\bar{X}}_r\right].$$

Then putting $\dfrac{1}{|z_j|}\left(D - \dfrac{z_j}{|z_j|}\right) = \omega_j, j = 1,...,r$, we obtain (3.100) and from

that moment the proof continues again with no changes.

Theorem is proven. This means that the MVN learning algorithm with either of the learning rules (3.92), (3.94)-(3.98) converges after a finite number of learning iterations. As well as the learning algorithm based on the rule (3.80), the learning algorithm based on the rules (3.92), (3.94)-(3.98) is derivative-free. As well as the algorithm based on the rule (3.80), it is not considered as the optimization problem of the minimization of the error functional. Therefore, a local minima problem (see Section 1.3, Fig. 1.11), which is typical for those learning rules that are based on the optimization technique, does not exist for the MVN learning algorithm based on the learning rules (3.92), (3.94)-(3.98), as well as for the learning algorithm based on the rule (3.80).

The error in the MVN learning algorithm based on the learning rules (3.92), (3.94)-(3.98) naturally decreases because of the same reasons that for the learning algorithm based on the learning rule (3.80). Each following weighting vector in the sequence $S_{\tilde{W}}$ should be closer to the "ideal" weighting vector $W$, which exists if the MVN input/output mapping is described by some $k$-valued threshold function and the learning subsets $A_0, A_1, ..., A_{k-1}$ are $k$-separable. According to (3.99), for the learning rule (3.92), and according to corresponding equations for the learning rules (3.94)-(3.98), the absolute value of the dot product of the vector $W$ and the weighting vector $W_{r+1}$ in the sequence $S_{\tilde{W}}$ should not decrease and moreover, as it follows from (3.103) and (3.104), it must be greater than or equal to the finite number proportional to $r$, which is the number of the correction. On the one hand, since $r$ increases, this means that $\left|\left(\tilde{W}_{r+1}, W\right)\right|$ should not decrease.

On the other hand, as we have proven, $\left|\left(\tilde{W}_{r+1}, W\right)\right|$ cannot increase to infinity. This means that the learning algorithm converges when $\left|\left(\tilde{W}_{r+1}, W\right)\right|$ reaches its maximum. This means that vectors $W_{r+1}$ and $W$ are as close to each other as it is possible. Ideally, they are collinear or close to collinearity. It follows from (3.99) and (3.100) that $\left|\left(\tilde{W}_{r+1}, W\right)\right|$ cannot decrease during the learning process. It may only increase or remain the same. If it increases, this means that the error decreases. This means that geometrically, the MVN learning algorithm based on either of the learning rules (3.92), (3.94)-(3.98) "rotates" the initial weighting vector and the intermediate weighting vectors $W_r$ such that $\left|\left(\tilde{W}_{r+1},\overline{W}\right)\right|$ should be maximized. It follows from this that the worst starting condition for the learning process is when the vectors $W_1$ (the starting weighting vector) and $W$ are orthogonal to each other and $\left|\left(\tilde{W}_{r+1},\overline{W}\right)\right|=0$, while the best starting condition is when the same vectors are about collinear. The closer they are to the collinearity, the smaller is the error and the shorter way is required for the convergence of the learning process.

Hence, the MVN learning algorithm based on the error-correction learning rule is another example of the "non-optimization" learning, which is based on the same idea that was developed in [12] by A. Novikoff for the threshold neuron and its error-correction learning.

It is important that a beautiful approach to the learning through the direct error-correction, which was proposed by F. Rosenblatt and developed by A. Novikoff for the threshold neuron about 50 years ago, was generalized for MVN. Unlike the threshold neuron, MVN employs this learning rule for multiple-valued and even continuous-valued input/output mappings.

We should recall that if the permutation $R=\left(\alpha_0,\alpha_1,...,\alpha_{k-1}\right)$ such that (3.73) holds for the entire learning set $A$ is not known, it is possible to find such a permutation by $k!$ means.

It is worth to mention that the MVN learning algorithm does not depend on the learning rate, unlike any learning algorithm for a real-valued neuron. As we saw, the learning rate $C_r$ in the learning rules (3.80), and (3.92), (3.94)-(3.98) can always be equal to 1. Evidently, that a variable learning rate $\dfrac{1}{|z_r|}$ in (3.94), (3.96), and (3.98) is self-adaptive.

### 3.3.3 Example of Error-Correction Learning

Let us consider how MVN learns. We will use the learning algorithm with the error-correction rule (3.92). We have already considered above (see Table 2.8 and Fig. 2.22a) how the discrete MVN implements in 3-valued logic the Post function $\max\left(y_1,y_2\right); y_i\in K; i=1,2$, which becomes in $k$-valued logic over the field of complex numbers $f_{\max}\left(x_1,x_2\right)=\max\left(x_1,x_2\right); x_i\in E_k; i=1,2$ Let us consider it again for $k=3$. We will obtain the weighting vector for this function using the learning algorithm with the error-correction learning rule (3.92).

The results are summarized in Table 3.10 and Fig. 3.38. The learning process starts from the random weighting vector $W_0=\left(0.96+0.32i,0.79+0.73i,0.59+0.5i\right)$ (all real and imaginary parts of the weights are random numbers in the interval [0, 1]). With the initial weighting vector, the actual outputs for the learning samples 1, 5, 6, 7, 8, 9 coincide with the desired outputs, while for the learning samples 2, 3, 4 the desired outputs are incorrect (see Fig. 3.38a and the column "Initial $W$" in Table 3.10). We have already shown in Section 2.1 that $f_{max}(x_1, x_2)$ is a threshold function, which can be implemented using MVN with the weighting vector $W = (-2 - 4\varepsilon_3, 4 + 5\varepsilon_3, 4 + 5\varepsilon_3)$.

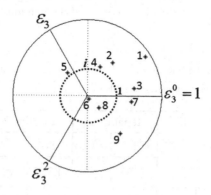

(a) with the initial random vector

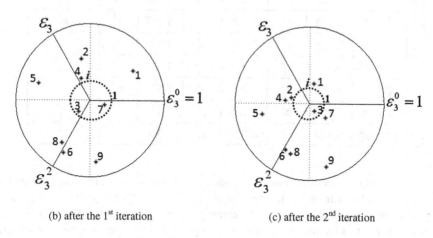

(b) after the 1$^{st}$ iteration　　　　　　　(c) after the 2$^{nd}$ iteration

**Fig. 3.38** Learning of the $f_{max}(x_1, x_2) = \max(x_1, x_2)$ for $k=3$, using the MVN learning algorithm with the rule (3.92).

Locations of the weighted sums corresponding to the learning samples 1-9 are shown

Thus, we may consider this weighting vector as the "ideal" one. According to Theorem 3.16 and Theorem 3.17, if the learning process starts from some arbitrary weighting vector whose components are chosen randomly, then this process should lead to the weighting vector, whose absolute dot product with the "ideal" weighting vector reaches its maximum. Moreover, the absolute value of this dot product should not decrease during the learning process.

For the starting weighting vector $W_0$ we obtain $|(W_0, W)| = 6.34$. After the first learning iteration the actual outputs for the learning samples 1, 3, 5, 6, 7, 9 coincide with the desired outputs, while for the learning samples 2, 4, 8 the desired outputs are incorrect (see Fig. 3.38b and the column "Iteration 1" in Table 3.10).

**Table 3.10** MVN learns the Post function $f_{\max}(x_1, x_2)$ in 3-valued logic using the learning algorithm with the error-correction learning rule (3.92)

| # | $x_1$ | $x_2$ | Initial $W$ | | Iteration 1 | | Iteration 2 | | $f_{\max}(x_1, x_2)$ |
|---|---|---|---|---|---|---|---|---|---|
| | | | arg(z) | P(z) | arg(z) | P(z) | arg(z) | P(z) | |
| 1 | $\varepsilon_3^0$ | $\varepsilon_3^0$ | 0.585 | $\varepsilon_3^0$ | 0.585 | $\varepsilon_3^0$ | 1.356 | $\varepsilon_3^0$ | $\varepsilon_3^0$ |
| 2 | $\varepsilon_3^0$ | $\varepsilon_3^1$ | 0.909 | $\varepsilon_3^0$ | 1.788 | $\varepsilon_3^0$ | 2.850 | $\varepsilon_3^1$ | $\varepsilon_3^1$ |
| 3 | $\varepsilon_3^0$ | $\varepsilon_3^2$ | 0.152 | $\varepsilon_3^0$ | 3.896 | $\varepsilon_3^1$ | 5.300 | $\varepsilon_3^2$ | $\varepsilon_3^2$ |
| 4 | $\varepsilon_3^1$ | $\varepsilon_3^0$ | 0.138 | $\varepsilon_3^0$ | 1.968 | $\varepsilon_3^0$ | 3.007 | $\varepsilon_3^1$ | $\varepsilon_3^1$ |
| 5 | $\varepsilon_3^1$ | $\varepsilon_3^1$ | 2.296 | $\varepsilon_3^1$ | 2.832 | $\varepsilon_3^1$ | 3.359 | $\varepsilon_3^1$ | $\varepsilon_3^1$ |
| 6 | $\varepsilon_3^1$ | $\varepsilon_3^2$ | 5.231 | $\varepsilon_3^2$ | 4.223 | $\varepsilon_3^2$ | 4.220 | $\varepsilon_3^2$ | $\varepsilon_3^2$ |
| 7 | $\varepsilon_3^2$ | $\varepsilon_3^0$ | 6.156 | $\varepsilon_3^2$ | 6.001 | $\varepsilon_3^2$ | 5.580 | $\varepsilon_3^2$ | $\varepsilon_3^2$ |
| 8 | $\varepsilon_3^2$ | $\varepsilon_3^1$ | 5.502 | $\varepsilon_3^2$ | 4.105 | $\varepsilon_3^1$ | 4.342 | $\varepsilon_3^2$ | $\varepsilon_3^2$ |
| 9 | $\varepsilon_3^2$ | $\varepsilon_3^2$ | 5.444 | $\varepsilon_3^2$ | 4.817 | $\varepsilon_3^2$ | 4.976 | $\varepsilon_3^2$ | $\varepsilon_3^2$ |

It should be mentioned that the weighted sums for the learning samples 2 and 4 have moved much closer to the desired sector 1 compared to the initial state, while the weighted sum for the learning sample 8 has moved a little bit in the incorrect direction from the correct sector 2 to the incorrect sector 1. After the first learning iteration, the updated weighting vector is

$$W_1 = (-1.04 - 0.84i, 0.79 + 1.31i, 0.59 + 1.08i) \text{ , and } |(W_1, W)| = 16.46.$$

After the second learning iteration, all actual outputs coincide with the desired outputs (see Fig. 3.38c and the column "Iteration 2" in Table 3.10) and therefore, the learning algorithm converges. The resulting weighting vector is

$W_2 = (-1.04 - 1.99i, 0.79 + 1.89i, 0.59 + 1.65i)$ , and $|(W_2, W)| = 24.66$. We

see that vectors $W$ and $W_2$ are not collinear, but they are close to collinearity (at least, all the real and imaginary parts of their components have the same sign) and this closeness is enough to ensure that both these vectors implement the same input/output mapping.

In Chapter 4, we will use the error-correction learning in the backpropagation learning algorithm for a feedforward neural network based on multi-valued neurons.

A level of growing of $|(W_r, W)|$ where $r$ is the number of the learning itera-

tion should be used as a measure of the *learning energy*, which the learning algorithm spends correcting the weights. We will see in Section 3.4 that the best choice for the starting weighting vector in the learning algorithm is the Hebbian weighting vector that is the vector obtained using the Hebb rule. The learning process, which starts from the Hebbian vector leads to fewer corrections of the weights than the learning process starting from the random vector.

## 3.4   Hebbian Learning and MVN

We have started consideration of different neural learning techniques from the Hebbian learning (see Section 1.2). Let us consider how this important learning technique works for MVN.

The mechanism of the Hebbian learning for MVN is the same as the one for the threshold neuron and as it was described by D. Hebb in his seminal book [8]. This is the mechanism of the association. The weight should pass the input signal or to enhance it or to weaken it depending on the correlation between the corresponding input and the output. As well as for the binary threshold neuron, the associations between the desired outputs and the given inputs should be developed through the dot product of the vector of all the desired outputs with the corresponding vectors of all the given inputs.

The Hebbian learning rule does not change for MVN, and equations (1.3) and (1.4) that describe this rule for the threshold neuron also work for MVN.

Just for the reader's convenience we will repeat these equations here.

Let us have $N$ $n$-dimensional learning samples $\left(x_1^j, ..., x_n^j\right)$, $j = 1, ..., N$. Let

$\mathbf{f} = \left(f_1, ..., f_N\right)^T$ be an $N$-dimensional vector-column of the desired outputs. Let

$\mathbf{X}_1, ..., \mathbf{X}_n$   be   $N$-dimensional   vectors   of   all   the   inputs

( $\mathbf{X}_1 = \left(x_1^1, x_1^2, ..., x_1^N\right)^T$ , $\mathbf{X}_2 = \left(x_2^1, x_2^2, ..., x_2^N\right)^T$ , ..., $\mathbf{X}_n = \left(x_n^1, x_n^2, ..., x_n^N\right)^T$ ).

Then according to the Hebbian learning rule (see (1.3)) the weights $w_1, ..., w_n$

are calculated as dot products of vector $\mathbf{f}$ and vectors $\mathbf{X}_1, ..., \mathbf{X}_n$, respectively.

Weight $w_0$ is calculated as a dot product of vector $\mathbf{f}$ and the $N$-dimensional vector-constant $\mathbf{x}_0 = (1,1,...,1)^T$ :

$$w_i = (\mathbf{f}, \mathbf{x}_i), i = 0,...,n,$$

where $(\mathbf{a}, \mathbf{b}) = a_1\bar{b}_1 + ... + a_n\bar{b}_n$ is the dot product of vector-columns $\mathbf{a} = (a_1,...,a_n)^T$ and $\mathbf{b} = (b_1,...,b_n)^T$ in the unitary space ("bar" is a symbol of complex conjugation), thus

$$w_i = (\mathbf{f}, \mathbf{x}_i) = f_1\bar{x}_i^1 + f_2\bar{x}_i^2 + ... + f_N\bar{x}_i^N, i = 0,1,...,n. \qquad (3.107)$$

Equation (1.4) determines the normalized version of the Hebbian learning rule

$$w_i = \frac{1}{N}(\mathbf{f}, \mathbf{x}_i), i = 0,...,n.$$

Let us now consider the following examples. Let $k=4$ in the MVN activation function (2.50). Thus, our MVN works in 4-valued logic whose values are encoded by the elements of the set $E_4 = \{1, i, -1, -i\}$ ($i = \varepsilon_4 = e^{i2\pi/4}$ is an imaginary unity and a primitive 4$^{th}$ root of a unity).

Let us first consider four examples illustrated in Fig. 3.39. In all these examples we calculate a weight for one of the MVN inputs and for a single learning sample.

In Fig. 3.39a, the desired MVN output is $i$ and the corresponding input is also $i$. According to (3.107) $w_0 = i, w_1 = f_1\bar{x}_1 = i \cdot (-i) = 1$, the weighted sum is $i + 1 \cdot i = 2i$, and according to (2.50) the neuron output is $P(2i) = i$. Thus, if the desired output coincides with the input, the weight just "passes" the input to the output.

In Fig. 3.39b, the desired MVN output is -$i$ and the corresponding input is $i$. According to (3.107) $w_0 = -i, w_1 = f_1\bar{x}_1 = -i \cdot (-i) = -1$, the weighted sum is $-i + (-1) \cdot i = -2i$, and according to (3.50) the neuron output is $P(-2i) = -i$. Thus, if the desired output is opposite to the input, the weight inverts the input passing it to the output.

The same situation is illustrated in Fig. 3.39c. Just the desired MVN output here is $i$, while the corresponding input is -$i$. According to (3.107) $w_0 = i, w_1 = f_1\bar{x}_1 = i \cdot i = -1$, the weighted sum is $i + (-1) \cdot (-i) = 2i$, and according to (3.50) $P(2i) = i$.

In Fig. 3.39d, the desired output and the input are neither the same nor opposite to each other. The desired MVN output is -1 and the corresponding input is $i$.

According to (3.107) $w_0 = -1, w_1 = f_1 \bar{x}_1 = -1 \cdot (-i) = i$, the weighted sum is $-1 + i \cdot i = -1 - 1 = -2$, and according to (2.50) the neuron output is $P(-2) = -1$. Thus, the weight "rotates" the input such that this input contributes to the desired output.

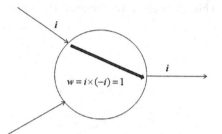

(a) the output coincides with the input, and the weight just pass the input to the output

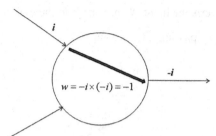

(b) the output is opposite to the input, and the weight inverts the input passing it to the output

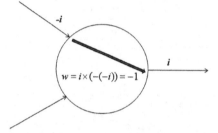

(c) the output is opposite to the input, and the weight inverts the input passing it to the output

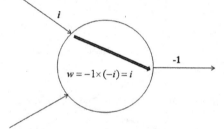

(d) the weight "rotates" the input such that this input contributes to the desired output

**Fig. 3.39** Calculation of the MVN weight using the Hebb rule for one of the neuron inputs and for a single learning sample: the weight is equal to the product of the desired output and the complex-conjugated input

Let us now consider calculation of the weights for the two MVN inputs using the Hebbian learning rule.

In Fig. 3.40a, the desired MVN output is $i$, while its two inputs are $i$ and -1, respectively. According to (3.107) $w_0 = i, w_1 = f_1 \bar{x}_1 = i \cdot (-i) = 1$, and $w_2 = f_1 \bar{x}_2 = i \cdot (-1) = -i$, the weighted sum is $i + 1 \cdot i + (-1) \cdot (-i) = 3i$ and according to (2.50) the neuron output is $P(3i) = i$. Thus, the weight $w_1$ passes the input $x_1$ to the output, while the weight $w_2$ "rotates" the input $x_2$ passing it to the output.

In Fig. 3.40b, the desired MVN output is -$i$, while its two inputs are $i$ and 1, respectively. According to (3.107) $w_0 = -i, w_1 = f_1\bar{x}_1 = -i \cdot (-i) = -1$ and $w_2 = f_1\bar{x}_2 = -i \cdot 1 = -i$, the weighted sum is $-i + (-1) \cdot i + (-i) \cdot 1 = -3i$, and according to (2.50) the neuron output is $P(-3i) = -i$. Thus, the weight $w_1$ inverts the input $x_1$ passing it to the output, while the weight $w_2$ "rotates" the input $x_2$ passing it to the output.

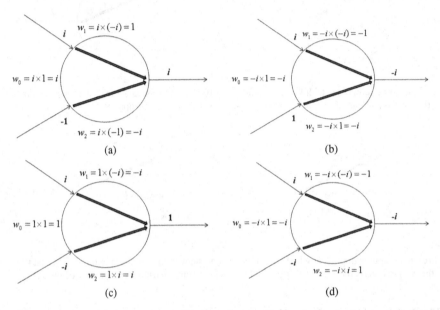

**Fig. 3.40** Calculation of the MVN weights using the Hebb rule for the two neuron inputs and for a single learning sample: the weight is equal to the product of the desired output and the complex-conjugated input

In Fig. 3.40c, the desired MVN output is 1, while its two inputs are $i$ and -$i$, respectively. According to (3.107) $w_0 = 1, w_1 = f_1\bar{x}_1 = 1 \cdot (-i) = -i$, and $w_2 = f_1\bar{x}_2 = 1 \cdot i = i$, the weighted sum is $1 + (-i) \cdot i + (i) \cdot (-i) = 3$, and according to (2.50) the neuron output is $P(3) = 1$. Thus, both weights $w_1$ and $w_2$ "rotate" the inputs $x_1$ and $x_2$ passing them to the output.

In Fig. 3.40d, the desired MVN output is -$i$, while its two inputs are $i$ and -$i$, respectively. According to (3.107) $w_0 = -i, w_1 = f_1\bar{x}_1 = -i \cdot (-i) = -1$ and $w_2 = f_1\bar{x}_2 = -i \cdot i = 1$, the weighted sum is $-i + (-1) \cdot i + 1 \cdot (-i) = -3i$, and according to (2.50) the neuron output is $P(-3i) = -i$. Thus, the weight $w_1$ inverts the input $x_1$ passing it to the output, while the weight $w_2$ passes the input $x_2$ to the output.

Evidently, the Hebbian learning can also be considered for the continuous MVN. In this case, nothing changes, and the Hebb rule still is still described by (1.3), (1.4), and (3.107).

It is important that the MVN Hebbian learning can be used for simulation of the associations that take place in biological neurons when they learn. We have already discussed above (Section 2.4) that the information transmitted by biological neurons to each other is completely contained in the frequency of the generated spikes. The phase, which determines the MVN state, is proportional to the frequency. Thus, the larger is phase, the higher is frequency. The reader may consider examples shown in Fig. 3.39 and Fig. 3.40 from the point of view of simulation of the biological neuron learning. In this case, the neuron states should be interpreted as follows: 1 – "inhibition" (phase 0), $i$ – slight excitation (phase $\pi/2$), -1 – moderate excitation (phase $\pi$), and $-i$ – maximal excitation (phase $3\pi/2$).

In all examples of the Hebbian learning, which we have considered above, the learning set has contained a single learning sample. When there are more learning samples in the learning set, the Hebbian learning rule usually does not lead to a weighting vector, which implements the corresponding input/output mapping. However, there is a wonderful property of the weighting vector obtained using the Hebbian learning rule. Although this vector usually does not implement the corresponding input/output mapping, the MVN learning algorithm based on the learning rules (3.80) and (3.92), (3.94)-(3.98) converges much faster when the learning process starts from this (Hebbian) vector than from a random vector.

We can illustrate this property using the example of learning the input/output mapping presented by the function $f_{\max}(x_1, x_2) = \max(x_1, x_2)$ for $k=3$, which we have already used several times. As it was shown in Section 2.1, this is a 3-valued threshold function, which can be implemented using MVN with the weighting vector $W = (-2 - 4\varepsilon_3, 4 + 5\varepsilon_3, 4 + 5\varepsilon_3)$. Thus, we may consider this weighting vector as the "ideal" one. According to Theorem 3.16 and Theorem 3.17, if the learning process starts from some arbitrary weighting vector whose components are chosen randomly, then this process should lead us to the weighting vector, whose absolute dot product with the "ideal" weighting vector reaches its maximum.

Let us find the Hebbian weights for $f_{\max}(x_1, x_2)$. According to (1.4) and taking into account (3.107), we obtain the following Hebbian weighting vector $W_H = (-0.33 + 0.19i, 0.5 - 0.096i, 0.5 - 0.096i)$. This weighting vector does not implement the function $f_{\max}(x_1, x_2)$. Distribution of the weighted sums with the weighting vector $W_H$ is shown in Fig. 3.41a. The outputs for five learning samples out of nine (samples 2, 4, 6, 8, 9) are incorrect (see the column "Hebbian Weights " in Table 3.11). However, they can easily be corrected using the learning algorithm, for example, with the error-correction rule (3.92). Moreover, the number of corrections of the weights is fewer than for the same learning algorithm when it starts from the random weighting vector.

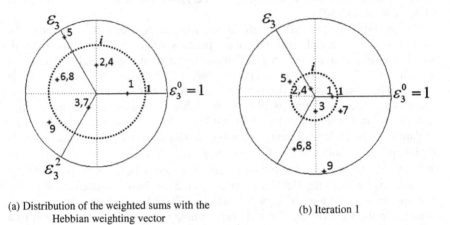

(a) Distribution of the weighted sums with the            (b) Iteration 1
     Hebbian weighting vector

**Fig. 3.41** Movement of the weighted sum $z$ after the correction of the weights according to (3.92) starting from the Hebbian weighting vector for the function

$$f_{\max}\left(x_1, x_2\right) = \max\left(x_1, x_2\right); k = 3$$

After a single learning iteration the actual outputs for all the learning samples coincide with the desired outputs (see Fig. 3.41b and the column "Iteration 1" in Table 3.11).

Let us evaluate the energy, which we have to spend for the learning, which starts from the Hebbian weights in terms of growing of the absolute value of the dot product $\left|\left(W_r, W\right)\right|$ of the current weighting vector $W_r$ and the "ideal" weighting vector $W$.

For the Hebbian weighting vector $W_H$ we obtain $\left|\left(W_H, W\right)\right| = 5.77$. After a single     learning     iteration,     for     the     weighting     vector $W_1 = \left(-0.33 - 1.35i, 0.5 + 0.67i, 0.5 + 0.67i\right)$ resulted from this iteration we obtain $\left|\left(W_1, W\right)\right| = 12.47$. Comparing this result to the one considered in Section 3.3 for the learning process started for the same input/output mapping from the random weighting vector, we see that not only a single iteration was enough for the convergence of the learning algorithm, but significantly smaller amount of the corrections of the weights is required for the learning process, which starts from the Hebbian weights. In fact, in the example with the learning algorithm started from the random weights (see Section 3.3) the absolute value of the dot product $\left|\left(W_r, W\right)\right|$ was 6.34 for the initial weights, 16.46 after the first iteration, and 24.66 after the second iteration. Hebbian weights ensure that correcting the weights for some learning sample, we do not corrupt the weights for other learning samples. Moreover, we may simultaneously improve the result for some other learning samples, which require correction of the weights.

**Table 3.11** MVN learns the Post function $f_{\max}\left(x_1, x_2\right)$ in 3-valued logic using the learning algorithm with the error-correction learning rule (3.92) and starting from the Hebbian weighting vector $W_H$

| # | $x_1$ | $x_2$ | Hebbian Weights | | Iteration 1 | | $f_{\max}\left(x_1, x_2\right)$ |
|---|---|---|---|---|---|---|---|
| | | | $\arg(z)$ | $P(z)$ | $\arg(z)$ | $P(z)$ | |
| 1 | $\varepsilon_3^0$ | $\varepsilon_3^0$ | 0.0 | $\varepsilon_3^0$ | 0.0 | $\varepsilon_3^0$ | $\varepsilon_3^0$ |
| 2 | $\varepsilon_3^0$ | $\varepsilon_3^1$ | 1.571 | $\varepsilon_3^0$ | 2.095 | $\varepsilon_3^1$ | $\varepsilon_3^1$ |
| 3 | $\varepsilon_3^0$ | $\varepsilon_3^2$ | 4.188 | $\varepsilon_3^2$ | 4.712 | $\varepsilon_3^2$ | $\varepsilon_3^2$ |
| 4 | $\varepsilon_3^1$ | $\varepsilon_3^0$ | 1.571 | $\varepsilon_3^0$ | 2.095 | $\varepsilon_3^1$ | $\varepsilon_3^1$ |
| 5 | $\varepsilon_3^1$ | $\varepsilon_3^1$ | 2.094 | $\varepsilon_3^1$ | 2.618 | $\varepsilon_3^1$ | $\varepsilon_3^1$ |
| 6 | $\varepsilon_3^1$ | $\varepsilon_3^2$ | 2.808 | $\varepsilon_3^1$ | 4.321 | $\varepsilon_3^2$ | $\varepsilon_3^2$ |
| 7 | $\varepsilon_3^2$ | $\varepsilon_3^0$ | 4.188 | $\varepsilon_3^2$ | 5.759 | $\varepsilon_3^2$ | $\varepsilon_3^2$ |
| 8 | $\varepsilon_3^2$ | $\varepsilon_3^1$ | 2.808 | $\varepsilon_3^1$ | 4.321 | $\varepsilon_3^2$ | $\varepsilon_3^2$ |
| 9 | $\varepsilon_3^2$ | $\varepsilon_3^2$ | 3.665 | $\varepsilon_3^1$ | 4.827 | $\varepsilon_3^2$ | $\varepsilon_3^2$ |

As we see comparing the vectors $W_H$ and $W_1$, only imaginary parts of the weights required correction.

Comparing the vectors $W$ and $W_1$, we see that they are not collinear, but close to collinearity. At least, the real and imaginary parts of their components have the same sign.

## 3.5 Concluding Remarks to Chapter 3

In this Chapter, we have considered fundamentals of MVN learning. If the input/output mapping is a $k$-valued threshold function, it can be learned by MVN. It was shown that in this case, a learning set consists of $k$ learning subsets, which are $k$-separable. The MVN learning algorithm is based on the sequential iterative examination of the learning samples and correction of the weights using a learning rule wherever it is necessary.

The learning process may continue until the zero error is reached or until the mean square error (or the root mean square error) drops below some reasonable predetermined value.

We have considered two learning rules. The first rule is based on the closeness of the actual output to the desired one in terms of angular distance. The second learning rule is the error-correction learning rule. The convergence theorems for the MVN learning algorithm based on both learning rules were proven. If the MVN input/output mapping is described by some $k$-valued threshold function, which means that a learning set corresponding to this mapping consists of $k$ disjoint $k$-separable subsets, then the MVN learning algorithm based on either of the considered learning rules converges.

It is fundamental that the MVN learning is based on the same principles as the perceptron learning in A. Novikoff's interpretation. It is not considered as the optimization problem of the error functional minimization. It is shown that each step of the learning process decreases the distance between the current weighting vector and the "ideal" weighting vector, which exists because the input/output mapping is a $k$-valued threshold function. The more this distance decreases, the more the error decreases too, and the iterative learning process always converges after a finite number of iterations.

We have also considered the Hebbian learning for MVN. It was shown that the Hebbian learning rule works for MVN in the same manner as for the threshold neuron. It builds associations between the inputs and desired outputs. We have also shown that Hebbian weights, even when they cannot implement the input/output mapping, should be optimal starting weights for the MVN learning algorithm, leading to fewer corrections of the weights rather than starting from the arbitrary random vector.

So we have considered all the MVN fundamentals, its mathematical background, its organization, and its learning rules.

Now we are ready to consider how MVN works in networks and first of all in the feedforward neural network.

# Chapter 4
# Multilayer Feedforward Neural Network Based on Multi-Valued Neurons (MLMVN)

"All truth passes through three stages. First, it is ridiculed. Second, it is violently opposed. Third, it is accepted as being self-evident."

Arthur Schopenhauer

In this Chapter, we consider one of the most interesting applications of MVN – its use as a basic neuron in a multilayer neural network based on multi-valued neurons (MLMVN). In Section 4.1, we consider basic ideas of the derivative-free backpropagation learning algorithm for MLMVN. In Section 4.2, we derive the error backpropagation rule for MLMVN with a single hidden layer and a single output neuron and then for the most general case of arbitrary number of layers and neurons in the layers. In Section 4.3, the Convergence Theorem for the MLMVN learning algorithm is proven. In Section 4.4, we consider in detail how MLMVN learns when solving a classification problem. We also consider in the same section how MLMVN solves a problem of the Mackey-Glass time series prediction. Concluding remarks are given in Section 4.5.

## 4.1   Introduction to Derivative-Free Backpropagation Learning

As it was shown in Chapters 2-3, MVN has a number of advantages over other neurons. MVN may learn multiple-valued discrete and continuous input/output mappings. Its learning algorithm is based on the same error-correction learning rule as the one of the threshold neuron. The MVN functionality is high. Nevertheless, it is limited. Of course, there are many multiple-valued functions that are not threshold, and therefore they cannot be implemented using a single MVN. There are also many real-world classification and prediction problems, which are described by multiple-valued and continuous functions that are not threshold. Therefore, they also cannot be solved with a single MVN. Hence, MVN-based neural networks should be used for solving such problems.

Perhaps, a multilayer feedforward neural network is the most popular type of neural networks. In Section 1.3, we have considered in detail MLF – a multilayer feedforward neural network based on sigmoidal neurons. As we have seen there, the MLF learning is based on the error backpropagation and on the minimization of the mean square error functional. The error is a composite function that is the

I. Aizenberg: Complex-Valued Neural Networks with Multi-Valued Neurons, SCI 353, pp. 133–172.
springerlink.com                                   © Springer-Verlag Berlin Heidelberg 2011

function of the activation function, which in turn is the function of the weights. According to the optimization principles, to minimize the error, we have to follow the direction shown by the derivative of the error function with respect to the weights. The error function decreases if its derivative is negative. However, since the error function is composite, its derivative with respect to the weights is the product of the its derivative with respect to the activation function and the derivative of the activation function with respect to the weights.

What happens if we want to design a feedforward neural network from multi-valued neurons? MVN activation functions, the discrete function (2.50) and the continuous function (2.54), are not differentiable as functions of a complex variable because the Cauchy–Riemann conditions [75] for the differentiability of a function of the complex variable[1] do not hold for them. Does it mean that it is not possible to design a backpropagation learning algorithm for a feedforward network based on multi-valued neurons? No, it does not! This only means that such an algorithm should be based on the different background, namely on the same error-correction rule that the learning algorithm for a single MVN is based. Just this rule should be generalized for the MLMVN case. This generalization was suggested by the author of this book together with C. Moraga and D. Paliy in [76].

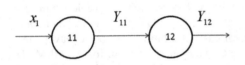

Let us consider the simplest possible multilayer feedforward neural network 1-1-1 containing only two neurons in two layers (one hidden neuron and one output neuron) with a single input (see Fig. 4.42). Let us use two indexes to number the neurons in the network. The $1^{st}$ index stands for the number of neuron in

**Fig. 4.42** The simplest multilayer feedforward neural network 1-1-1 containing two MVNs

the layer, and the second index stands for the number of layer. Thus, our network contains the neuron 11 in a single hidden layer and the output neuron 12. The network has a single input $x_1$, which is simultaneously a single input of the neuron 11. Let $Y_{11}$ and $Y_{12}$ be the actual outputs of the neurons 11 and 12, respectively. Let $D_{12}$ be the desired output of the neuron 12 and the entire network, respectively. Let us derive the error backpropagation rule for this network.

---

[1] The Cauchy–Riemann conditions state that the function $v = f(z)$ of the complex variable is differentiable at $z_0 = x_0 + iy_0$ if and only if its real and imaginary parts $u$ and $v$ are differentiable as functions of real variables $x$ and $y$ at $(x_0, y_0)$ and moreover,

$$\frac{\partial u}{\partial x} = \frac{\partial v}{\partial y} \text{ and } \frac{\partial u}{\partial y} = -\frac{\partial v}{\partial x}, \text{ or in other words } \frac{\partial f}{\partial x} + i\frac{\partial f}{\partial y} = 0.$$

Suppose that the actual output of the network does not coincide with its desired output. Hence, the error of the network is equal to $\delta^* = D_{12} - Y_{12}$.

Although, this error is evaluated for the output of the neuron 12, this is the error of the entire network. Since the network contains two neurons, both of them contribute to this global error. We need to understand, how we can obtain the local errors for each particular neuron, backpropagating the error $\delta^*$ form the right-hand side to the left-hand side.

Let us suppose that the errors for the neurons 11 and 12 are already known. Let us use the learning rule (3.92) for the correction of the weights. Let us suppose that the neuron 11 from the $1^{st}$ layer is already trained. Let us now correct the weights for the neuron 12 (the output neuron) and estimate its updated weighted sum $\tilde{z}_{12}$. Let $\left( w_0^{12}, w_1^{12} \right)$ be the current weighting vector of the neuron 12, which we have to correct, $z_{12}$ be the weighted sum of the neuron 12 before the correction, $\delta_{11}$ be the error of the neuron 11, and $\delta_{12}$ be the error of the neuron 12. Then we obtain the following

$$\tilde{z}_{12} = \left( w_0^{12} + \frac{1}{2}\delta_{12} \right) + \left( w_1^{12} + \frac{1}{2}\delta_{12}\overline{(Y_{11} + \delta_{11})} \right)(Y_{11} + \delta_{11}) =$$

$$= w_0^{12} + \frac{1}{2}\delta_{12} + w_1^{12}Y_{11} + w_1^{12}\delta_{11} + \frac{1}{2}\delta_{12} =$$

$$= \underbrace{w_0^{12} + w_1^{12}Y_{11}}_{z_{12}} + \frac{1}{2}\delta_{12} + \frac{1}{2}\delta_{12} + w_1^{12}\delta_{11} = \qquad (4.108)$$

$$= z_{12} + + \frac{1}{2}\delta_{12} + \frac{1}{2}\delta_{12} + w_1^{12}\delta_{11} = z_{12} + \delta_{12} + w_1^{12}\delta_{11}.$$

To ensure that after the correction procedure the weighted sum of the neuron 12 will be exactly equal to $z_{12} + \delta^*$, it is clear from (4.108) that we need to satisfy the following:

$$\delta_{12} + w_1^{12}\delta_{11} = \delta^*. \qquad (4.109)$$

Of course, if (4.109) is considered as a formal equation with respect to $\delta_{11}$ and $\delta_{12}$, we will not get something useful. This equation has an infinite number of solutions, while we have to find among them a single solution, which correctly represents the local errors of each neuron through the global error of the network.

Let us come back to the learning rule (3.92) for a single MVN. According to this rule, if $D$ be the desired neuron output and $Y$ be the actual neuron output, then

$\Delta W = \dfrac{C_r}{(n+1)}(D-Y)\bar{X}$. A factor $\dfrac{1}{(n+1)}$, as we have mentioned in Sections

3.2 and 3.3, is very important. Since we do not know, which of the weights contributes more to the error, this factor is used to share the error uniformly among all the inputs. Therefore the same factor is used to distribute $\Delta W$ uniformly among all the $n+1$ weights $w_0, w_1,..., w_n$. It is easy to see that if we would omit this factor then the weighted sum corrected according to the learning rule (3.92) would not be equal to $z+\delta$ (see (3.93) ), but to $z+(n+1)\delta$. However, all the inputs are equitable, and it is more natural that during the correction procedure $\Delta W$ should be shared among the weights. Thus $\Delta W$ is distributed among the weights

uniformly by using the factor $\dfrac{1}{(n+1)}$. This makes clear the important role of this

factor in the learning rule (3.92).

If we have not a single neuron, but a feedforward neural network, we have to take into account the same property. It has to be used, to generalize the error-correction learning rule, which we have derived for a single multi-valued neuron in Section 3.3, for a feedforward neural network based on multi-valued neurons. As we have seen, the error-correction learning algorithm for MVN is derivative-free. Let us generalize it for the feedforward MVN-based network.

Let us return to our simplest network (Fig. 4.42). The error $\delta^*$ of the entire network has to be shared among the neurons 12 and 11. We can assume that each neuron uniformly contributes to the global error $\delta^*$.

This important assumption, which is heuristic, but absolutely not contradictory, leads us to the following expression for the error $\delta_{12}$ of the output neuron

$$\delta_{12} = \frac{1}{2}\delta^*. \tag{4.110}$$

Thus, $\delta_{12}$ is the contribution of the neuron 12 to the global error $\delta^*$ of the entire network.

To find the error $\delta_{11}$ of the neuron 11, let us backpropagate the error $\delta_{12}$ of the output neuron, which we have just found, to the first hidden layer (to the neuron 11). From (4.109), and taking into account that for any non-zero complex

number $w \in \mathbb{C}, w^{-1} = \dfrac{\bar{w}}{|w|^2}$, we obtain

$$\delta_{11} = \frac{\delta^* - \delta_{12}}{w_1^{12}} = \left(\delta^* - \delta_{12}\right)\left(w_1^{12}\right)^{-1} = \frac{\left(\delta^* - \delta_{12}\right)\overline{w}_1^{12}}{\left|w_1^{12}\right|^2}. \tag{4.111}$$

But from (4.110) $\delta^* = 2\delta_{12}$ and therefore from (4.111) we finally obtain the following expression for $\delta_{11}$:

$$\delta_{11} = \frac{\left(\delta^* - \delta_{12}\right)\overline{w}_1^{12}}{\left|w_1^{12}\right|^2} = \frac{\left(2\delta_{12} - \delta_{12}\right)\overline{w}_1^{12}}{\left|w_1^{12}\right|^2} = \frac{\delta_{12}\overline{w}_1^{12}}{\left|w_1^{12}\right|^2} = \delta_{12}\left(w_1^{12}\right)^{-1}. \tag{4.112}$$

Equation (4.112) is not only a formal expression for $\delta_{11}$. It leads us to the following important conclusion: *during a backpropagation procedure the backpropagated error must be multiplied by the reciprocal values of the corresponding weights.* This is important distinction of the error backpropagation for a feedforward neural network based on multi-valued neurons from the error backpropagation for MLF.

Since now the errors of both neurons in the network are known, we can apply the error-correction learning rule (3.92) to adjust the weights. Equation (4.108) represents the updated weighted sum $\tilde{z}_{12}$ of the neuron 12 after the weights are corrected. Let us substitute there the errors $\delta_{11}$ and $\delta_{12}$, which we found in (4.110) and (4.112), respectively. Hence we obtain the following

$$\tilde{z}_{12} = z_{12} + \delta_{12} + w_1^{12}\delta_{11} + z_{12} + \frac{1}{2}w_1^{12}\left(\frac{\left(w_1^{12}\right)^{-1}\delta^*}{2}\right) = \tag{4.113}$$

$$= z_{12} + \frac{\delta^*}{2} + \frac{\delta^*}{2} = z_{12} + \delta^*$$

where $z_{12}$ is the weighted sum of the neuron 12 before the correction of the weights. Thus, we obtain exactly the result, which is our target $\tilde{z}_{12} = z_{12} + \delta^*$. This means that the updated weighted sum of the output neuron is equal to the previous value of the weighted sum plus the error of the network. Comparing (4.113) with the analogous expression (3.93) for the weighted sum of a single MVN, which was updated using the error-correction learning rule (3.92), we see that we obtained the same result, just this time not for a single neuron, but for the feedforward network.

This leads us to the following important conclusion. The error backpropagation for the MVN-based feedforward neural network has to be organized in such a way that the error, which has to be backpropagated from the current layer to the preceding layer, is shared among all the neurons contributed to this error. Then the weights of all the neurons should be adjusted using the error-correction learning rule. This ensures that the error of the entire network decreases after the correction of the weights in the same manner as for a single MVN. Therefore, the learning algorithm for the network can be based on the generalized error-correction learning rule.

Let us now consider the MVN-based feedforward neural network and its back-propagation learning in detail.

## 4.2  MVN-Based Multilayer Neural Network and Error Backpropagation

The multilayer neural network based on multi-valued neurons (MLMVN) was suggested by the author of this book, Claudio Moraga and Dmitriy Paliy in 2005, in the paper [76]. In that paper the network with one hidden layer and the output layer containing a single neuron was considered. Two years later, in 2007, the author of this book and C. Moraga presented MLMVN in detail and without restrictions on the number of layers and neurons in [62]. A number of MLMVN applications were also considered in the same paper. In 2008, the author of this book in co-authorship with Dmitriy Paliy, Jacek Zurada, and Jaakko Astola published the paper [77] where the MLMVN backpropagation learning algorithm was justified for the most general case.

Let us consider this algorithm in detail.

### 4.2.1  Simple MLMVN: A Single Hidden Layer and a Single Output Neuron

To make understanding of the MLMVN fundamentals easier, let us go step by step. In Section 4.1, we have already considered the simplest network (Fig. 4.42) with a single hidden neuron and a single output neuron. We have derived the error backpropagation rule for that network.

Let us now extend this rule to the network containing a single hidden layer with an arbitrary amount of neurons and a single output neuron.

Deriving the error backpropagation algorithm for MLMVN, let us use the same notations that we have used in Section 1.3 (see pp. 28-29) deriving the error backpropagation learning algorithm for MLF. Let our network has the topology $n\text{-}N\text{-}1$, which means that it has $n$ inputs $x_1, ..., x_n$, contains $N$ neurons in a single hidden layer (neurons 11, ..., $N1$) and a single output neuron 12 (see Fig. 4.43). Let $D_{12}$ be the desired output of the output neuron 12 and of the network. Let $Y_{12}$ be the actual output of the output neuron 12 and of the network and, $Y_{11}, ..., Y_{N1}$ be the actual outputs of the hidden neurons. Suppose the desired network output does not coincide with the actual output. According to (1.25) (see p. 29) the error of the network is $\delta^* = D_{21} - Y_{21}$.

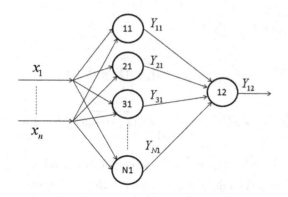

**Fig. 4.43** The $n$-$N$-1 MLMVN

To generalize the error-correction learning rule for the network, we have to ensure that after the weights of all the neurons are corrected, the updated weighted sum $\tilde{z}_{12}$ of the output neuron is equal to the previous its value plus the error $\tilde{z}_{12} = z_{12} + \delta^*$, like it was for a single MVN (see (3.93), p. 113) and for the simplest MLMVN (see (4.113) ).

Let $\left( w_0^{12}, w_1^{12}, ..., w_N^{12} \right)$ be an initial weighting vector of the output neuron 12. Since the inputs of this neuron are connected to the outputs of the corresponding neurons from the preceding layer and since there are exactly $N$ neurons in the layer 1, then the neuron 12 has exactly $N$ inputs. Let also $\delta_{i1}; i = 1, ..., N$ be the error of the neuron $i1$ from the hidden layer, and $\delta_{12}$ be the error of the neuron 12. Let us suppose that all neurons from the hidden layer are already trained and therefore their actual outputs $Y_{11}, ..., Y_{N1}$ are updated and equal to $Y_{11} + \delta_{11}, ..., Y_{N1} + \delta_{N1}$, respectively. So they are the inputs of the output neuron 12. Let us now correct the weights for the neuron 12 (the output neuron) using the error-correction rule (3.92) and estimate its updated weighted sum $\tilde{z}_{12}$. Taking into account our notations, according to (3.92) we obtain the following

$$\tilde{w}_0^{12} = w_0^{12} + \frac{1}{(N+1)} \delta_{12},$$

$$\tilde{w}_i^{12} = w_i^{12} + \frac{1}{(N+1)} \delta_{12} \overline{(Y_{i1} + \delta_{i1})}; i = 1, ..., N$$

where the bar sign under $Y_{i1} + \delta_{i1}$ means complex conjugation. Now we obtain the following equation for the updated weighted sum $\tilde{z}_{12}$

$$\tilde{z}_{12} = \left( w_0^{12} + \frac{1}{(N+1)}\delta_{12} \right) + \left( w_1^{12} + \frac{1}{(N+1)}\delta_{12}\overline{(Y_{11}+\delta_{11})} \right)(Y_{11}+\delta_{11}) + ... +$$

$$+ \left( w_N^{12} + \frac{1}{(N+1)}\delta_{12}\overline{(Y_{N1}+\delta_{N1})} \right)(Y_{N1}+\delta_{N1}) =$$

$$= \left( w_0^{12} + \frac{1}{(N+1)}\delta_{12} \right) + \sum_{i=1}^{N}\left( w_i^{12} + \frac{1}{(N+1)}\delta_{12}\overline{(Y_{i1}+\delta_{j1})} \right)(Y_{i1}+\delta_{i1}) =$$

$$= \left( w_0^{12} + \frac{1}{(N+1)}\delta_{12} \right) + \sum_{j=1}^{N}\left( w_i^{12}Y_{i1} + w_i^{12}\delta_{i1} + \frac{1}{(n+1)}\delta_{12}\overline{(Y_{i1}+\delta_{i1})}(Y_{i1}+\delta_{i1}) \right).$$

Taking into account that $\forall i = 1,...,N \;\; (Y_{i1}+\delta_{i1})$ is an output of the neuron, this complex number is always located on the unit circle and therefore its absolute value is always equal to 1. This means that $\forall i = 1,...,N \;\; \overline{(Y_{i1}+\delta_{i1})}(Y_{i1}+\delta_{i1}) = 1$. Taking this into account, we obtain the following:

$$\tilde{z}_{12} = \left( w_0^{12} + \frac{1}{(N+1)}\delta_{12} \right) +$$

$$+ \sum_{i=1}^{N}\left( w_i^{12}Y_{i1} + w_i^{12}\delta_{i1} + \frac{1}{(N+1)}\delta_{12}\overline{(Y_{i1}+\delta_{i1})}(Y_{i1}+\delta_{i1}) \right) =$$

$$= \left( w_0^{12} + \frac{1}{(N+1)}\delta_{12} \right) + \sum_{i=1}^{N}\left( w_i^{12}Y_{i1} + w_i^{12}\delta_{i1} + \frac{1}{(N+1)}\delta_{12} \right) =$$

$$= w_0^{12} + \frac{1}{(N+1)}\delta_{12} + \sum_{i=1}^{N}w_i^{12}Y_{i1} + \sum_{i=1}^{N}w_i^{12}\delta_{i1} + \frac{N}{(N+1)}\delta_{12} =$$

$$= w_0^{12} + \sum_{i=1}^{N}w_i^{12}Y_{i1} + \frac{1}{(N+1)}\delta_{12} + \frac{N}{(N+1)}\delta_{12} + \sum_{i=1}^{n}w_i^{12}\delta_{i1} =$$

$$= \underbrace{w_0^{12} + \sum_{i=1}^{N}w_i^{12}Y_{i1}}_{z_{12}} + \delta_{12} + \sum_{i=1}^{N}w_i^{12}\delta_{i1} = z_{12} + \delta_{12} + \sum_{i=1}^{N}w_i^{12}\delta_{i1}.$$

Thus finally

$$\tilde{z}_{12} = z_{12} + \delta_{12} + \sum_{i=1}^{N}w_i^{12}\delta_{i1} \qquad (4.114)$$

Evidently, (4.114) generalizes the analogous equation (4.108) for the updated weighted sum of the output neuron of the simplest network 1-1-1. To ensure that after the correction procedure the output of the network is equal exactly to $z_{12} + \delta^*$, it is clear from (4.114) that we need to satisfy the following condition for the network error $\delta^*$

$$\delta_{12} + \sum_{i=1}^{N} w_i^{12} \delta_{i1} = \delta^*. \tag{4.115}$$

One may see that (4.115) generalizes the analogous equation (4.109) for the simplest network. As well as (4.109), (4.115) should not be considered as a formal equation with respect to $\delta_{12}$ and $\delta_{i1}; i = 1, ..., N$. It has infinite amount of solutions, however we need to find a solution, which correctly represents the error $\delta_{12}$ and the errors $\delta_{i1}; i = 1, ..., N$ through the global error of the network $\delta^*$. Let us follow the same heuristic approach, which we used for the error-correction rule (3.92) for a single MVN and for the error-correction rule for the simplest MLMVN 1-1-1 in Section 4.1.

This approach is utilized in the *error sharing principle*.

**The Error Sharing Principle.** 1) The error of a single MVN must be shared among all the weights of the neuron. 2) The network error and the errors of each particular neuron in MLMVN must be shared among those neurons from the network that contribute to this error.

Particularly, for the network with a single hidden layer containing $N$ neurons and a single output neuron this means that the network error $\delta^*$ has to be distributed among the output neuron 12 and hidden neurons 11, ..., $N1$, and the error $\delta_{12}$ of the output neuron must be shared with the hidden neurons 11, ..., $N1$. We assume that *this sharing is uniform*: the contribution of each neuron to the global error is uniform accurate to the corresponding weight, which weights the error, passing it to the next layer neuron.

Taking these considerations into account, we obtain

$$\delta_{12} = \frac{1}{(N+1)} \delta^*, \tag{4.116}$$

because the network error $\delta^*$ includes the error $\delta_{12}$ of the output neuron and the errors $\delta_{11}, ..., \delta_{N1}$ of $N$ hidden neurons. Taking into account the latter considerations, (4.115) and (4.116), we obtain the following

$$w_i^{12} \delta_{i1} = \frac{1}{(N+1)} \delta^*; \ i = 1,...,N. \tag{4.117}$$

But from (4.116) $\delta^* = (N+1)\delta_{12}$, and substituting this expression to (4.117), we obtain

$$w_i^{12} \delta_{i1} = \delta_{12}; \ i = 1,...,N.$$

Hence for the errors $\delta_{i1}; i = 1,...,N$ of the neurons 11, ..., N1 we obtain the following error backpropagation rule

$$\delta_{i1} = \left(w_i^{12}\right)^{-1} \delta_{12}; \ i = 1,...,N. \tag{4.118}$$

This states that the errors of the hidden layer neurons equal to the output neuron error multiplied by the corresponding reciprocal weight of the output neuron.

Let us now substitute $\delta_{i1}; i = 1,...,N$ and $\delta_{12}$ from (4.118) and (4.116), respectively, into (4.114):

$$\tilde{z}_{12} = z_{12} + \delta_{12} + \sum_{i=1}^{N} w_i^{12} \delta_{j1} = z_{12} + \delta_{12} + \sum_{i=1}^{N} w_i^{12} \left(w_i^{12}\right)^{-1} \delta_{12} = \tag{4.119}$$

$$= z_{12} + \delta_{12} + N\delta_{12} = z_{12} + (N+1)\delta_{12} = z_{12} + \delta^*$$

So, we obtain exactly the result, which is our target $\tilde{z}_{12} = z_{12} + \delta^*$ - the weighted sum of the output neuron, which was updated after the weights of all the neurons were corrected according to the error-correction learning rule, is equal to the previous weighted sum plus the network error. We see that (4.119) exactly corresponds with (3.93), which demonstrates the same property for the updated weighted sum of a single MVN.

To summarize, we can formulate the learning algorithm for the two-layer MLMVN as follows.

1) The network error $\delta^*$ is backpropagated first to the output neuron according to (4.116).

2) The output neuron error $\delta_{12}$ is backpropagated to the hidden neurons according to (4.118).

3) After all the local errors $\delta_{i1}; i = 1,...,N$ and $\delta_{12}$ are obtained, the weights of all the neurons should be adjusted according to the error-correction learning rule (3.92).

This ensures that the weighted sum of the output neuron is updated according to (4.119), which exactly corresponds to the weighted sum of a single MVN (3.93), which in turn is updated according to the error-correction learning rule (3.92) (see p. 112).

This leads us to the very important conclusion that the mechanism of the MLMVN learning is exactly the same as the mechanism of the MVN learning based on the error-correction learning rule. The updated weighted sum should move exactly to the desired sector (for the discrete output neuron) or to the desired ray (for the continuous output neuron) or closer to them.

It is also important to mention that the learning algorithm and the error back-propagation rule, which we have just described, do not depend on the type of MVNs in the network. They all may have discrete activation function (2.50) or continuous activation function (2.54). Some of them may have the discrete activation function, while other may have the continuous activation function.

### 4.2.2 MLMVN and Backpropagation Learning:The General Case

Let us now consider MLMVN and its backpropagation learning algorithm for the most general case, where the network may have an arbitrary amount of hidden layers and neurons in these layers, and an arbitrary amount of output neurons.

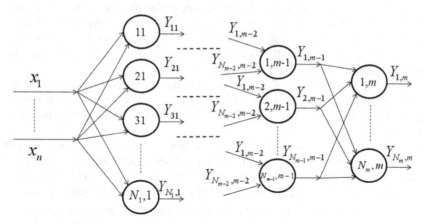

**Fig. 4.44** MLMVN $n - N_1 - ... - N_{m-1} - N_m$ (the most general case)

Let us consider the $m$-layer MLMVN $n - N_1 - ... - N_{m-1} - N_m$, which contains the input layer with $n$ inputs, $m$-1 hidden layers whose numbers are 1, $m$-1, and one output layer whose number is $m$. (see Fig. 4.44). Let us remind that according to our notations that we use for multilayer feedforward neural networks $N_j$ is the number of neurons in the layer $j$. Each layer including the output one contains an arbitrary amount of neurons.

Let us now generalize the learning algorithm and the error backpropagation rule, which we already derived for the networks 1-1-1 and $n$-$N$-1, for the most general case of MLMVN.

It follows from (1.116) and the *Error Sharing Principle* that if the error of a neuron in the layer $j$ is equal to $\delta$, this $\delta$ must contain a factor $\dfrac{1}{s_j}$, where

$s_j = N_{j-1} + 1$ is the number of neurons whose outputs are connected to the inputs of the considered neuron incremented by 1 (the considered neuron itself). Let us remind that the inputs of a neuron in the layer $j$ are connected to the outputs of all the corresponding neurons from the layer $j$-1.

The factor $\dfrac{1}{s_j}$ ensures sharing of the error of a particular neuron with all the neurons on whose errors the error of this neuron depends. In other words, *the error of each neuron is distributed among the neurons connected to it and itself.* It should be mentioned that for the 1$^{st}$ hidden layer $s_1 = 1$ because no hidden layer precedes to the first one and there are no neurons, which may share the error, respectively.

The next important rule of the MLMVN error backpropagation follows from (4.118). This rule states that *during a backpropagation procedure the backpropagated error must be multiplied by the reciprocal values of the corresponding weights.*[2] This is important distinction from the classical error backpropagation rule (1.33) (see p. 33) for MLF where during a backpropagation procedure the backpropagated error must be multiplied by the weights.

Now we are ready to formulate and to justify the learning algorithm and the error backpropagation rule for the most general case. We keep all those notations that were previously used for MLMVN. All additional notations will be introduced as necessary.

It is also important to mention that our considerations do not depend on the type of neuron activation function. It can be the discrete activation function (2.50) or the continuous activation function (2.54). Some of neurons may have the discrete activation function, while other may have the continuous activation function.

Let $D_{km}$ be a desired output of the $k$th neuron from the $m$th (output) layer; $Y_{km}$ be an actual output of the $k$th neuron from the $m$th (output) layer. Then the global error of the network taken from the $k$th neuron of the $m$th (output) layer is calculated as follows:

$$\delta^*_{km} = D_{km} - Y_{km}. \tag{4.120}$$

---

[2] Do not forget that our weights are complex numbers and therefore for any weight $w^{-1} = \dfrac{\bar{w}}{|w|^2}$, where the bar sign stands for the complex conjugation.

To find the errors of the hidden neurons, we have to sequentially backpropagate these errors to the hidden layers $m$-1, $m$-2, ..., 2, 1. We have to distinguish again the global errors of the network $\delta_{km}^*$ from the local errors $\delta_{km}$ of the particular output layer neurons. The global errors are obtained according to (4.120), while the local errors are obtained from the global ones through the backpropagation procedure. According to the *Error Sharing Principle* (see p. 141), it is essential that the global errors of the network consist not only from the output neurons errors, but from the local errors of all the output and hidden neurons. This means that in order to obtain the local errors for all neurons, the global errors must be distributed among these neurons through the error backpropagation. Hence, the local errors are represented in the following way. The errors of the $m$th (output) layer neurons:

$$\delta_{km} = \frac{1}{s_m} \delta_{km}^*$$

(4.121)

where $km$ specifies the $k$th neuron of the $m$th layer; $s_m + N_{m-1} + 1$, i.e. the number of all neurons on the preceding layer incremented by 1.

Let $w_i^{kj}$ be the weight corresponding to the $i$th input of the $kj$th neuron ($k$th neuron of the $j$th layer). Generalizing the backpropagation rule (4.118), which we obtained for the network $n$-$N$-1, let us backpropagate the errors of the output neurons $\delta_{im}; i = 1, ..., N_m$ to the neurons of the $m$-$1^{st}$ layer. The error $\delta_{kj}$ of the $k$th neuron in the $m$-$1^{st}$ layer is formed by the sum of the errors $\delta_{1m}, ..., \delta_{N_m, m}$ of all the output neurons weighted by the reciprocal weights $\left( w_k^{im} \right)^{-1}; i = 1, ..., N_m$ corresponding to the $k$th inputs of the output neurons. According to the *Error Sharing Principle* this sum must be multiplied by $\frac{1}{s_{m-1}}$ because this error must be shared with the neuron ($k$, $m$-1) and all the neurons from the $m$-$2^{nd}$ layer connected to the neuron ($k$, $m$-1). This is resulted in the following equation

$$\delta_{k,m-1} = \frac{1}{s_j} \sum_{i=1}^{N_m} \delta_{im} \left( w_k^{im} \right)^{-1}$$

Now the errors $\delta_{k,m-1}; k = 1, ..., N_{m-1}$ should be backpropagated in the same manner to the layer $m$-2, then the errors $\delta_{k,m-2}; k = 1, ..., N_{m-2}$ should be backpropagated in the same manner to the layer $m$-3, etc., up to the backpropagation to the first hidden layer.

Taking all these considerations into account, we obtain the following error backpropagation rule according to which the errors of all the hidden neurons are computed. The error $\delta_{kj}$ of the $k$th neuron in the $j$th layer is equal to

$$\delta_{kj} = \frac{1}{s_j} \sum_{i=1}^{N_{j+1}} \delta_{i,j+1} \left( w_k^{i,j+1} \right)^{-1},$$  (4.122)

where $kj$ specifies the $k$th neuron of the $j$th layer $(j=1,...,m\text{-}1)$; $s_j = N_{j-1} + 1; j = 2,...,m-1$ is the number of all neurons in the layer $j$-1 (the preceding layer $j$ where the error is backpropagated to) incremented by 1. Note again that $s_1 = 1$ because there is no preceding layer for the 1$^{st}$ hidden layer, that is $N_{1-1} = N_0 = 0$.

Hence, the MLMVN error backpropagation rule is determined by (4.121) for the output neurons errors and by (4.122) for the hidden layer neurons.

The main distinction of the MLMVN error backpropagation rule from the MLF backpropagation rule (1.30)-(1.32) (see p. 32) is that it is *derivative-free*. It is based on the generalization of the error-correction learning rule for MVN.

The MLMVN error backpropagation rule (4.121)-(4.122) can be justified if we will show that the weighted sums $\tilde{z}_{km}; k = 1,...,N_m$ of the output neurons updated after the weights of all the neurons in the network were corrected according to the MVN error-correction rule (3.92) or (3.94) will be equal to $z_{km} + \delta_{km}^*$ (to the previous weighted sum plus the network error taken from the $km$th neuron), like it is for a single MVN (see (3.93) ), for the simplest network 1-1-1 (see (4.113)), and for the network $n$-$N$-1 (see (4.119) ). Let us prove that this is definitely the case.

**Theorem 18** (about the MLMVN Error Backpropagation Rule). The updated weighted sums $\tilde{z}_{km}; k = 1,...,N_m$ of the MLMVN output neurons are equal to $z_{km} + \delta_{km}^*$ (where $\delta_{km}^*; k = 1,...,N_m$ are the network errors taken from all the output neurons) if the error backpropagation is determined by (4.121) for the output neurons errors and by (4.122) for the hidden layer neurons, and the weights of all the neurons are corrected according to the error-correction learning rule (3.92).

**Proof.** According to (4.120) the errors of the network that are taken from the $N_m$ output neurons are equal to $\delta_{km}^* = D_{km} - Y_{km}; k = 1,...,N_m$, where $N_m$ is the number of neurons in the output layer (the $m$th one). Let $\left( w_0^{km}, w_1^{km},...,w_{N_{m-1}}^{km} \right)$ be an initial weighting vector of the $km$th output neuron, $Y_{i,m-1}$ be an initial output of the neuron $i,m$-1 from the last hidden layer ($i$=1, ..., $N_{m-1}$), $z_{km}$ be the weighted sum on the $km$th output neuron before the correction of the weights, $\delta_{i,m-1}$ be the error of the neuron $i,m$-1 from the $m$-1$^{st}$ hidden layer ($i$=1, ..., $N_{m-1}$), and $\delta_{km}$ be the error of the $km$th output neuron. Without loss of generality, let us suppose that all hidden neurons are already trained. Thus, the outputs for all the neurons from the 1$^{st}$ hidden layer till the $m$-1$^{st}$ layer are already updated, and we have to train the neurons from the output layer. The MVN error-correction learning rule (3.92) is used for the correction of the weights. Let us suppose that the errors for all the neurons of the network are already known. Then, the weights for the $km$th output neuron are corrected according to (3.92). Let us take the learning rate $C = 1$. This means that the weights are updated as follows:

$$\tilde{w}_0^{km} = w_0^{km} + \frac{1}{(N_{m-1}+1)}\delta_{km};$$

$$\tilde{w}_i^{km} = w_i^{km} + \frac{1}{(N_{m-1}+1)}\delta_{km}\bar{\tilde{Y}}_{i,m-1}; i = 1,...,N_{m-1}$$

where $\tilde{Y}_{i,m-1}$ is the updated output of the $i$th neuron from the $m$-1$^{st}$ layer, that is the $i$th input of the $km$th output neuron, and the bar sign stands for complex conjugation.

Then the updated weighted sum for the $km$th output neuron is obtained as

$$\tilde{z}_{km} = \left(w_0^{km} + \frac{1}{(N_{m-1}+1)}\delta_{km}\right) +$$

$$+\left(w_1^{km} + \frac{1}{(N_{m-1}+1)}\delta_{km}\overline{(Y_{1,m-1}+\delta_{1,m-1})}\right)(Y_{1,m-1}+\delta_{1,m-1})+...+$$

$$+\left(w_{N_{m-1}}^{km} + \frac{1}{(N_{m-1}+1)}\delta_{km}\overline{(Y_{N_{m-1},m-1}+\delta_{N_{m-1},m-1})}\right)(Y_{N_{m-1},m-1}+\delta_{N_{m-1},m-1})=$$

$$\left(w_0^{km} + \frac{1}{(N_{m-1}+1)}\delta_{km}\right) + \qquad\qquad (4.123)$$

$$+\sum_{i=1}^{N_{m-1}}\left(w_i^{km} + \frac{1}{(N_{m-1}+1)}\delta_{km}\overline{(Y_{i,m-1}+\delta_{i,m-1})}\right)(Y_{i,m-1}+\delta_{i,m-1})=$$

$$=\left(w_0^{km} + \frac{1}{(N_{m-1}+1)}\delta_{km}\right) +$$

$$+\sum_{i=1}^{N_{m-1}}\left(w_i^{km}Y_{i,m-1}+w_i^{km}\delta_{i,m-1}+\frac{1}{(N_{m-1}+1)}\delta_{km}\overline{(Y_{i,m-1}+\delta_{i,m-1})}(Y_{i,m-1}+\delta_{i,m-1})\right)$$

where $\overline{Y}$ is the complex number complex conjugated to $Y$. It is important that $\forall i = 1,...,N_{m-1} \left| \left( Y_{i,m-1} + \delta_{i,m-1} \right) \right| = 1$ because since a complex number $\left( Y_{i,m-1} + \delta_{i,m-1} \right)$ is an output of MVN, it is always located on the unit circle, and therefore its absolute value is equal to 1. This means that $\forall i = 1,...,N_{m-1} \overline{\left( Y_{i,m-1} + \delta_{i,m-1} \right)} \left( Y_{i,m-1} + \delta_{i,m-1} \right) = 1$. Taking this into account, let us now simplify (4.123) as follows

$$\tilde{z}_{km} = \left( w_0^{km} + \frac{1}{(N_{m-1}+1)} \delta_{km} \right) +$$

$$+ \sum_{i=1}^{N_{m-1}} \left( w_i^{km} Y_{i,m-1} + w_i^{km} \delta_{i,m-1} + \frac{1}{(N_{m-1}+1)} \delta_{km} \overline{\left( Y_{i,m-1} + \delta_{i,m-1} \right)} \left( Y_{i,m-1} + \delta_{i,m-1} \right) \right) =$$

$$= \left( w_0^{km} + \frac{1}{(N_{m-1}+1)} \delta_{km} \right) + \sum_{i=1}^{N_{m-1}} \left( w_i^{km} Y_{i,m-1} + w_i^{km} \delta_{i,m-1} + \frac{1}{(N_{m-1}+1)} \delta_{km} \right) =$$

$$= w_0^{km} + \frac{1}{(N_{m-1}+1)} \delta_{km} + \sum_{i=1}^{N_{m-1}} w_i^{km} Y_{i,m-1} + \sum_{i=1}^{N_{m-1}} w_i^{km} \delta_{i,m-1} + \frac{N_{m-1}}{(N_{m-1}+1)} \delta_{km} =$$

$$= w_0^{km} + \sum_{i=1}^{N_{m-1}} w_i^{km} Y_{i,m-1} + \delta_{km} + \sum_{i=1}^{N_{m-1}} w_i^{km} \delta_{i,m-1} = z_{km} + \delta_{km} + \sum_{i=1}^{N_{m-1}} w_i^{km} \delta_{i,m-1}$$

Thus, finally, for the updated weighted sum $\tilde{z}_{km}$ of the $km$th output neuron we obtain

$$\tilde{z}_{km} = z_{km} + \delta_{km} + \sum_{i=1}^{N_{m-1}} w_i^{km} \delta_{i,m-1} . \tag{4.124}$$

To ensure that the weighted sum of the $km$th output neuron is exactly equal to $z_{km} + \delta_{km}^*$ after the correction procedure, it follows from (124) that we need to satisfy the following:

$$\delta_{km} + \sum_{i=1}^{N_{m-1}} w_i^{km} \delta_{i,m-1} = \delta_{km}^* . \tag{4.125}$$

If (4.125) is considered as a formal equation with respect to $\delta_{km}$ and $\delta_{i,m-1}; i = 1,...,N_{m-1}$, it has an infinite number of solutions. Let us apply the *Error Sharing Principle*, which requires that the network error $\delta_{km}^*$ taken from the

$km$th output neuron must be shared among this neuron and all the neurons from the preceding layer connected to this neuron (thus, the neurons $i,m$-1; $i=1,...,N_{m-1}$). According to the *Error Sharing Principle*

$$\delta_{km} = \frac{1}{(N_{m-1}+1)}\delta_{km}^*,$$
(4.126)

and according to the same principle and from (4.125)

$$w_i^{km}\delta_{i,m-1} = \frac{1}{(N_{m-1}+1)}\delta_{km}^*; i=1,...,N_{m-1}.$$
(4.127)

But from (4.126) $\delta_{km}^* = (N_{m-1}+1)\delta_{km}$, and substituting this expression into (4.127) we obtain the following

$$w_i^{km}\delta_{i,m-1} = \delta_{km}; i=1,...,N_{m-1}.$$
(4.128)

Hence, for the error $\delta_{i,m-1}$ of the neuron $(i,m$-1) we obtain the following

$$\delta_{i,m-1} = \left(w_i^{km}\right)^{-1}\delta_{km}; i=1,...,N_{m-1}.$$
(4.129)

Let us now substitute $\delta_{i,m-1}; i=1,...,N_{m-1}$ from (4.129) and $\delta_{km}$ from (4.126), respectively, into (4.124):

$$\tilde{z}_{km} = z_{km} + \delta_{km} + \sum_{i=1}^{N_{m-1}} w_i^{km}\delta_{i,m-1} =$$

$$= z_{km} + \delta_{km} + \sum_{i=1}^{N_{m-1}} w_i^{km}\left(w_i^{km}\right)^{-1}\delta_{km} =$$
(4.130)

$$z_{km} + \delta_{km} + N_{m-1}\delta_{km} = z_{km} + \left(N_{m-1}+1\right)\delta_{km} = z_{km} + \delta_{km}^*$$

So, we obtain exactly the result, which we were looking for: $\tilde{z}_{km} = z_{km} + \delta_{km}^*$.

To finalize the proof of the theorem we have to show that for the updated weighted sum $\tilde{z}_{kj}$ of any hidden neuron $kj, k=1,...,N_j; j=1,...,N_{m-1}$ the property analogous to (4.130) holds too. The errors of the hidden layer neurons are obtained according to (4.122). The correction of the weights according to the error-correction rule (3.92) is organized in the following way. For the 1st hidden layer neurons ($k$th neuron of the 1st layer, $k=1,...,N_1$)

$$\tilde{w}_0^{k1} = w_0^{k1} + \frac{1}{(n+1)}\delta_{k1};$$

$$\tilde{w}_i^{k1} = w_i^{k1} + \frac{1}{(n+1)}\delta_{k1}\bar{x}_i; i = 1,...,n,$$

where $x_1,...,x_n$ are the network inputs. For the rest of hidden neurons ($k$th neuron of the $j$th layer, $k = 1,...,N_j; j = 2,...,m-1$)

$$\tilde{w}_0^{kj} = w_0^{kj} + \frac{1}{(N_{j-1}+1)}\delta_{kj};$$

$$\tilde{w}_i^{kj} = w_i^{kj} + \frac{1}{(N_{j-1}+1)}\delta_{kj}\bar{\tilde{Y}}_{i,j-1}; i = 1,...,N_{j-1},$$

where $\tilde{Y}_{i,j-1}$ is the updated output of the $i$th neuron from the $j$-1$^{st}$ layer, that is the $i$th input of the $kj$th neuron, and the bar sign stands for the complex conjugation. Then the expressions for the updated weighted sums for the $kj$th hidden neuron ($k = 1,...,N_j; j = 2,...,m-1$) and for the $k$1st hidden neuron $(k = 1,...,N_1)$ are derived in the same way as (4.123):

$$\tilde{z}_{kj} = \left( w_0^{kj} + \frac{1}{(N_{j-1}+1)}\delta_{kj} \right) +$$

$$+ \sum_{i=1}^{N_{j-1}} \left( w_i^{kj}Y_{i,j-1} + w_i^{kj}\delta_{i,j-1} + \frac{1}{(N_{j-1}+1)}\delta_{kj}\overline{\left(Y_{i,j-1}+\delta_{i,j-1}\right)}\left(Y_{i,j-1}+\delta_{i,j-1}\right) \right); \quad (4.131)$$

$$\tilde{z}_{k1} = \left( w_0^{k1} + \frac{1}{(N_{j-1}+1)}\delta_{k1} \right) + \sum_{i=1}^{n} \left( w_i^{k1}x_i + \frac{1}{(n+1)}\delta_{k1}\bar{x}_ix_i \right).$$

Then, after simplifications we finally obtain for these updated weighted sums the following expressions

$$\tilde{z}_{kj} = z_{kj} + \frac{1}{(N_{j-1}+1)}\delta_{kj} + \sum_{i=1}^{N_{j-1}} w_i^{kj}\delta_{i,j-1};$$

$$\tilde{z}_{k1} = z_{k1} + \delta_{k1}.$$

$$(4.132)$$

It follows from the second equation of (4.132) that for all the neurons from the first hidden layer the updated weighted sums equal exactly to their previous values plus the corresponding errors, that is exactly to what is required.

Let us now prove that for all other hidden layers this is also true. The first equation from (4.132), which represents the updated weighted sum for the $kj$th hidden neuron $(k = 1,...,N_j; j = 2,...,m-1)$ is completely the same as the analogous equation (4.124) for the updated weighted sum of the $km$th output neuron $(k = 1,...,N_m)$. According to the *Error Sharing Principle* and in analogy with (4.125) we obtain for the output neurons errors

$$\delta_{kj} + \sum_{i=1}^{N_{j-1}} w_i^{kj} \delta_{i,j-1} = s_j \delta_{kj}, \tag{4.133}$$

where $s_j = N_{j-1} + 1$ is the number of neurons in the $j$-$1^{st}$ layer incremented by 1 (the number of neurons, which are directly involved in the formation of the error $\delta_{kj}$, and which have to share this error). From (4.133) we easily obtain

$$\sum_{i=1}^{N_{j-1}} w_i^{kj} \delta_{i,j-1} = (s_j - 1)\delta_{kj} = N_{j-1}\delta_{kj}$$

Substituting the last expression into the first equation of (4.132), we obtain

$$\tilde{z}_{kj} = z_{kj} + \frac{1}{(N_{j-1}+1)}\delta_{kj} + \sum_{i=1}^{N_{j-1}} w_i^{kj} \delta_{i,j-1} =$$

$$= z_{kj} + \frac{1}{(N_{j-1}+1)}\delta_{kj} + N_{j-1}\delta_{kj} = z_{kj} + \delta_{kj}$$

So, we obtain for all the neurons from our MLMVN that the updated weighted sum is equal to $\tilde{z}_{kj} = z_{kj} + \delta_{kj}; k = 1,...,N_j; j = 1,...,m$.

Theorem is proven.

It is very important that (4.130) as an expression for an updated weighted sum of an MLMVN output neuron exactly correspond to the analogous expressions for a single MVN (see (3.93) ), for the simplest network 1-1-1 (see (4.113) ), and for the network $n$-$N$-1 (see (4.119) ). This means that the MLMVN learning is completely based on the generalized error-correction learning rule.

Unlike the MLF error backpropagation rule (see (1.30) and (1.32) on p. 32), the MLMVN error backpropagation rule (see (4.121) and (4.122) on pp. 145-146) is derivative-free. While equations (1.30) and (1.32) contain the derivative of the activation function, equations (4.121) and (4.122) do not contain it.

This is not only a formal computational advantage of MLMVN. It is important to distinguish again that the MLMVN learning process is not considered as an optimization problem. It is based on the error-correction learning rule, as well as the learning process for a single MVN.

It is also worth to mention that the MLMVN error backpropagation rule does not depend on the number of layers and neurons in the network. It is equally efficient for the simplest network containing just two neurons and for a network containing tens of thousands neurons. This is another important advantage of MLMVN over MLF.

What is especially important: the MLMVN learning algorithm does not suffer from the local minima phenomenon as the MLF learning algorithm. If some input/output mapping, which we want to learn with MLMVN, can be implemented using this network, the MLMVN learning algorithm converges due to the same reasons as the MVN error-correction learning algorithm. Let us consider this in detail.

## 4.3  MLMVN Learning Algorithm and the Convergence Theorem

### 4.3.1  Basic Fundamentals

The MLMVN learning algorithm, as well as the MLF learning algorithm is a two-pass algorithm. First, the errors of all neurons must be calculated. To obtain these errors, the actual responses of the output neurons are subtracted from their desired responses to produce an error signal. This error signal is then propagated backward through the network, against the direction of synaptic connections and according to the error backpropagation rule (4.121)-(4.122). Then the weights should be adjusted so as to make the actual output of the network move closer to the desired output.

Let us formally describe the MLMVN learning algorithm. Let us consider MLMVN $n - N_1 - ... - N_{m-1} - N_m$ containing the input layer with $n$ inputs, $m$-1 hidden layers whose numbers are 1, ..., $m$-1, and one output layer whose number is $m$. (see Fig. 4.44, p. 143). It is important and interesting that this algorithm is very similar to the one for a single MVN (see Section 3.1).

It is worth to mention that the MLMVN learning algorithm does not depend on the type of MVN which from the network is consisted of (discrete or continuous). It should be empirically expected that if all the hidden neurons have the continuous activation function (2.54) (see p.68), the network may learn faster because of more flexibility. However, the type of the output neurons is no matter. They all may have the discrete activation function (2.50) (see p. 59) or the continuous activation function (2.54) (see p. 68), or some of them may have the discrete activation function (2.50), while some other may have the continuous activation function (2.54).

It is also important to mention that the type of network inputs is also no matter. Some of them can be discrete, while some other continuous or they all may be of the same type.

These advantages follow from the fact that regardless of their nature, MVN and MLMVN inputs and outputs are located on the unit circle.

Let us keep the same notations that we used in Sections 4.1 and 4.2. Let $D_{km}$ be a desired output of the $k$th neuron from the $m$th (output) layer; $Y_{km}$ be an actual output of the $k$th neuron from the $m$th (output) layer. The global error of the network taken from the $k$th neuron of the $m$th (output) layer is calculated according to (4.120) (see p. 144, we repeat here that equation for the reader's convenience):

$$\delta_{km}^* = D_{km} - Y_{km}; k = 1, ..., N_m.$$

The error backpropagation is organized as follows. The local errors of the output neurons are calculated according to (4.121) (see p. 145, we repeat here that equation for the reader's convenience):

$$\delta_{km} = \frac{1}{s_m} \delta_{km}^*; k = 1, ..., N_m.$$

Here and in what follows (as it was in the previous two sections) $s_j = N_{j-1} + 1$ is the number of neurons in the $j$-$1^{st}$ layer (the layer, which precedes to the layer $j$) incremented by one.

The errors of all the hidden neurons are calculated according to (4.122) (see p. 146, we repeat here that equation for the reader's convenience)

$$\delta_{kj} = \frac{1}{s_j} \sum_{i=1}^{N_{j+1}} \delta_{i,j+1} \left( w_k^{i,j+1} \right)^{-1}; k = 1, ..., N_j; j = 1, ..., m-1,$$

where $kj$ specifies the $k$th neuron of the $j$th layer.

As soon as all the errors are calculated, the weights of all the neurons can be adjusted according to the error-correction learning rule. First, the weights of the $1^{st}$ hidden layer neurons must be adjusted. Then, when the updated outputs of the $1^{st}$ hidden layer neurons are known, the weights of the second hidden layer neurons can be adjusted, etc.

Since the errors of the output layer neurons can be exactly calculated, the error-correction learning rule (3.92) (see p. 112) can be used to correct their weights. To correct the hidden neurons' weights, it should be better to use the error-correction learning rule (3.94) (see p. 114). This rule ensures more careful adjustment of the weights, and its use for the hidden neurons is reasonable because their errors are calculated according to the heuristic error backpropagation rule, and their exact errors unlike the exact errors of the output neurons are not known. As we saw (Theorem 3.17, p. 116) the MVN learning algorithm with the error-correction learning rule converges independently on which particular version of the learning rule is used. Thus, the correction of the MLMVN neurons' weights is determined by the following equations.

For the output layer neurons:

$$\tilde{w}_0^{km} = w_0^{km} + \frac{C_{km}}{(N_{m-1}+1)}\delta_{km};$$

$$\tilde{w}_i^{km} = w_i^{km} + \frac{C_{km}}{(N_{m-1}+1)}\delta_{km}\bar{\tilde{Y}}_{i,m-1}; i = 1,...,N_{m-1}. \qquad (4.134)$$

For the neurons from the hidden layers 2,...,$m$-1:

$$\tilde{w}_0^{kj} = w_0^{kj} + \frac{C_{kj}}{(N_{j-1}+1)}\delta_{kj};$$

$$\tilde{w}_i^{kj} = w_i^{kj} + \frac{C_{kj}}{(N_{j-1}+1)\,|\,z_{kj}\,|}\delta_{kj}\bar{\tilde{Y}}_{i,j-1}; \qquad (4.135)$$

$$i = 1,...,N_{j-1}; j = 2,...,m-1.$$

For the 1$^{st}$ hidden layer neurons:

$$\tilde{w}_0^{k1} = w_0^{k1} + \frac{C_{k1}}{(n+1)\,|\,z_{k1}\,|}\delta_{k1};$$

$$\tilde{w}_i^{k1} = w_i^{k1} + \frac{C_{k1}}{(n+1)\,|\,z_{k1}\,|}\delta_{k1}\bar{x}_i; i = 1,...,n, \qquad (4.136)$$

where $\tilde{Y}_{i,j-1}$ is the updated output of the $i$th neuron from the $j$-1$^{st}$ layer ($j$=1, ..., $m$-1), that is the $i$th input of the $kj$th neuron, the bar sign stands for the complex conjugation, $z_{kj}$ is the current value of the weighted sum of the $kj$th neuron (the $k$th neuron in the $j$th layer), $x_1,...,x_n$ are the network inputs, and $C_{kj}$ is a learning rate of the $kj$th neuron ($k$th neuron in the $j$th layer). We have already mentioned above that a learning rate can always be equal to 1 in the MVN error-correction rule. Let us also set $C_{kj} = 1, k = 1,...,N_j; j = 1,...,m$. Thus, for all the neurons in the MLMVN we set the learning rate equal to 1. At least in all applications considered in this book, we use a unitary learning rate.

### 4.3.2 MLMVN Learning Algorithm

Let now $A$ be a learning set, and $A_j \subset A, j = 0,1,..., K-1; A_s \cap A_t = \varnothing, s \neq t$ be $K$ disjoint learning subsets such that $\bigcup\limits_{i=0}^{K-1} A_i = A$.

Let $X_s = \left( x_1^s,..., x_n^s \right)$ be the $s$th element of the learning set $A$. Let $N$ be the cardinality of the set $A$, $|A| = N$, that is the total amount of the learning patterns.

The MLMVN learning may continue either until the zero-error is reached (which often is not possible) or either of the mean square error (MSE) (1.20) or the root mean square error (RMSE) (1.21) criteria is satisfied. We have already considered these stopping criteria for the learning algorithm for a single MVN (see Section 3.1, p. 99). For some applied problems the MLMVN learning with the zero error is not reasonable. It depends on the particular problem, which is necessary to learn. If all the output neurons in MLMVN are discrete, then both MSE and RMSE should be applied to the errors in terms of numbers of sectors (see Fig. 2.21, p. 59), thus not to the elements of the set $E_K = \left\{ \varepsilon_K^0, \varepsilon_K,..., \varepsilon_K^{K-1} \right\}$, but to the elements of the set $K = \{0,1,..., K-1\}$ or to their arguments $\left\{ \arg \varepsilon_k^0, \arg \varepsilon_k,..., \arg \varepsilon_k^{k-1} \right\}$ The local errors for the $s$th learning sample in these terms are calculated, respectively as

$$\gamma_s = \left( \alpha_{j_s} - \alpha_s \right) \bmod K ; \alpha_{j_s}, \alpha_s \in \{0,1,..., K-1\} , \tag{4.137}$$

$$\gamma_s = \left( \arg \varepsilon^{\alpha_{j_s}} - \arg \varepsilon^{\alpha_s} \right) \bmod 2\pi ; \varepsilon^{\alpha_{j_s}}, \varepsilon^{\alpha_s} \in E_K , \tag{4.138}$$

where $\varepsilon^{\alpha_{j_s}}$ is the desired output and $\varepsilon^{\alpha_s}$ is the actual output.

The functional of the MLMVN error is defined according to (1.26) (see p. 29) in the same way as it was defined for MLF:

$$E = \frac{1}{N} \sum_{s=1}^{N} E_s$$

where $E$ denotes *MSE*, and $E_s$ denotes the square error of the network for the $s$th pattern. It is

$$E_s = \gamma_s^2, s = 1,..., N , \tag{4.139}$$

($\gamma_s$ is the local error taken from (4.137) or (4.138)) for a single output neuron and

$$E_s = \frac{1}{N_m} \sum_{k=1}^{N_m} \gamma_{k_s}^2; s = 1, ..., N \qquad (4.140)$$

for $N_m$ output neurons, where $m$ is the output layer index.

The MLMVN learning process, as well as the MLF learning process (see p. 29) and the MVN learning process (see p. 99) continues until either of MSE or RMSE drops below some pre-determined acceptable minimal value $\lambda$. These MSE and RMSE stopping criteria are presented for the MVN learning by equations (3.76) and (3.77), respectively (see p. 99). These criteria are practically identical for MLMVN, accurate to the interpretation of a local error for multiple output neurons. They are

$$MSE = \frac{E_s}{N} = \frac{1}{N} \sum_{s=1}^{N} \gamma_s^2 < \lambda, \qquad (4.141)$$

$$RMSE = \sqrt{MSE} = \sqrt{\frac{E_s}{N}} = \sqrt{\frac{1}{N} \sum_{s=1}^{N} \gamma_s^2} < \lambda. \qquad (4.142)$$

It is also important to mention that to avoid *overfitting*, (it is known in machine learning as a phenomenon, which occurs when some learning set is "overlearned", and the results of this learning cannot be used for prediction because they produce mostly noise than something useful), an additional threshold, which determines the tolerance to a local error, can be specified. This means that we should check for $E_s; s = 1, ..., N$ obtained in (4.139) or (4.140) whether

$$E_s \le \mu; s = 1, ..., N, \qquad (4.143)$$

where $\mu$ is the mentioned additional threshold[3]. It is natural that it should be chosen such that $\mu \le \lambda$ if $\lambda$ is the acceptable MSE from (4.141) or $\mu \le \lambda^2$ if $\lambda$ is the acceptable RMSE from (4.142). If (4.143) holds, then the correction of the weights for the sth learning sample is not required.

Hence the learning algorithm consists of the following steps.

---

[3] In practical implementation, for MLMVN with a single output neurons it is more convenient to check whether $\sqrt{E_s} = \gamma_s \le \mu'$, where $\gamma_s$ is obtained according either to (4.137) or (4.138).

Let *Learning* be a flag, which is "True" if the weights adjustment is required and "False", if it is not required, and $r$ be the number of the weighting vector in the sequence $S_w^{kj}$, where $kj$ are the indexes of the $k$th neuron in the $j$th layer.

Step 1. The starting weighting vectors $W_0^{kj}; k = 1, ..., N_j; j = 1, ..., m$ are chosen arbitrarily (e.g., real and imaginary parts of their components can be the random numbers); $r=0$; $s=1$; *Learning* = 'False';

Step 2. Calculate the local error $\gamma_s$ according either to (4.137) if the network outputs are discrete or (4.138) if the network outputs are continuous or hybrid for the element $X_j^s$ from the learning set and, then calculate the square error $E_s$ according to (4.139) for a single output neuron or (4.140) for a network with more than one output neuron.
*if* (4.143) holds
*then go to* the Step 5
*else begin Learning* = 'True'; *go to* Step 3 *end*;

Step 3. To obtain the errors of all the network neurons, calculate the network errors according to (4.120), and backpropagate these errors to the output layer according to (4.121) and then backpropagate the obtained errors to the hidden layers according to (4.122).

Step 4. Sequentially, layer by layer, obtain the weighting vectors $W_{r+1}^{kj}$ from the ones $W_r^{kj}$ for all the neurons in the network ($j = 1, ..., m; k = 1, ..., N_j$) using the error-correction learning rule utilized in (4.136) for the 1ˢᵗ hidden layer neurons, in (4.135) for all other hidden neurons, and in (4.134) for the output layer neurons;

Step 5. $s = s+1$;  *if* $s \leq N$
 *then go to* Step 2
 *else*
  Calculate MSE according to (4.141) or RMSE according to (4.142) depending on which stopping criterion is used and check whether the error dropped below the pre-determined acceptable value $\lambda$. If so, then
  *Learning* = 'False'
  *if Learning* = 'False'
  *then* the learning process is finished successfully
  *else begin* $s=1$; *Learning* = 'False'; *go to* Step 2; *end*.

### 4.3.3 Convergence Theorem

Let us now prove that this learning algorithm converges. This proof is based on the same idea as the one of the convergence of the MVN learning algorithm with the error-correction learning rule (Theorem 3.17, p. 116). It is important to mention that this proof is published for the first time.

**Theorem 4.19** (About the convergence of the MLMVN learning algorithm with the error-correction learning rules (4.134)-(4.136) ). If there exist such a set of the non-zero weighting vectors $\Omega = \left\{ W^{kj}; j = 1,...,m; k = 1,...,N_m \right\}$, which guarantees that for the given MLMVN input/output mapping either of conditions (4.141) or (142) hold, then the MLMVN learning algorithm with the learning rules (4.134)-(4.136) converges to $\Omega = \left\{ W^{kj}; j = 1,...,m; k = 1,...,N_m \right\}$ or another set of the weighting vectors $\tilde{\Omega} = \left\{ \tilde{W}^{kj}; j = 1,...,m; k = 1,...,N_m \right\}$ for which either of conditions (4.141) or (4.142) also hold, after a finite number of steps.

**Proof.** We are given a condition that there exist such a set of the non-zero weighting vectors $\Omega = \left\{ W^{kj}; j = 1,...,m; k = 1,...,N_m \right\}$, which guarantees that for the given MLMVN input/output mapping either of conditions (4.141) or (4.142) hold.

Let us now train MLMVN applying the learning rules (4.134)-(4.136) according to our learning algorithm. For simplicity and without loss of generality, let us start learning process from the zero vectors $W_1^{kj} = \left( (0,0), (0,0),..., (0,0) \right); k = 1,...,N_m; j = 1,...,m$ for all the neurons of the network. Let $S_X = \left( X_1, X_2,..., X_N \right)$ be a learning sequence of input vectors $X_s = \left( 1, x_1^s,..., x_n^s \right), s = 1,..., N$ [4] that represent learning samples, and $S_W^{kj} = \left( W_r^{11},..., W_r^{N_1,1}, W_r^{12},..., W_r^{N_2,2},...,..., W_r^{1m},..., W_r^{N_m,m} \right); r = 1, 2,...$ be a sequence of the weighting vectors of all the neurons, which appear during the learning process. We have to prove that this sequence cannot be infinite. Let us remove from the learning sequence those vectors for which $\forall k = 1,...,N_j; j = 1,...,m \;\; W_{r+1}^{kj} = W_r^{kj}$, in other words, those input vectors, for which (4.143) hold without learning. Let $S_{\tilde{W}}$ be the reduced sequence of the weighting vectors. The Theorem will be proven if we will show that the sequence $S_{\tilde{W}}$ is finite. Let us suppose that the opposite is true: the sequence $S_{\tilde{W}}$ is infinite. Let $D_{km}; k = 1,...,N_m$ be the desired outputs for the input vector $X_1$ and the weighting vectors $W_1^{kj}$ and $Y_{km}; k = 1,...,N_m$ be the actual MLMVN outputs. If (4.143) does not hold, we first have to find the errors for all the neurons. We have to

---

[4] For simplicity, we will add the $0^{th}$ component equal to the constant 1 to the input vector, like we have done for a single neuron. This component corresponds to the weight $w_0$.

find the network errors according to (4.120), and then to backpropagate them according to (4.121)-(4.122). After all the errors are known, we have to apply the learning rules (4.134)-(4.136) to adjust the weights. According to (4.136) we obtain for 1$^{st}$ hidden layer neurons their updated weighted vectors

$$\tilde{W}_2^{11} = \frac{\delta_{11}^1}{(n+1)|z_{11}^1|}\bar{\tilde{X}}_1, ..., \tilde{W}_2^{N_1,1} = \frac{\delta_{N_1,1}^1}{(n+1)|z_{N_1,1}^1|}\bar{\tilde{X}}_1,$$

where $z_{kj}^1$ is the current weighted sum of the $kj$th neuron (the $k$th neuron from the $j$th layer) corresponding to the input vector $X_1$.

Then for the hidden layers' neurons we obtain according to (4.135)

$$\tilde{W}_2^{1j} = \frac{\delta_{1j}^1}{(N_{j-1}+1)|z_{1j}^1|}\bar{\tilde{V}}_{j-1}^1, ..., \tilde{W}_2^{N_1,1} = \frac{\delta_{N_j,j}^1}{(N_{j-1}+1)|z_{N_j,1}^1|}\bar{\tilde{V}}_{j-1}^1,$$

where $\bar{\tilde{V}}_{j-1} = \left(\bar{\tilde{Y}}_{1,j-1}, \bar{\tilde{Y}}_{2,j-1}, ..., \bar{\tilde{Y}}_{N_{j-1},j-1}\right)$ is the vector of the updated outputs of the neurons from $j$-1$^{st}$ layer (the one preceding to the layer $j$) taken complex-conjugated, and $\delta_{kj}^1$ is the error of the $kj$th neuron (the $k$th neuron from the $j$th layer) corresponding to the input vector $X_1$.

Then taking into account the last notation, we obtain according to (4.134) for the output layer neurons

$$\tilde{W}_2^{1m} = \frac{\delta_{1m}^1}{(N_{m-1}+1)}\bar{\tilde{V}}_{m-1}^1, ..., \tilde{W}_2^{N_m,m} = \frac{\delta_{N_m,1}^1}{(N_{m-1}+1)}\bar{\tilde{V}}_{m-1}^1.$$

Suppose that (4.143) does not hold for the input vector $X_2$, and the weights of all the neurons have to be corrected again. Calculating the errors according to (4.120), then backpropagating them according to (4.121)-(4.122), and applying the learning rules (4.134)-(4.136), we obtain the following.

For the neurons of the 1$^{st}$ hidden layer

$$\tilde{W}_3^{k1} = \tilde{W}_2^{k1} + \frac{\delta_{k1}^2}{(n+1)}\bar{\tilde{X}}_2 = \frac{1}{(n+1)}\left[\delta_{k1}^1\bar{\tilde{X}}_1 + \delta_{k1}^2\bar{\tilde{X}}_2\right]$$

$$k = 1, ..., N_1.$$

For the neurons of other hidden layers and the output neurons

$$\tilde{W}_3^{kj} = \tilde{W}_2^{kj} + \frac{\delta_{kj}^2}{\left(N_{j-1}+1\right)|z_{kj}^2|}\bar{\tilde{V}}_{j-1}^2 = \frac{1}{\left(N_{j-1}+1\right)}\left[\frac{\delta_{kj}^1}{|z_{kj}^1|}\bar{\tilde{Z}}_{j-1}^1 + \frac{\delta_{kj}^2}{|z_{kj}^2|}\bar{\tilde{V}}_{j-1}^2\right]$$

$$j = 2,...,m; k = 1,...,N_j.$$

Let us suppose without loss of generality that the $r$th correction of the weights is made for the $r$th learning sample that is for the input vector $X_r$. Calculating again the errors according to (4.120), then backpropagating them according to (4.121)-(4.122), and applying the learning rules (4.134)-(4.136), we obtain the following.

For the neurons of the 1$^{st}$ hidden layer

$$\tilde{W}_{r+1}^{k1} = \tilde{W}_r^{k1} + \frac{\delta_{k1}^r}{\left(n+1\right)}\bar{\tilde{X}}_r = \frac{1}{\left(n+1\right)}\left[\delta_{k1}^1\bar{\tilde{X}}_1 + ... + \delta_{k1}^r\bar{\tilde{X}}_r\right] \quad (4.144)$$

$$k = 1,...,N_1.$$

For the neurons of other hidden layers and the output neurons

$$\tilde{W}_{r+1}^{kj} = \tilde{W}_r^{kj} + \frac{\delta_{kj}^r}{\left(N_{j-1}+1\right)|z_{kj}^r|}\bar{\tilde{V}}_{j-1}^r =$$

$$= \frac{1}{\left(N_{j-1}+1\right)}\left[\frac{\delta_{kj}^1}{|z_{kj}^1|}\bar{\tilde{V}}_{j-1}^1 + ... + \frac{\delta_{kj}^r}{|z_{kj}^r|}\bar{\tilde{V}}_{j-1}^r\right]; j = 2,...,m; k = 1,...,N_j. \quad (4.145)$$

Let us set $\dfrac{\delta_{kj}^t}{|z_{kj}^t|} = \omega_{kj}^t$ in (4.145). We are given a condition that weighting vec-

tors $\Omega = \left\{W^{kj}; j = 1,...,m; k = 1,...,N_m\right\}$ exist such that for the given MLMVN input/output mapping either of conditions (4.141) or (4.142) hold. Let us find the dot products of both parts of (4.144) for each $k = 1,...,N_1$ with the corresponding weighting vector $W^{k1}; k = 1,...,N_1$, and the dot products of both parts of (4.145) for each $j = 2,...,m; k = 1,...,N_j$ with the corresponding weighting vector $W^{kj}; j = 2,...,m; k = 1,...,N_j$. We obtain the following, respectively:

$$\left(\tilde{W}_{r+1}^{k1}, \bar{W}^{k1}\right) = \frac{1}{n+1}\left[\left(\delta_{k1}^1\bar{\tilde{X}}_1, \bar{W}^{k1}\right) + ... + \left(\delta_{k1}^r\bar{\tilde{X}}_r, \bar{W}^{k1}\right)\right] \quad (4.146)$$

$$k = 1,...,N_1$$

$$\left(\tilde{W}_{r+1}^{kj}, \bar{W}^{kj}\right) = \frac{1}{\left(N_{j-1}+1\right)} \left[\left(\omega_{kj}^1 \bar{\tilde{V}}_{j-1}^1, \bar{W}^{kj}\right) + \ldots + \left(\omega_{kj}^r \bar{\tilde{V}}_{j-1}^r, \bar{W}^{kj}\right)\right]$$

(4.147)

$$k = 1, \ldots, N_j; \; j = 2, \ldots, m.$$

Let us estimate the absolute values of these dot products $\left|\left(\tilde{W}_{r+1}^{kj}, \bar{W}^{kj}\right)\right|; k = 1, \ldots, N_j; \; j = 1, \ldots, m:$

$$\left|\left(\tilde{W}_{r+1}^{k1}, \bar{W}^{k1}\right)\right| = \frac{1}{n+1} \left\|\left[\left(\delta_{k1}^1 \bar{\tilde{X}}_1, \bar{W}^{k1}\right) + \ldots + \left(\delta_{k1}^r \bar{\tilde{X}}_r, \bar{W}^{k1}\right)\right]\right\|$$

$$k = 1, \ldots, N_1;$$

$$\left|\left(\tilde{W}_{r+1}^{kj}, \bar{W}^{kj}\right)\right| = \frac{1}{\left(N_{j-1}+1\right)} \left\|\left[\left(\omega_{kj}^1 \bar{\tilde{V}}_{j-1}^1, \bar{W}^{kj}\right) + \ldots + \left(\omega_{kj}^r \bar{\tilde{V}}_{j-1}^r, \bar{W}^{kj}\right)\right]\right\|$$

(4.148)

$$k = 1, \ldots, N_j; \; j = 2, \ldots, m.$$

Since for any complex number $\beta$ $|\beta| \geq |\text{Re}\,\beta|$ and $|\beta| \geq |\text{Im}\,\beta|$, then the absolute value of the sum in the right-hand side of all the equalities in (4.148) is always greater than or equal to the absolute values of the real and imaginary parts of this sum. Let us set $a_{k1} = \min_{t=1,\ldots,r} \left|\text{Re}\left(\delta_{k1}^t \bar{\tilde{X}}_t, \bar{W}^{k1}\right)\right|$ and also $a_{kj} = \min_{t=1,\ldots,r} \left|\text{Re}\left(\omega_{kj}^t \bar{\tilde{Z}}_{j-1}^t, \bar{W}^{kj}\right)\right|; \; j = 1, \ldots, m$. Then it follows from (4.148) that

$$\left|\left(\tilde{W}_{r+1}^{kj}, \bar{W}^{kj}\right)\right| \geq \frac{r a_{kj}}{M}; k = 1, \ldots, N_j; \; j = 1, \ldots, m,$$

(4.149)

where $M = \max\left\{n+1, N_1+1, \ldots, N_m+1\right\}$.

According to the fundamental Schwarz inequality [74] the squared dot product of the two vectors does not exceed the product of the squared norms of these vectors or in other words, the norm of the dot product of the two vectors does not exceed the product of the norms of these vectors $\left\|(V_1, V_2)\right\| \leq \left\|V_1\right\| \cdot \left\|V_2\right\|$. Thus, according to the Schwartz inequality

$$\left|\left(\tilde{W}_{r+1}^{kj}, \bar{W}^{kj}\right)\right| \leq \left\|\tilde{W}_{r+1}^{kj}\right\| \cdot \left\|W^{kj}\right\|; k = 1, \ldots, N_j; \; j = 1, \ldots, m.$$

(4.150)

Taking into account (4.149), we obtain from (4.150) the following

$$\frac{ra_{kj}}{M} \le \left|\left(\tilde{W}_{r+1}^{kj}, \bar{W}^{kj}\right)\right| \le \left\|\tilde{W}_{r+1}^{kj}\right\| \cdot \left\|W^{kj}\right\|; k = 1,...,N_j; j = 1,...,m..$$

Then it follows from the last inequality that

$$\left\|\tilde{W}_{r+1}^{kj}\right\| \ge \frac{ra_{kj}}{M \left\|W^{kj}\right\|}; k = 1,...,N_j; j = 1,...,m. \tag{4.151}$$

Let for simplicity $\dfrac{a_{kj}}{M} = \tilde{a}_{kj}; k = 1,...,N_j; j = 1,...,m$. Then (4.151) is transformed as follows:

$$\left\|\tilde{W}_{r+1}^{kj}\right\| \ge \frac{r\tilde{a}_{kj}}{\left\|W^{kj}\right\|}; k = 1,...,N_j; j = 1,...,m. \tag{4.152}$$

The weighting vectors $W^{kj}; k = 1,...,N_j; j = 1,...,m$ exist according to the condition of the Theorem. According to our assumption, the sequence $S_{\tilde{W}}$ of the weighting vectors is infinite. Since $r$ is the number of the learning step, let us consider (4.152) when $r \to \infty$. $\left\|\tilde{W}_{r+1}^{kj}\right\|$ is a non-negative finite real number, $\left\|W^{kj}\right\|$ is a finite positive real number ($\left\|W^{kj}\right\| > 0$ because according to the condition of the Theorem vectors $W^{kj}$ are non-zero) $\tilde{a}_{kj}$ is a finite positive real number. It follows from this analysis that

$$\frac{r\tilde{a}_{kj}}{\left\|W^{kj}\right\|} \xrightarrow[r \to \infty]{} \infty; k = 1,...,N_j; j = 1,...,m.$$

However, this means that from (4.152) we obtain

$$\left\|\tilde{W}_{r+1}^{kj}\right\| \ge \frac{r\tilde{a}_{kj}}{\left\|W^{kj}\right\|} \to \infty; k = 1,...,N_j; j = 1,...,m. \tag{4.153}$$

All the inequalities (4.153) are contradictory. Indeed, the norm of a vector, which is in the left-hand side, is a finite non-negative real number. However, it has to be greater than or equal to the infinity in the right-hand side of (4.153), which is impossible. This means that (4.153) is contradictory. This means in turn that either it is impossible that $r \to \infty$ or the vectors $W^{kj}; k = 1,...,N_j; j = 1,...,m$ do not exist. The latter contradicts to the condition of the Theorem. Hence, $r \nrightarrow \infty$ and it is always a finite integer number. Thus, our assumption that the sequence $S_{\tilde{w}}$ of the weighting vectors is infinite, is false, which means that it is always finite. Hence, the learning algorithm converges after a finite number of steps.

Theorem is proven.

### 4.3.4 Important Advantages of the MLMVN Learning Algorithm

Thus, the MLMVN learning algorithm, which is based on the error backpropagation rule (4.121)-(4.122) and on the generalized error-correction learning rule (4.134)-(4.136), converges after a finite number of learning epochs. It is important that the MLMVN error-correction learning algorithm is based on the same principles as the error-correction learning algorithm for a single MVN (see Sections 3.1 and 3.3). The mechanism of the MLMVN error-correction learning is the same as the one for the MVN learning. The MLMVN learning process consists of the iterative approaching the "ideal" weights of all the neurons. Since it is based on the error-correction learning rule, the network error decreases due to the same reasons as for a single MVN.

This algorithm has a number of important advantages over the MLF learning algorithm. Let us distinguish how they compensate those disadvantages that we have found considering the MLF learning algorithm (see Section 1.3, p. 34).

1) The MLMVN learning algorithm is based on the same error-correction learning rule as the MVN learning algorithm. Since it is not reduced to solving an optimization problem, it does not suffer from the local minima phenomenon. If some input/output mapping is learnable accurate a specified mean-square error or a root mean-square error, then according to Theorem 4.19 the learning algorithm converges after a finite number of steps (epochs).

2) The MLMVN learning algorithm is derivative-free. It works equally well for a network based on the discrete MVNs, and for a network based on the continuous MVNs. It also perfectly works for a network containing MVNs with both discrete and continuous activation functions. There is no matter of the type of the network inputs and outputs (discrete or continuous, or hybrid). Thus, this learning algorithm can be used to train, for example, a network whose input/output mapping simulates some hybrid dynamical system and therefore it is hybrid.

3) Since a single MVN has a higher functionality than a sigmoidal neuron, MLMVN has also significantly higher functionality than MLF. Typically, smaller MLMVN solves any problem much better than larger MLF.

4) The MLMVN learning algorithm does not depend on the number of hidden layers in the network and on the number of neurons in a layer. It works equally well for the simplest network containing just two neurons (Fig. 4.42) and for the MLMVN containing tens of thousands hidden neurons.

Let us now consider how this learning algorithm can be used for solving some benchmark problems. We will also consider in Chapter 6 how it can be successfully used for solving real-world problems.

## 4.4   Examples of MLMVN Applications[5]

We will start from a simple example, which will be considered in detail, and then we will consider how MLMVN solves a problem of Mackey-Glass time series prediction.

### *4.4.1   Simple Example: 3-Class Classification Problem*

Let us consider the following simple classification problem. We take this solution from [77], where it was considered by the author of this book together with Dmitriy Paliy, Jacek Zurada, and Jaakko Astola.

Suppose we wish to classify three training vectors $X_1, X_2, X_3$ of the length 2 to be members of three different classes $\tilde{T}_1, \tilde{T}_2, \tilde{T}_3$. Our vectors are as follows

$$X_1 = \left( e^{4.23i}, e^{2.10i} \right) \to \tilde{T}_1,$$

$$X_2 = \left( e^{5.34i}, e^{1.24i} \right) \to \tilde{T}_2,$$

$$X_3 = \left( e^{2.10i}, e^{0i} \right) \to \tilde{T}_3,$$

where $i$ is an imaginary unity. Classes $\tilde{T}_1, \tilde{T}_2, \tilde{T}_3$ are determined in such a way that the argument of the desired output of the network must belong to the interval

$$[\arg T_j - 0.05, \arg T_j + 0.05], \, j = 1, 2, 3, . \tag{4.154}$$

where $T_1 = e^{0.76i}, \, T_2 = e^{2.56i}, \, T_3 = e^{5.35i}$.

This means that the desired outputs of the network should be equal to $T_1$ for the input vector $X_1$, to $T_2$ for the input vector $X_2$, and to $T_3$ for the input vector $X_3$. Hence, our desired outputs are continuous. Taking into account (4.154), we may easily conclude that the learning can be stopped according either to the MSE criterion (4.141) or the RMSE criterion (4.142). In (141), $\lambda = \left( 0.05 \right)^2 = 0.0025$ should be taken, while in (4.142) $\lambda = 0.05$ should be taken, respectively.

Let the MLMVN has a topology 2-2-1 (two inputs, two hidden neurons in a single hidden layer, and the output layer with a single output neuron). All the neurons have continuous inputs and outputs, and the activation function (2.54).

---

[5] MLMVN software simulator (executable code) is available at
http://www.freewebs.com/igora/Downloads.htm

The initial weights (both real and imaginary parts) of all the MLMVN neurons are taken as random numbers from the interval $[0,1]$:

| Neuron 11 | Neuron 21 | Neuron 12 |
|---|---|---|
| $w_0 = 0.23 - 0.38i$ | $w_0 = 0.23 - 0.38i$ | $w_0 = 0.23 - 0.38i$ |
| $w_1 = 0.19 - 0.46i$ | $w_1 = 0.19 - 0.46i$ | $w_1 = 0.19 - 0.46i$ |
| $w_2 = 0.36 - 0.33i$ | $w_2 = 0.36 - 0.33i$ | $w_2 = 0.36 - 0.33i$ |

Let $t$ be the index of training epoch and $j$ be the index of a training vector. So, initially $t=1, j=1$. The MLMVN training is performed as follows.

Iteration 1:

1) *Compute the weighting sum*: $z_{k1} = w_0^{k1} + w_1^{k1}x_1 + w_2^{k1}x_2; k = 1,2$ for each neuron from the hidden layer for the first $(j=1)$ pattern vector $X_j = (x_1, x_2)$, where $x_1 = \exp(4.23i) = -0.46 - 0.88i$ and $x_2 = \exp(2.10i) = -0.51 + 0.85i$ :

$$z_{11} = (0.23 - 0.38i) + (0.19 - 0.46i)(-0.46 - 0.88i) +$$
$$+(0.36 - 0.33i)(-0.51 + 0.85i) = -0.16 + 0.13i.$$

It is easy to see that $z_{21} = z_{11} = -0.16 + 0.13i$ in this particular case.

2) *Compute the outputs for the hidden layer neurons*: According to (2.51) and (2.54) the actual outputs are $Y_{11} = z_{11}/|z_{11}| = -0.78 + 0.62i$ and $Y_{21} = z_{21}/|z_{21}| = -0.78 + 0.62i$ . They are used as inputs to the output neuron (21). Then, the output $Y_{12}$ for the neuron (12) is computed according to (2.54) as follows:

$$z_{12} = (0.23 - 0.38i) + (0.19 - 0.46i)(-0.78 + 0.62i) +$$
$$+(0.36 - 0.33i)(-0.78 + 0.62i) = 0.29 + 0.57i,$$

and finally $Y_{12} = z_{12}/|z_{12}| = 0.45 + 0.89i$ .

3) *Find the error* (4.138) and *check condition* (4.142) for MSE or (4.143) for RMSE. The angular error is the difference between the argument of the output $Y_{12}$ and the argument of the desired output $T_{12} = T_j$ . This difference must be compared with the maximal acceptable error $\mu = 0.05$, which is determined in (4.154) by the angular size of the class

$$\left|\arg(Y_{12}) - \arg(T_{12})\right| = \left|1.10 - 0.76\right| = 0.34 > 0.05 .$$

The condition (4.143) does not hold and the weights must be corrected.

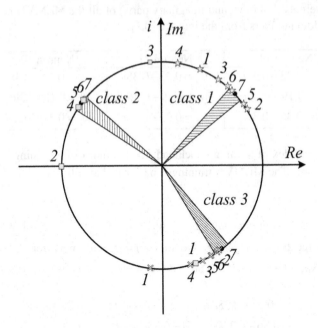

**Fig. 4.45** Illustration to the simple example of MLMVN 2-2-1 training (see also Table 4.12). The network outputs and the corresponding number of epochs from 1 to 7 are shown

☆- Class 1; □ - Class 2; x – Class3. Shaded sectors correspond to the classes

4) *Weights update*: The network error calculated at the output neuron according to (4.120) is $\delta_{12}^* = 0.27 - 0.21i$. After this error is backpropagated, we get the error $\delta_{12} = 0.09 - 0.07i$ for the output neuron according to (4.121), and the errors $\delta_{11} = 0.88 + 0.53i$ and $\delta_{21} = 1.04 + 0.11i$ for the hidden layer neurons according to (4.122).

Weights corrections are performed using (4.136) for the hidden layer neurons and using (4.134) for the output layer neuron and result in the following updated weights:

$$W = (0.52 - 0.21i, -0.10 - 0.28i, 0.36 - 0.67i) \text{ for } (11),$$
$$W = (0.58 - 0.35i, 0.00 - 0.17i, 0.21 - 0.65i) \text{ for } (21),$$
$$W = (0.26 - 0.41i, 0.20 - 0.50i, 0.38 - 0.36i) \text{ for } (12).$$

5) *Completion of the training epoch*: Repeat steps 1-4 for all the pattern vectors (till $j=3$).

**Table 4.12** The example of the 2-2-1 MLMVN training

| Epoch | Pattern vector $j$ | Weights $W^{12} = (w_0^{12}, w_1^{12}, w_2^{12})$ for the output neuron (21) | Actual Output $Y_{12}$ | Error (4.138) $\leq 0.05$ | MSE (4.142) $\leq 0.0025$ |
|---|---|---|---|---|---|
| 1 | 1 | $w$=(0.23-0.38$i$, 0.19-0.46$i$, 0.36-0.33$i$) | 0.45+0.89$i$ | 0.342 | |
|   | 2 | $w$=(0.26-0.41$i$,0.2-0.5$i$,0.38-0.36$i$) | 0.39-0.92$i$ | 2.553 | 2.4213 |
|   | 3 | $w$=(0.12-0.24$i$, 0.07-0.67$i$, 0.18-0.43$i$) | -0.19-0.98$i$ | 0.829 | |
| 2 | 1 | $w$=(0.21-0.22$i$, 0.16-0.63$i$, 0.27-0.4$i$) | 0.83+0.55$i$ | 0.174 | |
|   | 2 | $w$=(0.2-0.21$i$, 0.18-0.64$i$, 0.29-0.41$i$) | -1.0+0.0$i$ | 0.581 | 0.1208 |
|   | **3** | **$w$=(0.21-0.15$i$, 0.13-0.68$i$, 0.3-0.47$i$)** | **0.57-0.82$i$** | **0.030** | |
| 3 | 1 | $w$=(0.21-0.15$i$, 0.13-0.68$i$, 0.3-0.47$i$) | 0.57+0.83$i$ | 0.209 | |
|   | 2 | $w$=(0.23-0.16$i$, 0.11-0.69$i$, 0.28-0.47$i$) | -0.1+0.99$i$ | 0.888 | 0.2872 |
|   | 3 | $w$=(0.15-0.21$i$, 0.19-0.73$i$, 0.36-0.41$i$) | 0.46-0.89$i$ | 0.160 | |
| 4 | 1 | $w$=(0.16-0.2$i$, 0.21-0.72$i$, 0.38-0.41$i$) | 0.19+0.98$i$ | 0.619 | |
|   | **2** | **$w$=(0.22-0.24$i$, 0.16-0.77$i$, 0.32-0.44$i$)** | **-0.85+0.53$i$** | **0.024** | 0.1486 |
|   | 3 | $w$=(0.22-0.24$i$, 0.16-0.77$i$, 0.32-0.44$i$) | 0.37-0.93$i$ | 0.259 | |
| 5 | 1 | $w$=(0.25-0.22$i$, 0.18-0.75$i$, 0.35-0.44$i$) | 0.78+0.63$i$ | 0.080 | |
|   | **2** | **$w$=(0.24-0.21$i$, 0.19-0.75$i$, 0.35-0.44$i$)** | **-0.82+0.57$i$** | **0.025** | 0.0049 |
|   | 3 | $w$=(0.24-0.21$i$, 0.19-0.75$i$, 0.35-0.44$i$) | 0.53-0.85$i$ | 0.080 | |
| 6 | 1 | $w$=(0.25-0.21$i$, 0.2-0.74$i$, 0.36-0.43$i$) | 0.68+0.73$i$ | 0.060 | |
|   | **2** | **$w$=(0.15-0.21$i$, 0.19-0.75$i$, 0.36-0.44$i$)** | **-0.82+0.58$i$** | **0.034** | 0.0026 |
|   | 3 | $w$=(0.25-0.21$i$, 0.19-0.75$i$, 0.36-0.44$i$) | 0.55-0.83$i$ | 0.052 | |
| 7 | **1** | **$w$=(0.26-0.21$i$, 0.19-0.74$i$, 0.36-0.43$i$)** | **0.71+0.71$i$** | **0.025** | |
|   | **2** | **$w$=(0.26-0.21$i$, 0.19-0.74$i$, 0.36-0.43$i$)** | **-0.81+0.58$i$** | **0.039** | **0.0009** |
|   | **3** | **$w$=(0.26-0.21$i$, 0.19-0.74$i$, 0.36-0.43$i$)** | **0.58-0.82$i$** | **0.022** | |

6) *Termination of training*: Compute MSE according to (4.142) and check the condition determined by the same equation (4.142). If it holds, the learning process has converged. If it does not hold then increment $t$ and perform steps 1-5 again. This process should continue until the condition (4.142) is satisfied.

The process is illustrated in detail in Table 4.12 and in Fig. 4.45. The table rows highlighted by the bold font indicate that the MLMVN has reached the desired output for the particular training vector $X_j$. Fig. 4.45 shows how the outputs of the network are distributed for all 7 iterations. It is clearly visible that the training error is decreasing very quickly and the actual outputs approach the desired ones with the quickly increasing precision starting from the 4$^{th}$ training epoch. The learning process converges after the 7$^{th}$ epoch, as soon as MSE drops below 0.0025.

## 4.4.2 Mackey-Glass Time Series Prediction

In [62], where MLMVN was explicitly presented by the author of this book and C. Moraga, it was demonstrated that this network outperforms MLF, different kernel-based and neuro-fuzzy techniques in terms of learning speed, a number of parameters employed, and in terms of generalization capability, when solving a number of popular benchmark problems. One of the most convincible examples is a famous problem of Mackey-Glass time-series prediction. Thus, we would like to demonstrate this important example here.

The Mackey-Glass time series is generated by the chaotic Mackey-Glass differential delay equation defined as follows [78]:

$$\frac{dx(t)}{dt} = \frac{0.2x(t-\tau)}{1+x^{10}(t-\tau)} - 0.1x(t) + n(t), \tag{4.155}$$

where $n(t)$ is a uniform noise (it is possible that $n(t)=0$). $x(t)$ is quasi-periodic, and choosing $\tau = 17$ it becomes chaotic [78, 79]. This means that only short term forecasts are feasible. Exactly $\tau = 17$ was used in our experiment. To integrate (4.155) and to generate the data, we used an initial condition $x(0) = 1.2$ and a time step $\Delta t = 1$. The Runge-Kutta method was used for the integration of (4.155). The data was sampled every 6 points, as it is usually recommended for the Mackey Glass time-series (see e.g., [80, 81]).

Thus, we use the Mackey-Glass time series generated with the same parameters and in the same way as in the recently published papers [80-82]. The task of prediction is to predict $x(t+6)$ from $x(t)$, $x(t-6)$, $x(t-12)$, $x(t-18)$. Thus, each member of the series was considered as a function from the four preceding members taken with the step 6

$$x(t+6) = f\left(x(t), x(t-6), x(t-12), x(t-18)\right).$$

We generated 1000 points data set. The first 500 points were used for training (a fragment of the first 250 points is shown in Fig. 4.46a) and the next 500 points were used for testing (a fragment of the last 250 points is shown in Fig. 4.46b). The true values of $x(t+6)$ were used as the target values during training.

Since the root mean square error (RMSE) is a usual estimator of the training control and the quality for the Mackey-Glass time series prediction [79-83], we also used it here. Thus, the RMSE criterion (4.142)

$$RMSE = \sqrt{MSE} = \sqrt{\frac{E_s}{N}} = \sqrt{\frac{1}{N}\sum_{s=1}^{N}\gamma_s^2} < \lambda$$

was used to determine the convergence of the learning process. $\lambda$ determines a maximum possible RMSE for the training data. The local error $\gamma$ was measured in terms of output's argument (see (4.138) ). Wherever the actual MLMVN output did not coincide with the desired one, the weights were corrected (thus, we took $\mu = 0$ in (4.143)).

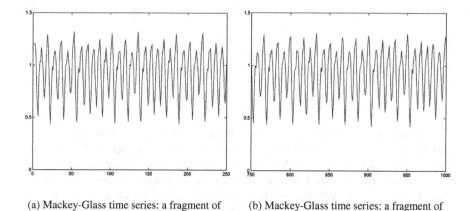

(a) Mackey-Glass time series: a fragment of the first 250 points of the training data

(b) Mackey-Glass time series: a fragment of the last 250 points of the testing data

**Fig. 4.46** Mackey-Glass Time series

In all our experiments, we used MLMVN 4-$n$-1 (four inputs, one hidden layer containing $n$ neurons, and a single output neuron). The four inputs were taken, since we have to predict a series member from its four predecessors taken with the step 6. To transform the generated data into the complex numbers located on the unit circle and to transform the predicted data back into a real-valued form, transformations (2.55) and (2.56) (see p. 69) were used, respectively. We have chosen $\alpha = \beta = 7\pi/8$.

The results of our experiments are summarized in Table 4.13. For each of the three series of experiments we made 30 independent runs of training and prediction, like it was done in [81]. Our experiments show that choosing a smaller $\lambda$ in (4.142) it is possible to decrease the RMSE for the testing data significantly. The results of training and prediction are illustrated in Fig. 4.47a and Fig. 4.47b, respectively. Since both the testing and prediction errors are very small and it is practically impossible to show the difference among the actual and predicted values at the same graph, Fig. 4.47a and Fig. 4.47b show not the actual and predicted data, but the error among them. Fig. 4.47a presents the error on the training set after the convergence of the training algorithm for the network containing 50 hidden neurons. $\lambda = 0.0035$ was used as a maximum acceptable RMSE in (4.142).

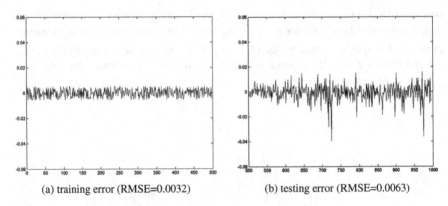

(a) training error (RMSE=0.0032)            (b) testing error (RMSE=0.0063)

**Fig. 4.47** Mackey-Glass time series prediction

An actual RMSE on the training set at the moment, when training was stopped, was equal to 0.0032. Fig. 4.47b presents the error on the testing set (RMSE=0.0063, which is a median value of the 30 independent experiments). To estimate a training time, one can base on the following data for the networks containing 50 and 40 hidden neurons, respectively. 100,000 epochs require 25 minutes for the first network and 15 minutes for the second one on a PC with the Intel® Core™2 Duo 2.4 GHz CPU.

Comparing the results of the Mackey-Glass time series prediction using MLMVN to the results that were obtained using other techniques, we have to conclude that MLMVN outperforms all of them (see Table 4.14). The results comparative with the ones obtained using MLMVN are obtained only using GEFREX [84] and ANFIS [83]. However, it is not mentioned in [83] and [84] whether the reported RMSE is the result of averaging over the series of experiments or it is the best result of this series.

In any case, MLMVN with the 50 hidden neurons steadily outperforms ANFIS and shows at least a comparative result with GEFREX. However, the implementation of MLMVN is incompatibly simpler than the one of GEFREX (for example, referring to the difficulty of the GEFREX implementation, the SuPFuNIS model was proposed in [80]).

It is important to mention that classical MLF with its backpropagation learning [86] loses to MLMVN dramatically. It is interesting that the best results of our predecessors were obtained using different fuzzy techniques [80, 83, 84]. The only pure neural solution, which shows a comparative result (which is not as good as the one obtained using MLMVN), is the recently proposed cooperative neural networks ensemble (CNNE) [81].

This is an ensemble of several feedforward neural networks. Its average result obtained with a greater number of the hidden neurons (56 in average) is worse than the one obtained using MLMVN with the 50 hidden neurons. It should also be mentioned that a training procedure of the networks ensemble (CNNE) is more

**Table 4.13** The results of Mackey-Glass time series prediction using MLMVN

| # of neurons on the hidden layer | | **50** | **50** | **40** |
|---|---|---|---|---|
| $\lambda$ - a maximum possible **RMSE** in (142) | | 0.0035 | 0.0056 | 0.0056 |
| Actual RMSE for the training set (min - max) | | 0.0032 - 0.0035 | 0.0053 – 0.0056 | 0.0053 – 0.0056 |
| **RMSE** for the **testing** **set** | Min | **0.0056** | 0.0083 | 0.0086 |
| | Max | **0.0083** | 0.0101 | 0.0125 |
| | Median | **0.0063** | 0.0089 | 0.0097 |
| | Average | **0.0066** | 0.0089 | 0.0098 |
| | SD | 0.0009 | 0.0005 | 0.0011 |
| Number of training epochs | Min | 95381 | 24754 | 34406 |
| | Max | 272660 | 116690 | 137860 |
| | Median | 145137 | 56295 | 62056 |
| | Average | 162180 | 58903 | 70051 |

**Table 4.14** Comparison of Mackey-Glass time series prediction using MLMVN with other techniques

| MLMVN min | MLMVN average | GEFREX [84] | EPNet [85] | ANFIS [83] | CNNE [81] | SuPFuNIS [80] | MLF (taken from [86]) |
|---|---|---|---|---|---|---|---|
| **0.0056** | **0.0066** | 0.0061 | 0.02 | 0.0074 | 0.009 | 0.014 | 0.02 |

complicated than the one of MLMVN. The additional important advantage of MLMVN is an opportunity to control a level of the prediction error by choosing an appropriate value of the maximum training error $\lambda$ in (4.142).

We hope that this example shows many advantages of MLMVN compared to many other machine learning techniques. It learns faster, generalizes better, and employs fewer parameters.

In Chapter 6, we will consider applications of MLMVN in solving some real-world problems. We will consider how MLMVN can be used for solving a problem of blur identification (for image deblurring) and for a long-term financial time-series prediction.

## 4.5 Concluding Remarks to Chapter 4

In this Chapter, we have considered a multilayer neural network with multi-valued neurons (MLMVN). This is a neural network with a standard feedforward topology, but built from the multi-valued neurons.

The MLMVN has a number of very important advantages over competitive techniques and first of all over MLF.

The most important advantage is that the MLMVN learning algorithm is based on the generalization of the error-correction learning rule for a single MVN. It is not reduced to solving an optimization problem, and respectively, it does not suffer from the local minima phenomenon. If some input/output mapping is learnable accurate a specified mean-square error or a root mean-square error, then according to Theorem 4.19 the learning algorithm converges after a finite number of steps (epochs).

Both the MLMVN error beckpropagation rule and the MLMVN learning algorithm, which we have derived in this Chapter, are derivative-free. It was shown that after the weights are corrected, the updated weighted sums of the network output neurons are equal to the previous weighted sums plus the corresponding network errors that were taken from the output neurons. This MLMVN property is the same as the one of a single MVN.

The MLMVN learning algorithm works equally well for a network based on the discrete MVNs, and for a network based on the continuous MVNs. It also perfectly works for a network containing both discrete and continuous MVNs. There is no matter of the type of the network inputs and outputs (discrete or continuous, or hybrid). Thus, this learning algorithm can be used to train, for example, a network whose input/output mapping simulates some hybrid dynamical system and therefore it is hybrid.

Since a single MVN has a higher functionality than a sigmoidal neuron, MLMVN has also significantly higher functionality than MLF. Typically, smaller MLMVN solves any problem much better than larger MLF. MLMVN also employs fewer parameters than many other machine learning techniques and shows better generalization capability (we will show even more such examples in Chapter 6).

The MLMVN learning algorithm does not depend on the number of hidden layers in the network and on the number of neurons in a layer. It works equally well for a small network, which contains just a few neurons and for a network, which contains tens of thousands hidden neurons.

All these wonderful properties make MLMVN a very powerful and efficient machine learning tool for solving various problems of classification, pattern recognition, prediction, and simulation of complex hybrid dynamical systems depending on hundreds and thousands parameters.

# Chapter 5
# Multi-Valued Neuron with a Periodic Activation Function

"The opposite of a profound truth may well be another profound truth"

Niels Bohr

In this Chapter, we consider MVN with a periodic activation function. As we have already seen, MVN's functionality is higher than the one of, for example, sigmoidal neurons. In this Chapter, we will consider how a single MVN may learn non-linearly separable input/output mappings in that initial $n$-dimensional space where they are defined. In Section 5.1, we consider a universal binary neuron (UBN), which in fact is the discrete MVN with a periodic activation function for $k=2$. We show how this neuron may learn non-linearly separable Boolean functions, for example, XOR and Parity $n$, projecting them into larger valued logic. In Section 5.2, we generalize that approach, which is used in UBN, and introduce a periodic activation function for the discrete MVN. We also consider the learning algorithm for MVN with a periodic activation functions. In Section 5.3, we show how a number of non-linearly separable benchmark classification problems can be solved using a single MVN with a periodic activation function. Concluding remarks are given in Section 5.4.

## 5.1 Universal Binary Neuron (UBN): Two-Valued MVN with a Periodic Activation Function

Introducing complex-valued neurons in Section 1.4, we have shown (see p. 41) that a single complex-valued neuron can learn a classical non-linearly separable problem, the XOR problem. We have mentioned there that a neuron, which can learn the XOR problem, is called the universal binary neuron (UBN).

Actually, UBN is nothing else than the two-valued MVN with a periodic activation function. This periodicity is a main idea behind UBN.

### 5.1.1 A Periodic Activation Function for k=2

We have considered earlier the $k$-valued activation function (2.50) of the discrete MVN for $k=2$ (see p. 60). It is transformed to the two-valued function (2.52)

I. Aizenberg: Complex-Valued Neural Networks with Multi-Valued Neurons, SCI 353, pp. 173–206.
springerlink.com                                            © Springer-Verlag Berlin Heidelberg 2011

$$P(z) = \begin{cases} 1; 0 \le \arg z < \pi \\ -1; \pi \le \arg z < 2\pi, \end{cases}$$

which divides the complex plane into two sectors, the top and the bottom half-planes (see Fig. 5.48a). However, it was shown in [33, 35, 37, 60] that the functionality of a neuron with this activation function exactly coincides with the functionality of the threshold neuron with the activation function (1.1).

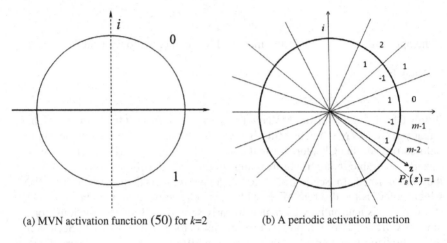

(a) MVN activation function (50) for $k$=2          (b) A periodic activation function

**Fig. 5.48** MVN activation function for $k$=2 and a periodic activation function for $k$=2

This means that the discrete MVN with the activation function (2.50) with $k$=2 (or with the activation function (2.52), which is (2.50) with $k$=2) can learn only linearly separable input/output mappings. Thus, such binary non-linearly separable problems as XOR, Parity $n$ and others cannot be learned using MVN with the activation function (2.52).

An idea to modify the activation function (2.52) in such a way that it should have ensured non-linearly separated Boolean functions to be learned by a single neuron was developed by the author of this book in his Ph.D. dissertation in 1986. This idea was as follows. We have to use complex weights with the binary inputs and outputs taken from the set $E_2 = \{1, -1\}$. Hence, our weighted sum is a complex number. If, dividing the complex plane into two sectors, we cannot learn non-linearly separable Boolean functions, will we be able to do so dividing the complex plane into more than two sectors and determining a periodic activation function? In the paper [29], which was published in 1985, the author of this book suggested the activation function (1.40), which divides the complex plane into four sectors, and ensures the implementation of the XOR function using a single neuron with complex-valued weights (this example we have already considered in detail in Section 1.4, see p. 41). The activation function (1.40)

$$\varphi(z) = \begin{cases} 1, & \text{if } 0 \le \arg z < \pi/2 \text{ or } \pi \le \arg z < 3\pi/2 \\ -1, & \text{if } \pi/2 \le \arg z < \pi \text{ or } 3\pi/2 \le \arg z < 2\pi \end{cases}$$

divides the complex plane into four sectors (see Fig. 1.14, p. 41), and its value (which is the neuron output) is determined as the alternating sequence 1, -1, 1, -1. Let us number sectors in the natural order (0, 1, 2, 3). If a complex weighted sum is located in an even sector, then the activation function is equal to 1, while if the weighted sum is located in an odd sector, then the activation function is equal to -1. In fact, the activation function (1.40) is periodic. Its period is 2, and its two values 1 and -1 are repeated two times each. In his Ph.D. dissertation, the author of this book also suggested the following generalization of the activation function (1.40), which was explicitly presented in [30].

Let us have a neuron with $n$ binary inputs taken from the set $E_2 = \{1, -1\}$. Let weights of this neuron are arbitrary complex numbers $w_i \in \mathbb{C}; i = 0, 1, ..., n$. Thus, the weighted sum $z = w_0 + w_1 x_1 + ... + w_n x_n$ is also a complex number. Let us choose some even positive integer $m = 2l$ where $l \ge n$. Let us consider now the following $l$-multiple activation function, which was suggested in [30]

$$P_B(z) = (-1)^j, \text{if } 2\pi j / m \le \arg(z) < 2\pi(j+1)/m; \\ m = 2l, l \ge n, \tag{5.156}$$

where $j$ is a non-negative integer $0 \le j < m$.

The activation function (5.156) is illustrated in Fig. 5.48b. It divides the complex plane into $m = 2l$ equal sectors. It determines the neuron output by the alternating periodic sequence of 1, -1, 1, -1,..., depending on the parity of the sector's number. The activation function (5.156) is equal to 1 for the complex numbers located in the even sectors 0, 2, 4, ..., $m$-2 and to -1 for the complex numbers located in the odd sectors 1, 3, 5, ..., $m$-1. Similarly to the MVN activation function (2.50), function (5.156) also depends only on the argument of the weighted sum and does not depend on its magnitude. The activation function (5.156) is a periodic and $l$-multiple continuation of the activation function (2.52) or, which is the same, of the discrete MVN activation function (2.50) for $k$=2. This is clearly illustrated in Fig. 5.48. This periodicity of the activation function (5.156) is its main property. It is also easy to check, that when $l = 2, m = 4$ in (5.156), we obtain (1.40).

A neuron with the activation function (5.156) was called in [30] the universal logical element over the field of complex numbers. A bit later, it was suggested to call it the *universal binary neuron* (UBN) [87]. Its universality is determined by its ability to learn and implement non-linearly separable input/output mappings (along with the linearly separable ones). Let us illustrate this by the following example.

### 5.1.2 Implementation of the Parity n Function Using a Single Neuron

We have already considered (see p. 41) how a single UBN with the activation function (1.40) (which is the same as (4.156) with $l = 2, m = 4$) implements the XOR function. Let us consider now how a single UBN with the activation function (5.156) with $l = 3, m = 6$ (see Fig. 5.49) implements the Parity 3 function.

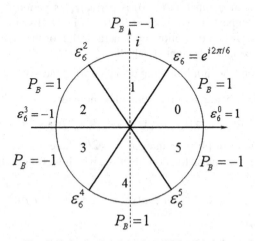

**Fig. 5.49** Activation function (156) with $l=3$, $m=6$

**Table 5.15** Solution of the Parity 3 problem using a single UBN with the activation function (5.156) with $l=2$, $m=6$ and with the weighting vector $W = (0, \varepsilon_6, 1, 1)$

| $x_1$ | $x_2$ | $x_3$ | $z = w_0 + w_1 x_1 + w_2 x_2 + w_3 x_3$ | arg $z$ | Number of sector | $P_B(z)$ | $f(x_1, x_2, x_3) = x_1 \oplus x_2 \oplus x_3$ |
|---|---|---|---|---|---|---|---|
| 1 | 1 | 1 | $\varepsilon_6 + 2$ | 0.335 | 0 | 1 | 1 |
| 1 | 1 | -1 | $\varepsilon_6$ | $\pi/3$ | 1 | -1 | -1 |
| 1 | -1 | 1 | $\varepsilon_6$ | $\pi/3$ | 1 | -1 | -1 |
| 1 | -1 | -1 | $\varepsilon_6 - 2$ | $2.618 = 5\pi/6$ | 2 | 1 | 1 |
| -1 | 1 | 1 | $-\varepsilon_6 + 2$ | $11\pi/6$ | 5 | -1 | -1 |
| -1 | 1 | -1 | $-\varepsilon_6 = \varepsilon_6^4$ | $4\pi/3$ | 4 | 1 | 1 |
| -1 | -1 | 1 | $-\varepsilon_6 = \varepsilon_6^4$ | $4\pi/3$ | 4 | 1 | 1 |
| -1 | -1 | -1 | $-\varepsilon_6 - 2$ | 3.475 | 3 | -1 | -1 |

While the XOR function is a mod 2 addition of two Boolean variables, the Parity $n$ function is a mod 2 addition of $n$ Boolean variables. The Parity $n$ function is a non-linearly separable Boolean function for any $n$. Let us consider a single UBN with the activation function (5.156) with $l = 3, m = 6$. It is easy to check that the weighting vector $W = (0, \varepsilon_6, 1, 1)$ (where $\varepsilon_6 = e^{i2\pi/6}$ is the primitive 6$^{\text{th}}$ root of a unity) implements the Parity 3 function $f(x_1, x_2, x_3) = x_1 \oplus x_2 \oplus x_3$. This is illustrated in Table 5.15.

In [88], it was experimentally shown by the author of this book that a single UBN easily solves the Parity $n$ problem up to $n=14$. This will be considered in detail in Section 5.3.5. The ability of a single UBN, a neuron with complex-valued weights, to implement non-linearly separable input/output mappings one more time shows that the functionality of a single neuron with complex-valued weights is higher than the functionality of real-valued neurons.

## 5.1.3  Projection of a Two-Valued Non-linearly Separable Function into an m-Valued Threshold Function

Let us now consider in detail that mechanism, which makes it possible implementation of non-linearly separable input/output mappings by a single UBN. The following theorem is very important.

**Theorem 5.20.** If the input/output mapping $f(x_1,...,x_n): E_2^n \to E_2$ can be implemented using a single UBN with the activation function (5.156) and the weighting vector $W = (w_0, w_1,..., w_n)$, then there exist a partially defined $m$-valued threshold function $\tilde{f}(x_1,...,x_n): E_2^n \to E_m$, which can be implemented using a single discrete MVN with the activation function (2.50) (where $k=m$) and the same weighting vector $W = (w_0, w_1,..., w_n)$ as the function $f$.

**Proof.** Since a single UBN implements the input/output mapping $f(x_1,...,x_n): E_2^n \to E_2$ with the weighting vector $W = (w_0, w_1,..., w_n)$, then

$$\forall (x_1,...,x_n) \in E_2 \qquad w_0 + w_1 x_1 +...+ w_n x_n = z \qquad \text{such} \qquad \text{that}$$

$P_B(z) = f(x_1,...,x_n)$. This means that if $f(x_1,...,x_n) = 1$, then $z$ is located in one of the "even" sectors (0, 2, ..., $m$-2) in which the activation function (5.156) divides the complex plane (see Fig. 5.48b and Fig. 5.49). If $f(x_1,...,x_n) = -1$, then $z$ is located in one of the "odd" sectors (1, 3, ..., $m$-1) in which the activation function (5.156) divides the complex plane (see again Fig. 5.48b and Fig. 5.49). This means that the number of a sector where the weighted sum $z$ can be located, belongs to the set $M = \{0, 1,..., m-1\}$.

Let us apply the discrete MVN activation function (2.50) to $z$. Then we obtain $P(z) = \varepsilon_m^j$, ( $j \in M$ is the number of the sector on the complex plane where $z$ is located). Let us build a partially defined $m$-valued function $\tilde{f}(x_1,...,x_n): E_2^n \to E_m$ in the following way (it is partially defined because $E_2^n \subset E_m^n$, thus, it is defined only on the binary inputs). Let us set

$$\tilde{f}(x_1,...,x_m) = P(z) = P(w_0 + w_1 x_1 + ... + w_n x_n) = \varepsilon_m^j \in E_m;$$
$$j \in \{0,1,...,m-1\}$$

From the composition of the function $\tilde{f}$ it is clear that it is an $m$-valued threshold function with the weighting vector $W = (w_0, w_1,..., w_n)$ according to Definition 2.5. Theorem is proven.

On the one hand, the function $\tilde{f}(x_1,...,x_n): E_2^n \to E_m$ is a partially defined $m$-valued function because $E_2^n \subset E_m^n$ (its domain is a subset of $E_m^n$). On the other hand, Theorem 5.20 can easily be generalized for any function $f: T \to E_2$, where $T \subset O^n$, and $O$ is the set of points located on the unit circle. This generalization leads us to the following statement.

If the input/output mapping $f(x_1,...,x_n): T \to E_2$ (where $T \subset O^n$) can be implemented using a single UBN with the activation function (5.156) and the weighting vector $W = (w_0, w_1,..., w_n)$, then there exist a partially defined $m$-valued threshold function $\tilde{f}(x_1,...,x_n): T \to E_m$, which can be implemented using a single discrete MVN with the activation function (2.50) (where $k=m$) and the same weighting vector $W = (w_0, w_1,..., w_n)$ as the function $f$.

Theorem 5.20 and its generalization establish the mechanism that projects a two-valued function $f(x_1,...,x_n)$, which can be implemented using a single UBN, into an $m$-valued threshold function $\tilde{f}(x_1,...,x_n)$. The most important here is the following.

If the two-valued function $f(x_1,...,x_n)$ is a non-linearly separable function in the real domain and cannot be implemented using a single real-valued neuron, but can be implemented using a single UBN, then its projection $\tilde{f}(x_1,...,x_n)$, is an $m$-valued threshold function, which can be learned using a single MVN.

For example, the Parity 3 function $f(x_1, x_2, x_3) = x_1 \oplus x_2 \oplus x_3$ is non-linearly separable in the real domain. It follows from Theorem 5.20 that there exists its projection $\tilde{f}(x_1, x_2, x_3)$, which is built using the weighting vector $W = (0, \varepsilon_6, 1, 1)$, as it is shown in Table 5.16.

Table 5.16 Projection of the non-linearly separable Parity 3 function into 6-valued multiple-valued threshold function using a single UBN with the activation function (5.156) with $l=2$, $m=6$ and with the weighting vector $W = (0, \varepsilon_6, 1, 1)$

| $x_1$ | $x_2$ | $x_3$ | $z = w_0 + w_1 x_1 + w_2 x_2 + w_3 x_3$ | arg $z$ | $P_B(z)$ | $f(x_1, x_2, x_3) = x_1 \oplus x_2 \oplus x_3$ | $\tilde{f}(x_1, x_2, x_3)$ |
|---|---|---|---|---|---|---|---|
| 1 | 1 | 1 | $\varepsilon_6 + 2$ | 0.335 | 1 | 1 | $\varepsilon_6^0$ |
| 1 | 1 | -1 | $\varepsilon_6$ | $\pi/3$ | -1 | -1 | $\varepsilon_6$ |
| 1 | -1 | 1 | $\varepsilon_6$ | $\pi/3$ | -1 | -1 | $\varepsilon_6$ |
| 1 | -1 | -1 | $\varepsilon_6 - 2$ | $2.618 = 5\pi/6$ | 1 | 1 | $\varepsilon_6^2$ |
| -1 | 1 | 1 | $-\varepsilon_6 + 2$ | $11\pi/6$ | -1 | -1 | $\varepsilon_6^5$ |
| -1 | 1 | -1 | $-\varepsilon_6 = \varepsilon_6^4$ | $4\pi/3$ | 1 | 1 | $\varepsilon_6^4$ |
| -1 | -1 | 1 | $-\varepsilon_6 = \varepsilon_6^4$ | $4\pi/3$ | 1 | 1 | $\varepsilon_6^4$ |
| -1 | -1 | -1 | $-\varepsilon_6 - 2$ | 3.475 | -1 | -1 | $\varepsilon_6^3$ |

It follows from Table 5.16 that the function $\tilde{f}(x_1, x_2, x_3)$ is a partially defined 6-valued threshold function with the weighting vector $W = (0, \varepsilon_6, 1, 1)$.

According to Definition 2.9 (see p. 72) it is also a complex-valued threshold function (we can always set $P(0) = \varepsilon_6^0 = 1$, for example). Then according to Definition 2.10 (see p. 82) such a 6-edge $\overline{Q} = \{Q_0, Q_1, ..., Q_5\}$ exists that $\forall \alpha = (\alpha_1, \alpha_2, \alpha_3) \in E_2^3 \cap Q_j \; P(\tilde{f}(\alpha)) = \varepsilon_6^j; j = 0, 1, ..., 5$. The last equation means that this 6-edge separates a 3-dimensional space where the function $\tilde{f}(x_1, x_2, x_3)$ is defined, into six edges (subspaces) $Q_0, Q_1, ..., Q_5$, where our function takes the values $\varepsilon_6^0, \varepsilon_6^1, ..., \varepsilon_6^5$, respectively. However, the same 6-edge also separates the 1s of the Parity 3 function from its -1s! This is illustrated

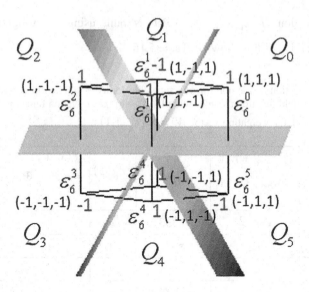

**Fig. 5.50** 6-edge separates a 3-dimensional space where the Parity 3 function and its 6-valued projection are defined

in Fig. 5.50. The planes, which create the edges of the 6-edge are shown in color. Since the Parity 3 function is defined on the set $E_2^3$, its 8 values are located in the vertices of the cube, which is also shown in Fig. 5.50. The values of the Parity 3 function are shown in red, while the values of the function $\tilde{f}(x_1, x_2, x_3)$ located in the same cube vertices are shown in blue. The labels $Q_0, Q_1, ..., Q_5$ of the edges of the 6-edge are also shown in blue. As we see, the cube vertices $(1, 1, 1)$ and $(1, -1, -1)$ where the Parity 3 function takes the same value 1 are located in the different edges - $Q_0$ and $Q_2$, respectively (the function $\tilde{f}(x_1, x_2, x_3)$ takes there the values $\varepsilon_6^0$ and $\varepsilon_6^2$, respectively). The cube vertices $(-1, 1, 1)$ and $(-1, -1, -1)$ where the Parity 3 function takes the same value -1 are also located in the different edges - $Q_5$ and $Q_3$, respectively (the function $\tilde{f}(x_1, x_2, x_3)$ takes there the values $\varepsilon_6^5$ and $\varepsilon_6^3$, respectively). At the same time, the cube vertices $(1, -1, 1)$ and $(1, 1, -1)$ where the Parity 3 function takes the same value -1 are located in the same edge $Q_1$ where the function $\tilde{f}(x_1, x_2, x_3)$ takes the value $\varepsilon_6^1$. The cube vertices $(-1, -1, 1)$ and $(-1, 1, -1)$ where the Parity 3 function takes the same value 1 are also located in the same edge $Q_4$ where the function $\tilde{f}(x_1, x_2, x_3)$ takes the value $\varepsilon_6^4$.

While in the real domain the Parity 3 function is not linearly separable and cannot be implemented using a single real-valued neuron, it becomes linearly separable in the complex domain. This separation is utilized by the 6-edge. As we have seen, the Parity 3 function can be implemented using a single UBN. We have also seen that this implementation is equivalent to the implementation of the partially defined 6-valued threshold function $\tilde{f}(x_1, x_2, x_3)$ using a single MVN.

## 5.1.4  UBN Learning

According to Theorem 5.20, if a Boolean function $E_2^n \rightarrow E_2$ can be implemented using a single UBN with the activation function (5.156), then there exist a partially defined $m$-valued threshold function $E_2^n \rightarrow E_m$, which can be implemented using a single MVN with the activation function (2.50). Thus the UBN learning can be reduced to the MVN learning. This means that the same learning rules (3.81) or (3.92) or (3.94)-(3.98) that are used for the MVN learning can be used for the UBN learning. The only special moment is specification of the desired output for either of these learning rules. If we have to learn any multiple-valued function, the desired output for each element of the learning set is always unambiguous. However, if we have to learn a Boolean function or a mapping like $O \rightarrow E_2$, the desired output in terms of multiple-valued logic is ambiguous. Indeed, the activation function (5.156) divides the complex plane into $m$ sectors (see Fig. 5.48b). In a half of them the UBN output is equal to 1, while in another half it is equal to -1. Where we have to direct the weighted sum during the learning process? How we can specify the desired output, to be able to use the MVN learning rules?

It was suggested in [60] by the author of this book, Naum Aizenberg, and Joos Vandewalle to resolve this problem in the following way. The choice of the desired sector $q$ in either of (3.81) or (3.92) or (3.94)-(3.98) should be based on the closeness of the current weighted sum to the right or left adjacent sector. Indeed, if the current UBN output is incorrect, this means that it should become corrected if the weighted sum is moved to either left or right adjacent sector. This follows from the construction of the activation function (5.156) (see also Fig. 5.48b). The adjacent sector, which is closer to the current value of the weighted sum in terms of angular distance, is chosen as the "correct" one. The number $q$ of this sector determines the desired output in either of the learning rules (3.81) or (3.92) or (3.94)-(3.98).

Hence, the UBN learning algorithm can be described as follows.

Let $X_j^t$ be the $t$th element of the learning set $A$ belonging to the learning subset $A_j$. Let $N$ be the cardinality of the set $A$, $|A| = N$. Let $Y_t \in E_2$ and

$D_t \in E_2$ be the actual and the desired UBN outputs, respectively, corresponding to the $t$th element of the learning set.

Let *Learning* be a flag, which is "True" if the weights adjustment is required and "False", if it not required, and $r$ be the number of the weighting vector in the sequence $S_w$ of weighting vectors obtained during the learning process. Let $z_t$ be the weighted sum corresponding to the $t$th element of the learning set.

Step 1. The starting weighting vector $W_0$ is chosen arbitrarily (e.g., real and imaginary parts of its components can be random numbers); $m=0$; $t=1$; *Learning* = "False";

Step 2. Check for $X_j^t$ :

$$\text{if } Y_t = D_t$$

*then go to* the step 5
*else begin Learning* = "True"; *go to* Step 3 *end;*

Step 3. Find $P(z_t) = \varepsilon_m^s$ (where $P$ is the MVN activation function (2.50)).

Find $q_1 \in M = \{0,1,...,m-1\}$, which determines the adjacent sector from the right (to the $s$th one), and find $q_2 \in M$, which determines the adjacent sector from the left (to the $s$th one) one, where the output is correct.

*If*

$$\left(\arg z_t - \arg\left(e^{i(q_1+1)2\pi/m}\right)\right) \bmod 2\pi \leq \left(\arg\left(e^{iq_2 2\pi/m}\right) - \arg z_t\right) \bmod 2\pi$$

*then* $q = q_1$

*else* $q = q_2$, where $q$ is the number of the desired sector.

Step 4. Obtain the vector $W_{r+1}$ from the vector $W_r$ by setting the desired output to $\varepsilon_m^q$ and applying either of the learning rules (3.81) or (3.92) or (3.94)-(3.98);

Step 5. $t = t+1$;         *if* $t \leq N$
*then go to* Step 2
*else if Learning* = "False"
*then* the learning process is finished successfully
*else begin* $t=1$; *Learning* = "False"; *go to* Step 2; *end.*

Since this UBN learning algorithm is reduced to the MVN learning algorithm, its convergence directly follows from the convergence of the MVN learning algorithm (see Theorem 3.16 and Theorem 3.17).

## 5.2  *k*-Valued MVN with a Periodic Activation Function

### 5.2.1  *Some Important Fundamentals*

We have just considered UBN – the universal binary neuron. We have shown that UBN is nothing else than the discrete MVN with the activation function (2.50) with $k=2$ (or simply with the activation function (2.52), which is (2.50) for $k=2$), periodically extended. This periodic extension transforms (2.52) to the binary periodic activation function (5.156). While the activation function (2.52) divides the complex plane into two sectors (top and bottom half-planes), the periodic activation function (5.156) divides the complex plane into $m=2l$ equal sectors. In this case, the neuron output is determined by the alternating periodic sequence of 1, -1, 1, -1,..., depending on the parity of the ordinal sector's number.

As we have seen, this approach leads to one very important advantage. A single UBN may implement those input/output mappings that are non-linearly separable in the real domain. Perhaps, the most convincible examples, which illustrate this advantage of UBN, are XOR and Parity $n$ that can easily be learned by a single UBN, without any network. Actually, this is achieved by the projection of 2-valued logic, where the initial non-linearly separable problem is defined, to $m$-valued logic. While in 2-valued logic our input/output mapping is not linearly-separable, in $m$-valued logic it becomes linearly separable.

A natural question is whether it is possible to generalize this approach for multiple-valued input/output mappings that is for the activation function (2.50) with $k > 2$? In other words, if there is some $k$-valued input/output mapping $T \rightarrow E_k$ (where $T = E_k^n$ or $T \subseteq O^n$), which is not a $k$-valued threshold function (and therefore it cannot be learned using a single MVN with the $k$-valued activation function (2.50)), can the same input/output mapping be a partially defined $m$-valued threshold function $T \rightarrow E_m$ for $m > k$?

This question is very important because there is a great practical sense behind it. Suppose we have to solve some $n$-dimensional $k$-class classification problem and the corresponding classes are non-linearly separable. The commonly used approach for solving such a problem, as we already know from this book, is its consideration in the larger dimensional space. One of the ways to utilize this approach is a neural network, where hidden neurons form a new space, and a problem becomes linearly separable. Another popular machine learning approach to solving non-linearly separable problems projecting them into a higher dimensional space is the support vector machine (SVM) introduced in [25]. In SVM, a larger dimensional space is formed using the kernels and a problem becomes linearly separable in this new space. We would like to approach the same problem from a different angle, that is, to consider an $n$-dimensional $k$-class classification problem as an $n$-dimensional $m$-class classification problem (where $m > k$ and each of $k$ initial classes is a union of some of $t$ disjoint subclasses (clusters) of an initial class):

$$C_j = \bigcup_{i=1}^{t_j} \tilde{C}_i^j, j = 1,...,k; 1 \le t_j < m; \tilde{C}_t^j \cap \tilde{C}_s^j = \varnothing, t \ne s,$$

where $C_j, j = 1,...,k$ is an initial class and each $\tilde{C}_i^j, i = 1,...,m$ is a new subclass). Thus, we would like to modify the formation of a decision rule instead of increasing the dimensionality. In terms of neurons and neural networks this means increasing the functionality of a single neuron by modification of its activation function.

Recently, this problem was comprehensively considered by the author of this book in his paper [61]. Let us present these considerations here adding more details.

### 5.2.2  Periodic Activation Function for Discrete MVN

Let us consider an MVN input/output mapping described by some $k$-valued function

$$f(x_1,...,x_n): T \to E_k$$

where $T = E_k^n$ or $T \subseteq O^n$).

It is important to mention that since there exists a one-to-one correspondence between the sets $K = \{0,1,...,k-1\}$ and $E_k = \{\varepsilon_k^0, \varepsilon_k,..., \varepsilon_k^{k-1}\}$ (see p. 49), our function $f$ can also be easily re-defined as $f_K: T \to K$. These both definitions are equivalent.

Suppose that the function $f(x_1,...,x_n)$ is not a $k$-valued threshold function. This means that it cannot be learned by a single MVN with the activation function (2.50).

Let us now project the $k$-valued function $f(x_1,...,x_n)$ into $m$-valued logic, where $m = kl$, and $l \ge 2$ similarly to what we have done in Section 5.1 for 2-valued functions projecting them into $m$-valued logic where we used $m = 2l$. To do this, let us define the following new discrete activation function for MVN

$$P_l(z) = j \bmod k, \text{ if } 2\pi j / m \le \arg z < 2\pi(j+1) / m,$$
$$j = 0,1,...,m-1; \ m = kl, l \ge 2. \tag{5.157}$$

This definition is illustrated in Fig. 5.51. The activation function (5.157) divides the complex plane into $m$ equal sectors and $\forall d \in K$ there are exactly $l$ sectors, in which the activation function (5.157) equals $d$.

This means that the activation function (5.157) establishes mappings from $K$ into $M = \{0,1,...,k-1,k,k+1,...,m-1\}$, and from $E_k$ into $E_m = \{1, \varepsilon_m, \varepsilon_m^2, ..., \varepsilon_m^{m-1}\}$, respectively.

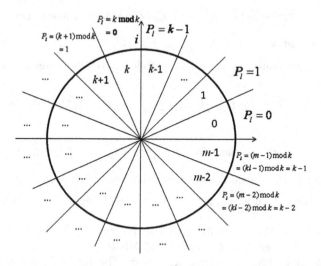

**Fig. 5.51** Geometrical interpretation of the $k$-periodic $l$- multiple discrete-valued MVN activation function (5.157)

Since $m = kl$, then each element from $M$ and $E_m$ has exactly $l$ prototypes in $K$ and $E_k$, respectively. In turn, this means that the neuron's output determined by (5.157) is equal to

$$\underbrace{\underbrace{0,1,...,k-1}_{0},\underbrace{0,1,...,k-1}_{1},...,\underbrace{0,1,...,k-1}_{l-1},}_{lk=m} \tag{5.158}$$

depending on which one of the $m$ sectors (whose ordinal numbers are determined by the elements of the set $M$) the weighted sum is located in.

Hence, the MVN activation function in this case becomes $k$-periodic and $l$-multiple.

In terms of multiple-valued logic, the activation function (5.157) projects a $k$-valued function $f(x_1,...,x_n)$ into an $m$-valued function $\tilde{f}(x_1, ..., x_n)$. Evidently, $\tilde{f}(x_1,...,x_n)$ is a partially defined function in $m$-valued logic because $K \subset M, E_k \subset E_m$, and $E_k^n \subset E_m^n$.

If $f(x_1,...,x_n):T \to E_k$ is not a $k$-valued threshold function, then its domain does not allow the *edged decomposition* (see Section 2.3) $T = [C_0, C_1, ..., C_{k-1}]$. Projecting $f(x_1,...,x_n)$ into $kl = m$-valued logic using the activation function (5.157), we create in this $m$-valued logic the function $\tilde{f}(x_1,...,x_n):T \to E_m; T \subseteq E_k^n \vee T \subseteq O^n$ whose domain may have the edged decomposition $T = [\tilde{C}_0, \tilde{C}_1, ..., \tilde{C}_{k-1}, \tilde{C}_k, \tilde{C}_{k+1},..., \tilde{C}_{m-1}]$. Moreover, if this edged decomposition exists, it exactly follows from its existence that

$$C_j = \bigcup_{i=1}^{t_j} \tilde{C}_i^j, j=0,...,k-1; 1 \le t_j < m; \tilde{C}_t^j \cap \tilde{C}_s^j = \varnothing, t \ne s.$$

It is important that both functions $f(x_1,...,x_n)$ and $\tilde{f}(x_1,...,x_n)$ have the same domain $T$. This means that the initial function $f(x_1,...,x_n)$, not being a $k$-valued threshold function, is projected to a partially defined $m$-valued threshold function.

Let us refer the MVN with the activation function (5.157) as the *multi-valued neuron with a periodic activation function* (MVN-P).

The following theorem, which generalizes Theorem 5.20 for $k$-valued input/output mappings, is proven by the last considerations.

**Theorem 5.21.** If the input/output mapping $f(x_1,...,x_n):T \to E_k$ can be implemented using a single MVN-P with the activation function (5.157) and the weighting vector $W = (w_0, w_1,..., w_n)$, then there exist a partially defined $m$-valued threshold function $\tilde{f}(x_1,...,x_n):T \to E_m$, which can be implemented using a single discrete MVN with the activation function (2.50) (where $k=m$) and the same weighting vector $W = (w_0, w_1,..., w_n)$.

It is important to mention that if $l=1$ in (5.157) then $m=k$ and the activation function (5.157) coincides with the activation function (2.50) accurate within the interpretation of the neuron's output (if the weighted sum is located in the $j$th sector, then according to (2.50) the neuron's output is equal to $e^{ij2\pi/k} = \varepsilon^j \in E_k$, which is the $j$th of $k$th roots of unity, while in (5.157) it is equal to $j \in K$), and MVN-P becomes regular MVN. It is also important to mention that if $k=2$ in (5.157), then the activation function (5.157) coincides with the UBN activation function (5.156), sequence (5.158) becomes an alternating sequence 1, -1, 1, -1, ..., and MVN-P becomes UBN. Hence, the MVN-P is a neuron, for which both MVN and UBN are its particular cases.

MVN-P may have a great practical sense if $f(x_1,...,x_n)$, being a non-threshold function of $k$-valued logic, could be projected into a partially defined threshold function of $m$-valued logic and therefore it will be possible to learn it using a single MVN-P with the activation function (5.157).

When we told that the activation function (5.157) projects a $k$-valued function $f(x_1,...,x_n)$ into $m$-valued logic, we kept in mind that the weighted sum $z$, on which the activation functions depends, is known. But in turn, the weighted sum is a function of the neuron weights for the fixed inputs $x_1,...,x_n$

$$z = w_0 + w_1 x_1 + ... + w_n x_n .$$

This means that to establish the projection determined by the activation function (5.157), we have to find the corresponding weights. To find them, a learning algorithm should be used.

### 5.2.3 Learning Algorithm for MVN-P

A learning algorithm, which we are going to present here, was recently developed by the author of this book and comprehensively described in the paper [61]. On the one hand, this learning algorithm is based on the modification of the MVN learning algorithm considered above in Section 3.1. On the other hand, this learning algorithm is based on the same idea as the UBN learning algorithm, which we have just presented in Section 5.1. The latter becomes clear when we take into account that UBN is nothing else than a particular case of MVN-P, just for $k=2$.

Let us take the MVN learning algorithm described in Section 3.1 and based on either of the error-correction learning rules (3.92) or (3.94)-(3.96) as the initial point for out MVN-P learning algorithm. Let us adapt this MVN learning algorithm to MVN-P. Thus, we have to modify the standard MVN learning algorithm based on the error-correction rule in such a way that it will work for MVN with the periodic activation function (5.157) (for the MVN-P). Let us assume for simplicity, but without loss of generality that the learning rule (3.92) will be used. It is important to mention that the learning rules (3.94)-(3.96) can also be used because, as we have seen (Theorem 3.17), the convergence of the MVN learning algorithm with the error-correction learning rule does not depend on its modification ((3.92) or (3.94)-(3.96)).

Let the MVN input/output mapping is described by the $k$-valued function $f(x_1,...,x_n)$, which is not a threshold function of $k$-valued logic. Since this function is non-threshold, there is no way to learn it using a single MVN with the activation function (2.50). Let us try to learn $f(x_1,...,x_n)$ in $m$-valued logic using a single MVN-P with the activation function (5.157).

Thus, the expected result of this learning process is the representation of $f(x_1,...,x_n)$ according to (2.51), where the activation function $P_l$ determined by (5.157) substitutes for the activation function $P$ determined by (2.50)

$$f(x_1,...,x_n) = P_l(w_0 + w_1 x_1 + ... + w_n x_n)$$     (5.159)

To organize this learning process, we will use the same learning rule (3.92).

To determine a desired output in the error-correction learning rule (3.92), we may use the same approach, which we used for the UBN learning algorithm in Section 5.1.

So, the learning rule (3.92) requires that a desired neuron output is pre-determined. Unlike the case of regular MVN with the activation function (2.50), a desired output in terms of $m$-valued logic cannot be determined unambiguously for MVN-P with the activation function (3.157) for $l \geq 2$. According to (3.157), there are exactly $l$ sectors out of $m$ on the complex plane, where this activation function is equal to the given desired output $d \in K$ (see (5.158) and Fig. 5.51). Therefore, there are exactly $l$ out of $m$ $m$th roots of unity that can be used as the desired outputs in the learning rule (3.92). Which one of them should we choose?

Let us make this choice using two self-adaptive learning strategies, which will make it possible to determine a desired output during the learning process every time, when the neuron's output is incorrect.

The first strategy is based on the same idea, which was used for UBN. Since MVN-P is a generalization of UBN for $k > 2$, we suggest using here the same learning strategy that was used in the UBN error-correction learning algorithm in Section 5.1. There is the following idea behind this approach. The UBN activation function (5.156) determines an alternating sequence 1, -1, 1, -1, ... with respect to sectors on the complex plane. Hence, if the actual output of UBN is not correct, in order to make the correction, we can "move" the weighted sum into either of the sectors adjacent to the one where the current weighted sum is located. It was suggested to always move it to the sector, which is the closest one to the current weighted sum (in terms of angular distance).

Let us employ the same approach here for MVN-P. Let $l \geq 2$ in (5.157) and $d \in \{0,1,...,k-1\}$ be the desired output. The activation function (5.157) determines the $k$-periodic and $l$-multiple sequence (5.158) with respect to sectors on the complex plane. Suppose that the current MVN-P output is not correct and the current weighted sum is located in the sector $s \in M = \{0,1,...,m-1\}$, where $m = kl$.

Since $l \geq 2$ in (5.157), there are exactly $l$ sectors on the complex plane, where function (5.157) takes a correct value (see also Fig. 5.51). Two of these $l$ sectors are the closest ones to sector $s$ (from right and left sides, respectively). From these two sectors, we choose sector $q$ whose border is closer to the current weighted sum $z$ in terms of the angular distance. Then the learning rule (3.92) can be applied. Hence, the first learning strategy for the MVN-P with the activation function (5.157) is as follows. Let a learning set for the function (input/output mapping) $f(x_1,...,x_n)$ to be learned contains $N$ learning samples and $j \in \{1,...,N\}$ be the number of the current learning sample, $r$ be the number of the learning iteration, and *Learning* is a flag, which is "True" if the weights adjustment is required and "False" otherwise.

The iterative learning process consists of the following steps:

**Learning Strategy 1.**
1) Set $r=1, j=1$, and Learning='False'.
2) Check (5.159) for the learning sample $j$.
3) *If* (5.159) holds
   *then* set $j = j+1$, otherwise set Learning='True' and
   *go to* Step 5.
4) If $j \leq N$ then go to Step 2, otherwise go to Step 9.
5) Let $z$ be the current value of the weighted sum and $P(z) = \varepsilon_m^s, s \in M$,
   $P(z)$ is the activation function (2.50), where $m$ is substituted for $k$. Hence
   the MVN-P actual output is $P_l(z) = s \bmod k$. Find
   $q_1 \in M = \{0,1,...,m-1\}$, which determines the closest sector to the $s$th
   one, where the output is correct, from the right, and find $q_2 \in M$, which determines the closest sector to the $s$th one, where the output is correct, from
   the left (this means that $q_1 \bmod k = d$ and $q_2 \bmod k = d$).
6) *If* $\left( \arg z - \arg\left(e^{i(q_1+1)2\pi/m}\right)\right) \bmod 2\pi \leq \left(\arg\left(e^{iq_2 2\pi/m}\right) - \arg z\right) \bmod 2\pi$
   *then* $q = q_1$
   *else* $q = q_2$.
7) Set the desired output for the learning rule (3.92) equal $\varepsilon_m^q$.
8) Apply the learning rule (3.92) to adjust the weights.
9) Set $j = j+1$ and return to Step 4.
10) If *Learning*='False'
    *then* go to Step 10,
    *else* set $r=r+1, j=1$, *Learning*='False' and go to Step 2.
11) End.

Let us now consider the second learning strategy, which is somewhat different. The activation function (5.157) divides the complex plane into $l$ domains, and each of them consists of $k$ sectors (Fig. 5.51). Since a function $f$ to be learned as a partially defined function of $m$-valued logic ($m=lk$) is in fact a $k$-valued function, then each of $l$ domains contains those $k$ values, which may be used as the desired outputs of the MVN-P. Suppose that the the current MVN-P output is not correct, and the current weighted sum is located in the sector $s \in M = \{0,1,...,m-1\}$. This sector in turn is located in the $t$th $l$-domain (out of $l$, $t = [s/k]$). Since there are $l$ $l$-domains and each of them contains a potential correct output, we have $l$ options to choose the desired output. Let us choose it in the same $t$th $l$-domain, where the current actual output is located. Hence, $q = tk + f_K(x_1,...,x_n)$, where $f_K(x_1,...,x_n)$ is a desired value of the function to be learned in terms of traditional multiple-valued logic ($f_K(x_1,...,x_n) \in K = \{0,1,...,k-1\}$                    and                    respectively, $f(x_1,...,x_n) \in E_k = \{1, \varepsilon_k, \varepsilon_k^2,..., \varepsilon_k^{k-1}\}$). Once $q$ is determined, this means that $\varepsilon_m^q$ be the desired output and the learning rule (3.92) can be applied.

Let again a learning set for the function (input/output mapping) $f(x_1,...,x_n)$ to be learned contains $N$ learning samples, $j \in \{1,...,N\}$ be the number of the current learning sample, $r$ be the number of the learning iteration, and *Learning* is a flag, which is "True" if the weights adjustment is required and "False" otherwise. The iterative learning process for the second strategy consists of the following steps:

**Learning Strategy 2.**
1) Set $r=1, j=1$, and Learning='False'.
2) Check (5.159) for the learning sample $j$.
3) *If* (5.159) holds
   *then* set $j = j+1$,
   *else* set Learning ='True' and go to Step 5.
4) *If* $j \leq N$
   *then* go to Step 2,
   *else* go to Step 8.
5) Let the actual neuron output is located in the sector $s \in M = \{0,1,...,m-1\}$. Then $t = [s/k] \in \{0,1,...,l-1\}$ is the number of that $l$-domain, where sector $s$ is located. Set $q = tk + f_K(x_1,...,x_n)$.
6) Apply the learning rule (3.92) to adjust the weights.
7) Set $j = j+1$ and return to Step 4.
8) *If* Learning='False'
   *then* go to Step 9,
   *else* set $r=r+1, j=1$, Learning='False' and go to Step 2.
9) End.

The learning strategies 1 and 2 determine two variants of the same MVN-P learning algorithm, which can be based on either of the learning rules (3.92), (3.94)-(3.98). The convergence of this learning algorithm follows from the convergence of the regular MVN learning algorithm with the error-correction learning rule (see Theorem 3.17). Indeed, if our input/output mapping $f(x_1,...,x_n)$ is a non-threshold function of $k$-valued logic, but it can be projected to a partially defined threshold function $\tilde{f}(x_1,...,x_n)$ of $m$-valued logic (where $m = kl, l \geq 2$), then the MVN learning algorithm has to converge for the last function according to Theorem 3.17. The MVN-P learning algorithm based on the either of learning rules (3.92), (3.94)-(3.96) differs from the MVN learning algorithm only at one point. While for the regular MVN learning a desired output is pre-determined, for the MVN-P learning a desired output in terms of $m$-valued logic should be determined during the learning process. If the function $\tilde{f}(x_1,...,x_n)$ obtained using either of Learning Strategies 1 or 2 is a partially defined $m$-valued threshold function, its learning has to converge to a weighting vector of this function (a weighting vector of this function can always be obtained after a finite number of learning iterations).

Thus, in other words, if a non-threshold $k$-valued function $f(x_1,...,x_n)$ can be projected to and associated with a partially defined $m$-valued threshold function $\tilde{f}(x_1,...,x_n)$, then its learning by a single MVN-P is reduced to the learning of the function $\tilde{f}(x_1,...,x_n)$ by a single MVN.

We have to mention that we do not consider here any general mechanism of such a projection of a $k$-valued function into $m$-valued logic that the resulting $m$-valued function will be threshold and therefore it will be possible to learn it by a single neuron. It is a separate problem, which is still open and can be a good and interesting subject for the further work.

It is interesting that in terms of learning a $k$-valued function, the learning algorithm presented here is supervised. However, in terms of learning an $m$-valued function, this learning algorithm is unsupervised. We do not have a prior knowledge about those $m$-valued output values, which will be assigned to the input samples. A process of this assignment is self-adaptive, and this adaptation is reached by the learning procedure (Strategies 1 and 2), if a corresponding function is a partially defined $m$-valued threshold function.

It should be mentioned that for $k=2$ in (5.157) the MVN-P learning algorithm (Strategy 1) coincides with the UBN learning algorithm based on the error-correction rule (see Section 5.1). On the other hand, for $k>2$ and $l=1$ in (5.157) the MVN-P learning algorithm (both Strategy 1 and Strategy 2) coincides with the MVN learning algorithm based on the error-correction rule (see Sections 3.1 and 3.3).

This means that a concept of the MVN-P generalizes and includes the corresponding MVN and UBN concepts.

## 5.3 Simulation Results for $k$-Valued MVN with a Periodic Activation Function

As it was shown above, MVN-P can learn input/output mappings that are non-linearly separable in the real domain. We would like to consider here a number of non-linearly separable benchmark classification problems and a non-linearly separable $\bmod k$ addition problem, which can be learned using a single MVN-P. Moreover, we would like to show that a single MVN-P not only formally learns non-linearly separable problems, but it can really be successfully used for solving non-linearly separable classification problems, showing very good results that are better or comparable with the solutions obtained using neural networks or support vector machines. However, it is very important to mention that MVN-P is just a single neuron, and it employs fewer parameters than any network or SVM.

So, let us consider some examples. Most of them were presented by the author of this book in his recently published paper [61], some of them will be presented here for the first time, but even those published earlier will be presented here in more detail. In all simulations, we used the MVN-P software simulator written in Borland Delphi 5.0 environment, running on a PC with the Intel® Core™2 Duo CPU.

### 5.3.1 Iris

This famous benchmark database was downloaded from the UC Irvine Machine Learning Repository [89]. The data set contains 3 classes of 50 instances each, where each class refers to a type of iris plant. Four real-valued (continuous) features are used to describe the data instances. Thus, we have here 4-dimensional 3-class classification problem. It is known [89] that the first class is linearly separable from the other two but the latter are not linearly separable from each other. Thus, a regular single MVN with the discrete activation function (2.50), as well as any other single artificial neuron cannot learn this problem completely.

However, a single MVN-P with the activation function (5.157) ( $l = 3, k = 3, m = 9$ ) learns the Iris problem completely with no errors. To transform the input features into the numbers located on the unit circle, we used the linear transformation (2.53) with $\alpha = 2\pi / 3$ (this choice of α is based on the consideration that there are exactly 3 classes in this problem). It is necessary to say that the problem is really complicated and it is not so easy to learn it. For example, the learning algorithm based on the Learning Strategy 1 does not converge even after 55,000,000 iterations independently from the learning rule, which is applied, although the error decreases very quickly and after 50-100 iterations there are stably not more than 1-7 samples, which still require the weights adjustment. However, the learning algorithm based on the Learning Strategy 2 and the learning rule (3.96) converges with the zero error. Seven independent runs of the learning algorithm starting from the different random weights[1] converged after 9,379,027 – 43,878,728 iterations. Every time the error decreases very quickly and after 50-100 iterations there are stably 1 or just a few more samples, which still require the weights adjustment, but their final adjustment takes time (5-12 hours). Nevertheless, this result is very interesting, because to our best knowledge this is the first time when the "Iris" problem was learned using just a single neuron.

The results of the one of the learning sessions are shown in Fig. 5.52. In Fig. 5.52a, the normalized weighted sums are plotted for all the 150 samples from the data set. It is interesting that after the learning process converges, for the Class "0" (known and referred to as "Iris Setosa" [89]), the weighted sums for all instances appear in the same single sector on the complex plane (sector 6, see Fig. 5.52a). By the way, according to the activation function (5.157), the MVN-P output for all the samples from this class is equal to $6 \bmod 3 = 0$. Thus, Class "0" is a single cluster class.

Each of two other classes contains two different clusters (this is why they cannot be linearly separated from each other in the real domain!). For the second class ("Iris Versicolour"), 45 out of 50 learning samples appear in the sector 7, but other 5 learning samples appear in the sector 1 located in the different "$l$-domain" (cluster) (see Fig. 5.52a). According to the activation function (5.157), the MVN-P output for all the samples from this class is equal to 1 ($7 \bmod 3 = 1$ and $1 \bmod 3 = 1$). For the third class ("Iris Virginica"), the weighted sums for all the instances except one appear in the same single sector on the complex plane (sector 2), but for the one instance (every time the same) the weighted sum appears in the different sector (sector 8) belonging to the different "$l$-domain" (cluster). According to the activation function (5.157), the MVN-P output for all the samples from this class is equal to 2 ($2 \bmod 3 = 2$ and $8 \bmod 3 = 2$). For the reader's convenience, a fragment showing where exactly five "special" elements from the Class 1 and one "special" element from the Class 2 are located is enlarged in Fig. 5.52b. Hence, the second and the third classes, which initially are known as non-linearly separable (in the real domain), become linearly separable in the complex domain. This means that while there is no 3-edged decomposition $T = \begin{bmatrix} C_0, C_1, C_2 \end{bmatrix}$ (where $C_0, C_1, C_2$ are our three classes) for the Iris problem, there exists the 9-edged decomposition $T = \begin{bmatrix} A_0, A_1, ..., A_8 \end{bmatrix}$. Subsets $A_0, A_3, A_4, A_5$ are empty. Other subsets of the edged decomposition contain all the elements of the Iris dataset as follows (see Fig. 5.52a)

$$C_0 = A_6; C_1 = A_1 \cup A_7; C_2 = A_2 \cup A_8.$$

---

[1] Here and further the initial weights (both real and imaginary parts) are random numbers from the interval [0, 1] generated using a standard generator.

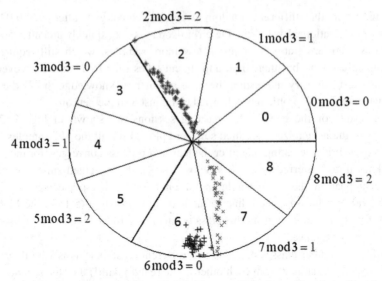

(a) The results of learning of the "Iris" problem. + - Class 0, x – Class 1, * - Class 2
While Class 0 contains a single cluster (sector 6), Class 1 and Class 2 contain two
clusters each (Class 1 - sectors 1 and 7, Class 2 – sectors 2 and 8)

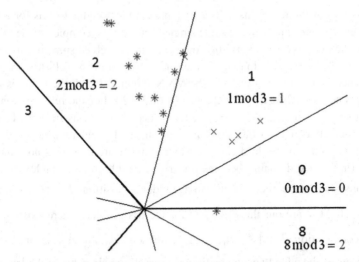

(b) 5 out of 50 representatives of Class 1 belong to the cluster located in the sector1,
and a single representative of Class 2 belong to the cluster located in the sector 8

**Fig. 5.52** Learning of the "Iris" problem. Three "Iris" classes are linearly separated in
9-valued logic

It is interesting that this effect is achieved by the self adaptation of the MVN-P learning algorithm.

Another important experiment with the "Iris" data set was checking the MVN-P's ability to solve a classification problem. We used 5-fold cross validation. The data set was every time randomly separated into a learning set containing 75 samples (25 from each class) and a testing set also containing 75 samples. The best results are obtained for the activation function (5.157) with $l = 2, k = 3, m = 6$. The Learning Strategy 1 and the learning rule (3.92) were used. The learning algorithm requires for its convergence with the zero error 10-288 iterations (which takes just a few seconds). The classification results are absolutely stable: 73 out of 75 instances are classified correctly (the classification rate is 97.33%). All instances from the first class are always classified correctly and there is one classification error in each of other two classes. These results practically coincide with the best known results for this benchmark data set [90]: (97.33 for the one-against-one SVM and 97.62 for the dendogram-based SVM). However, it is important to mention that the one-against-one SVM for 3 classes contains 3 binary decision SVMs, the dendogram-based SVM for 3 classes contains 5 binary decision SVMs, while we solved the Iris problem using just a single MVN-P.

### 5.3.2  Two Spirals

The two spirals problem is a well known non-linearly separable classification problem, where the two spirals points (see Fig. 5.53) must be classified as belonging to the 1st or to the 2nd spiral. Thus, this is 2-dimensional, 2-class classification problem. The standard two spirals data set usually consists of 194 points (97 belong to the 1st spiral and other 97 points belong to the 2nd spiral). The following results are known as the best for this problem so far. The two spirals problem can be learned completely with no errors by the

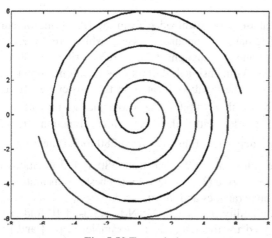

**Fig. 5.53** Two spirals

MLMVN [62] containing 30 hidden neurons in a single hidden layer and a single output neuron. This learning process requires about 800,000 iterations. The best known result for the standard backpropagation network (MLF) with the same topology is 14% errors after 150,000 learning iterations [91]. For the cross-validation testing, where each second point of each spiral goes to the learning set and each other second point goes to the testing set, one of the best known results is reported in [92]. The classification accuracy up to 94.2% is shown there by BDKSVM, which employs along with a traditional SVM the merits of the kNN classifier. A fuzzy kernel perceptron shows the accuracy up to 74.5%. [93]. The MLMVN shows the accuracy of about 70% [62].

A single MVN-P with the activation function (5.157) $\left(l = 2, k = 2, m = 4\right)$ significantly outperforms all mentioned techniques. Just 2-3 learning iterations are required to learn the two spirals problem completely with no errors using the Learning Strategy 1 and learning rule (3.92). Just 3-6 iterations are required to achieve the same result using the Learning Strategy 1 and learning rule (3.94). These results are based on the ten independent runs of the learning algorithm for each of the learning rules. We also used ten independent runs to check the classification ability of a single MVN-P with the activation function (5.157) $\left(l = 2, k = 2, m = 4\right)$ using the cross-validation. The two spirals data were divided into the learning set (98 samples) and testing set (96 samples). We reached the absolute success in this testing: 100% classification accuracy is achieved in all our experiments. Just 2-3 iterations were needed to learn the learning set using the Learning Strategy 1 and the learning rule (3.92), and 3-5 iterations were needed to do the same using the Learning Strategy 1 and the learning rule (3.94).

### 5.3.3  Breast Cancer Wisconsin (Diagnostic)

This famous benchmark database was downloaded from the UC Irvine Machine Learning Repository [89]. The data set contains 2 classes, which are represented by 569 instances (357 benign and 212 malignant) that are described by 30 real-valued features. Thus, this is 30-dimensional, 2-class classification problem. To transform the input features into the numbers located on the unit circle, we used the linear transformation (2.53) with $\alpha = 6.0$. The whole data set may be easily learned by a single MVN-P with the activation function (5.157) $\left(l = 2, k = 2, m = 4\right)$. Ten independent runs give from 280 to 370 iterations for the Learning Strategy 1 – learning rule (3.92) and from 380 to 423 iterations for the Learning Strategy 2 – learning rule (3.92) (there are the best results among different combinations of learning strategies and rules).

To check the classification ability of a single MVN-P, we used 10-fold cross-validation as it is recommended for this data set, for example, in [89] and [94]. The entire data set was randomly divided into the 10 subsets, 9 of them contained 57 samples and the last one contained 56 samples. The learning set every time was formed from 9 of 10 subsets and the remaining subset was used as the testing set. We used the same parameters in the activation function (5.157) $\left(l = 2, k = 2, m = 4\right)$. The best average classification accuracy (97.68%) was achieved with the Learning Strategy 2 and learning rule (3.92). The learning process required from 74 to 258 iterations. The classification accuracy is comparable with the results reported for SVM in [94] (98.56% for the regular SVM and 99.29% for the SVM with an additional "majority decision" tool) and a bit better than 97.5% reported as the estimated accuracy for the different classification methods in [89]. However, a single MVN-P use fewer parameters than, for example, SVM (the average amount of support vectors for different kernels used in [94] is 54.6, whereas the MVN-P uses 31 weights).

### 5.3.4   Sonar

This famous benchmark database was also downloaded from the UC Irvine Machine Learning Repository [89]. It contains 208 samples that are described by 60 real-valued features. Thus, this is 30-dimensional, 2-class classification problem, which is non-linearly separable. To transform the input features into the numbers located on the unit circle, we used (2.53) with $\alpha = 6.0$. There are two classes ("mine" and "rock") to which these samples belong. The whole data set may be easily learned by a single MVN-P with the activation function (5.157) $\left( l = 2, k = 2, m = 4 \right)$. Ten independent runs give from 75 to 156 iterations for the Learning Strategy 1 – learning rule (3.92) and from 59 to 78 iterations for the Learning Strategy 2 – learning rule (3.92) (there are the best results among different combinations of learning strategies and rules).

   To check the classification ability of a single MVN-P, we divided the data set into a learning set and a testing set (104 samples in each), as it is recommended by the developers of this data set [89]. The same parameters were used in the activation function (5.157) $\left( l = 2, k = 2, m = 4 \right)$. The best classification results were achieved using the Learning Strategy 2 - learning rule (3.92). The learning process required from 24 to 31 iterations. The average classification accuracy for 10 independent runs is 86.63% and the best achieved accuracy is 91.3%. This is comparable to the best known results reported in [93] – 94% (Fuzzy Kernel Perceptron), 89.5% (SVM), and in [62] - 88%-93% (MLMVN). It is important to mention that all mentioned competitive techniques employ more parameters than a single MVN-P. It is also necessary to take into account that a Fuzzy Kernel Perceptron is much more sophisticated tool than a single neuron.

### 5.3.5   Parity N Problem

Parity $n$ problem is a mod 2 addition of $n$ variables. Its particular case for $n=2$ is XOR, perhaps the most popular non-linearly separable problem considered in the literature. We have already convinced that the XOR problem can easily be solved using a single UBN (see Table 1.7, p. 42). We have also seen (see Table 5.15, p. 176) that the Parity 3 problem can be solved using a single UBN. As we have mentioned, it was experimentally shown by the author of this book in [88] that the Parity $n$ problem is easily solvable using a single UBN up to $n=14$ (this does not mean that for larger $n$ it is not solvable, simply experimental testing was not performed for $n>14$). Let us summarize the results of this experimental testing here.

Actually, as we have seen, UBN is nothing else than MVN-P for $k=2$. So, we used MVN-P with the activation function (5.157) and the learning algorithm with the Learning Strategy 1, which we have just described. The learning rule (3.92) was used in this learning algorithm. The results are summarized in Table 5.17. Everywhere we show the results for such minimal $l$ in (5.157), for which the learning process converged after not more than 200,000 iterations.

**Table 5.17** The results of solving Parity n problem for $3 \leq n \leq 14$ using a single MVN-P

| Number of variables, $n$ in the Parity $n$ problem | $l$ in (157) | Number of sectors ($m$ in (157)) | Number of learning iterations (average of 5 independent runs) |
|---|---|---|---|
| 3 | 3 | 6 | 8 |
| 4 | 4 | 8 | 23 |
| 5 | 5 | 10 | 37 |
| 6 | 6 | 12 | 52 |
| 7 | 7 | 14 | 55 |
| 8 | 8 | 16 | 24312 |
| 9 | 11 | 22 | 57 |
| 10 | 14 | 28 | 428 |
| 11 | 15 | 30 | 1383 |
| 12 | 18 | 36 | 1525 |
| 13 | 19 | 38 | 16975 |
| 14 | 22 | 44 | 3098 |

### 5.3.6 Mod k Addition of n k-Valued Variables

This problem may be considered as a generalization of the famous and popular Parity $n$ problem for the $k$-valued case. In fact, Parity $n$ problem is a mod 2 addition of $n$ variables. mod $k$ addition of $n$ variables is a non-threshold $k$-valued function for any $k$ and any $n$ and therefore it cannot be learned by a single MVN. To our best knowledge there is no evidence that this function can be learned by any other single neuron. However, as we will see now, it is not a problem to learn this problem using a single MVN-P with the activation function (5.157).

We do not have a universal solution of the problem of $\mod k$ addition of $n$ variables in terms of the relationship between $k$ and $n$ on the one side and $l$ in (5.157) on the other side. However, we can show here that this multiple-valued problem is really solvable at least for those $k$ and $n$, for which we have performed experimental testing [61, 95].

The experimental results are summarized in Table 5.18. Since the Learning Strategy 1 showed better performance for this problem (fewer learning iterations and time), all results are given for this strategy only.

Our goal was to find a minimal $l$ in (5.157), for which the learning process converges. For each combination of $k$ and $n$ on the one hand, and for each learning rule, on the other hand, such a minimal value of $l$ is presented. The average number of iterations for seven independent runs of the learning process and its standard deviation are also presented for each combination of $k$ and $n$ for each learning rule. We considered the learning process non-converging if there was no convergence after 200,000 iterations. The learning error for all learning rules drops very quickly, but fine adjustment of the weights takes more time. For some $k$ and $n$ and for some of learning rules we used a staggered learning technique. This means that unlike a regular learning technique, where all learning samples participate in the learning process from the beginning, the staggered method expands a learning set step by step. For example, let $A$ be a learning set and its cardinality is $N$. The set $A$ can be represented as a union of non-overlapping subsets $A_1, A_2, ..., A_s$. Then the learning process starts from the subset $A_1$. Once it converged, it has to continue for the extended learning set $A_1 \cup A_2$ starting from the weights that were obtained for $A_1$. Once it converges, the learning set has to be extended to $A_1 \cup A_2 \cup A_3$, adding one more subset after the previous learning session converged. Finally, we obtain the learning set $A_1 \cup A_2 \cup ... \cup A_{s-1} \cup A_s = A$. This approach is efficient when the function to be learned has a number of high jumps.

If there are exactly $s$ high jumps, then $s$ sequential learning sessions with $s$ expanding learning sets $A_1, A_1 \cup A_2, ..., A_1 \cup ... \cup A_s = A$ lead to faster convergence of the learning algorithm. For example, the function mod $k$ addition of $n$ variables has exactly $k^n$ learning samples. For any $k$ and $n$, this function has multiple high jumps from $k$-1 to 0. These jumps can be used to determine partitioning of the corresponding learning set into $s$ non-overlapping subsets $A_1, A_2, ..., A_s$. First, the learning process has to be run for $A_1$. Once it converges, it has to be run for $A_1 \cup A_2$. This set contains one high jump, but since the starting weighting vector, which already works for $A_1$, better approaches the resulting weighting vector, the learning process for $A_1 \cup A_2$ converges better starting from this weighting vector than starting from a random one. Then this process has to be continued up to the learning set $A_1 \cup ... \cup A_s = A$. We used this staggered learning technique, for example, for $k = 5, n = 2$, for $k = 6, n = 2$ and some other $k$ and $n$ (see the footnote in Table 5.18.). While neither of learning rules (3.92), (3.94)-(3.96) leads to the convergence of the standard learning algorithm after 200,000 iterations for these specific $k$ and $n$, the staggered technique leads to very quick convergence of the learning process for all four learning rules.

Thus, for all the experiments, we show in Table 5.18 the smallest $l$ in (5.157), for which the convergence was reached.

**Table 5.18** Simulation Results for mod $k$ addition of $n$ $k$-valued variables

| k | n | Average number of learning iterations (Iter.) for 7 independent runs, its standard deviation (SD), and minimal value of $l$ in (5.157), for which the learning process converged. | | | | | | | | | | | |
|---|---|---|---|---|---|---|---|---|---|---|---|---|---|
| | | Learning rule (3.92) | | | Learning rule (3.94) | | | Learning rule (3.95) | | | Learning rule (3.96) | | |
| | | Iter. | SD | l | Iter. | SD | l | Iter. | SD | l | Iter. | SD | l |
| 3 | 2 | 14 | 5 | 2 | 18048 | 44957 | 2 | 54 | 14 | 2 | 1005 | 1384 | 2 |
| 3 | 3 | 2466 | 2268 | 10 | 3773 | 1721 | 10 | 2568 | 1988 | 10 | 2862 | 676 | 12 |
| 3 | 4 | 4296 | 2921 | 11 | 391404 | 158688 | 7 | 2002 | 1272 | 14 | 1728 | 767 | 11 |
| 3 | 5 | 78596 | 87158 | 18 | 344440 | 308044 | 22 | 236242 | 188276 | 18 | 23372 | 8255 | 24 |
| 3 | 6 | 237202 | 172100 | 36 | 50292 | 91260 | 39 | 291950 | 346862 | 27 | 41083 | 23117 | 30 |
| 3 | 7 | 518313 | 395671 | 41 | 1556379 | 798841 | 41 | 489366 | 229706 | 22 | 390786 | 260953 | 25 |
| 4 | 2 | 2693 | 3385 | 3 | 135 | 163 | 3 | 94[1] | 23 | 3 | 660[1] | 567 | 3 |
| 4 | 3 | 2571 | 1772 | 7 | 12175 | 5407 | 7 | 411 | 190 | 7 | 602 | 436 | 7 |
| 4 | 4 | 50151 | 35314 | 10 | 140850 | 88118 | 10 | 47818 | 53349 | 13 | 3797 | 2756 | 13 |
| 4 | 5 | 734691[1] | 231353 | 13 | 355910[1] | 98208 | 13 | 174649[1] | 148655 | 16 | 209629[1] | 189481 | 15 |
| 4 | 6 | 131139[1] | 185316 | 42 | 171269[1] | 104685 | 15 | 59807[1] | 57060 | 34 | 306672[1] | 312548 | 30 |
| 4 | 7 | 108050 | 30309 | 39 | 110809 | 37286 | 38 | 90734 | 35474 | 37 | 95055 | 39581 | 38 |
| 5 | 2 | 96[1] | 22 | 4 | 81[1] | 16 | 4 | 82[1] | 23 | 4 | 197[1] | 82 | 5 |
| 5 | 3 | 1202[1] | 193 | 9 | 1419[1] | 264 | 9 | 1460[1] | 308 | 9 | 1470[1] | 191 | 9 |
| 5 | 4 | 4604[1] | 393 | 13 | 4893[1] | 211 | 13 | 5606[1] | 374 | 13 | 6182[1] | 616 | 13 |
| 5 | 5 | 22812[1] | 3977 | 22 | 17274[1] | 1682 | 18 | 17402[1] | 3415 | 18 | 21470[1] | 1959 | 18 |
| 5 | 6 | 105672[1] | 20071 | 26 | 196441[1] | 5635 | 22 | 66609[1] | 9888 | 33 | 2693[1] | 310 | 24 |
| 5 | 7 | 630490 | 192494 | 52 | 557635 | 305579 | 49 | 16192 | 24618 | 35 | 5280 | 2433 | 31 |
| 6 | 2 | 272[1] | 110 | 4 | 120[1] | 33 | 4 | 95[1] | 35 | 4 | 295[1] | 65 | 4 |
| 6 | 3 | 109733 | 2400 | 14 | 4131 | 4039 | 10 | 861 | 150 | 11 | 834 | 221 | 10 |
| 6 | 4 | 118128 | 15596 | 14 | 48002 | 11652 | 14 | 9615[1] | 1159 | 16 | 10951[1] | 1265 | 16 |
| 6 | 5 | 71241 | 74463 | 21 | 15986 | 14835 | 21 | 128550 | 105115 | 20 | 42122 | 26631 | 20 |
| 6 | 6 | 92257[1] | 4773 | 33 | 194544[1] | 203936 | 26 | 130405[1] | 25104 | 27 | 122814[1] | 9447 | 27 |
| 6 | 7 | 247232[1] | 64747 | 31 | 262564[1] | 13614 | 31 | 260654[1] | 32977 | 37 | 257298[1] | 2291 | 37 |

1 - staggered learning technique used. This means that a learning set was extended step by step. Initially first $k$ samples were learned, then starting from the obtained weights $2k$ samples were learned, then $3k$, etc. up to $k^n$ samples in a whole learning set

As we have discovered above (Theorem 5.21), if some $k$-valued input/output mapping is a non-threshold $k$-valued function, but it can be learned using a single MVN-P, this means that this non-threshold $k$-valued function is projected to a partially defined $kl = m$-valued function. In practice, this partially-defined $m$-valued threshold function, which is not known prior to the learning session, is generated by the learning process.

As it follows from our experiments mod $k$ addition of $n$ variables functions for any $k \geq 2$ and any $n$ are projected into partially defined $kl = m$-valued threshold functions, which are so-called minimal monotonic functions.

This means the following. Let $X_i = \left( x_1^i, ..., x_n^i \right)$ and $X_j = \left( x_1^j, ..., x_n^j \right)$. Vector $X_i$ precedes to vector $X_j$ ($X_i \prec X_j$) if $x_s^i \leq x_s^j$, $\forall s = 1, ..., n$. Function $f \left( x_1, ..., x_n \right)$ is called *monotonic* if for any two sets of variables $X_i$ and $X_j$, such that $X^i \prec X^j$, the following holds

$$f \left( X_i \right) = f \left( x_1^i, ..., x_n^i \right) \leq f \left( X_j \right) = f \left( x_1^j, ..., x_n^j \right).$$

An $m$-valued function $f \left( x_1, ..., x_n \right)$ is called *minimal monotonic* [88], if it is monotonic and for any two closest comparable sets of variables $X_i = \left( x_1^i, ..., x_n^i \right)$ and $X_j = \left( x_1^j, ..., x_n^j \right)$, if $X_i \prec X_j$, then $f \left( x_1^j, ..., x_n^j \right) - f \left( x_1^i, ..., x_n^i \right) \leq 1$, that is $f \left( x_1^j, ..., x_n^j \right)$ is either equal to $f \left( x_1^i, ..., x_n^i \right)$ or is greater than $f \left( x_1^i, ..., x_n^i \right)$ by exactly 1.

A very interesting experimental fact is that all partially defined $m$-valued functions, to which mod $k$ additions of $n$ variables were projected by the learning process are minimal monotonic $m$-valued functions. Let us consider several examples for different $k$ and $n$ (see Table 5.19 - Table 5.24).

**Table 5.19** XOR – mod 2 addition of 2 variables, $l=2$, $m=4$ in (5.157)

| $x_1$ | $x_2$ | $f(x_1, x_2) =$ $= (x_1 + x_2) \bmod 2$ | $j \in M = \{0, 1, 2, 3\}$ 2x2=4-valued function $\tilde{f}(x_1, x_2)$ |
|---|---|---|---|
| 0 | 0 | 0 | 0 |
| 0 | 1 | 1 | 1 |
| 1 | 0 | 1 | 1 |
| 1 | 1 | 0 | 2 |

In all these tables, the first columns contain values of the corresponding inputs (input variables) $x_1, ..., x_n$. The second to the last column contains values of the $k$-valued function $f(x_1, ..., x_n)$, which we learn, and the last column contains values of that partially defined $kl=m$-valued function $\tilde{f}(x_1, ..., x_n)$, to which the initial function was projected by the learning algorithm. For simplicity, we show the values of the input variables and functions $f$ and $\tilde{f}$ in the regular multiple-valued alphabets $K = \{0,1,...,k-1\}$ and $M = \{0,1,...,k,...,m-1\}$. The reader may easily convert these values to such that belong to $E_k$ and $E_m$, respectively (if $j \in K$, then $e^{i2\pi j/k} \in E_k$).

**Table 5.20** Parity 3 – mod 2 addition of 3 variables, $l$=3, $m$=6 in (5.157)

| $x_1$ | $x_2$ | $x_3$ | $f(x_1,x_2,x_3) =$ $=(x_1+x_2+x_3)\bmod 2$ | $j \in M = \{0,1,...,5\}$ 2x3=6-valued function $\tilde{f}(x_1,x_2,x_3)$ |
|---|---|---|---|---|
| 0 | 0 | 0 | 0 | 0 |
| 0 | 0 | 1 | 1 | 1 |
| 0 | 1 | 0 | 1 | 1 |
| 0 | 1 | 1 | 0 | 2 |
| 1 | 0 | 0 | 1 | 1 |
| 1 | 0 | 1 | 0 | 2 |
| 1 | 1 | 0 | 0 | 2 |
| 1 | 1 | 1 | 1 | 3 |

**Table 5.21** Parity 4 – mod 2 addition of 4 variables, $l$=3, $m$=6 in (5.157)

| $x_1$ | $x_2$ | $x_3$ | $x_4$ | $f(x_1,x_2,x_3,x_4) =$ $=(x_1+x_2+x_3+x_4)\bmod 2$ | $j \in M = \{0,1,...,7\}$ 2x4=8-valued function $\tilde{f}(x_1,x_2,x_3,x_4)$ |
|---|---|---|---|---|---|
| 0 | 0 | 0 | 0 | 0 | 0 |
| 0 | 0 | 0 | 1 | 1 | 1 |
| 0 | 0 | 1 | 0 | 1 | 1 |
| 0 | 0 | 1 | 1 | 0 | 2 |
| 0 | 1 | 0 | 0 | 1 | 1 |
| 0 | 1 | 0 | 1 | 0 | 2 |
| 0 | 1 | 1 | 0 | 0 | 2 |
| 0 | 1 | 1 | 1 | 1 | 3 |
| 1 | 0 | 0 | 0 | 1 | 1 |
| 1 | 0 | 0 | 1 | 0 | 2 |
| 1 | 0 | 1 | 0 | 0 | 2 |
| 1 | 0 | 1 | 1 | 1 | 3 |
| 1 | 1 | 0 | 0 | 0 | 2 |
| 1 | 1 | 0 | 1 | 1 | 3 |
| 1 | 1 | 1 | 0 | 1 | 3 |
| 1 | 1 | 1 | 1 | 0 | 4 |

**Table 5.22** mod 3 addition of 3 variables, $l=3$, $m=9$ in (5.157)

| $x_1$ | $x_2$ | $x_3$ | $f(x_1,x_2,x_3) =$ $=(x_1+x_2+x_3)\bmod 3$ | $j \in M = \{0,1,...,9\}$ 3x3=9-valued function $\tilde{f}(x_1,x_2,x_3)$ |
|---|---|---|---|---|
| 0 | 0 | 0 | 0 | 21 |
| 0 | 0 | 1 | 1 | 22 |
| 0 | 0 | 2 | 2 | 23 |
| 0 | 1 | 0 | 1 | 22 |
| 0 | 1 | 1 | 2 | 23 |
| 0 | 1 | 2 | 0 | 24 |
| 0 | 2 | 0 | 2 | 23 |
| 0 | 2 | 1 | 0 | 24 |
| 0 | 2 | 2 | 1 | 25 |
| 1 | 0 | 0 | 1 | 22 |
| 1 | 0 | 1 | 2 | 23 |
| 1 | 0 | 2 | 0 | 24 |
| 1 | 1 | 0 | 2 | 23 |
| 1 | 1 | 1 | 0 | 24 |
| 1 | 1 | 2 | 1 | 25 |
| 1 | 2 | 0 | 0 | 24 |
| 1 | 2 | 1 | 1 | 25 |
| 1 | 2 | 2 | 2 | 26 |
| 2 | 0 | 0 | 2 | 23 |
| 2 | 0 | 1 | 0 | 24 |
| 2 | 0 | 2 | 1 | 25 |
| 2 | 1 | 0 | 0 | 24 |
| 2 | 1 | 1 | 1 | 25 |
| 2 | 1 | 2 | 2 | 26 |
| 2 | 2 | 0 | 1 | 25 |
| 2 | 2 | 1 | 2 | 26 |
| 2 | 2 | 2 | 0 | 27 |

As it is clearly seen from all three examples, the corresponding $m$-valued functions are minimal monotonic functions. All these functions can be learned using a single MVN with the activation function (2.50) because they are partially defined $kl=m$-valued threshold functions.

**Table 5.23** mod 5 addition of 2 variables, $l=4$, $m=20$ in (5.157)

| $x_1$ | $x_2$ | $f(x_1,x_2) =$ $=(x_1+x_2)\bmod 5$ | $j\in M=\{0,1,...,9\}$ 5x4=20-valued function $\tilde{f}(x_1,x_2)$ |
|---|---|---|---|
| 0 | 0 | 0 | 0 |
| 0 | 1 | 1 | 1 |
| 0 | 2 | 2 | 2 |
| 0 | 3 | 3 | 3 |
| 0 | 4 | 4 | 4 |
| 1 | 0 | 1 | 1 |
| 1 | 1 | 2 | 2 |
| 1 | 2 | 3 | 3 |
| 1 | 3 | 4 | 4 |
| 1 | 4 | 0 | 5 |
| 2 | 0 | 2 | 2 |
| 2 | 1 | 3 | 3 |
| 2 | 2 | 4 | 4 |
| 2 | 3 | 0 | 5 |
| 2 | 4 | 1 | 6 |
| 3 | 0 | 3 | 3 |
| 3 | 1 | 4 | 4 |
| 3 | 2 | 0 | 5 |
| 3 | 3 | 1 | 6 |
| 3 | 4 | 2 | 7 |
| 4 | 0 | 4 | 4 |
| 4 | 1 | 0 | 5 |
| 4 | 2 | 1 | 6 |
| 4 | 3 | 2 | 7 |
| 4 | 4 | 3 | 8 |

**Table 5.24** mod 6 addition of 2 variables, $l=4$, $m=24$ in (5.157)

| $x_1$ | $x_2$ | $f(x_1, x_2) =$ $= (x_1 + x_2) \bmod 6$ | $j \in M = \{0,1,...,23\}$ 6x4=24-valued function $\tilde{f}(x_1, x_2)$ |
|---|---|---|---|
| 0 | 0 | 0 | 6 |
| 0 | 1 | 1 | 7 |
| 0 | 2 | 2 | 8 |
| 0 | 3 | 3 | 9 |
| 0 | 4 | 4 | 10 |
| 0 | 5 | 5 | 11 |
| 1 | 0 | 1 | 7 |
| 1 | 1 | 2 | 8 |
| 1 | 2 | 3 | 9 |
| 1 | 3 | 4 | 10 |
| 1 | 4 | 5 | 11 |
| 1 | 5 | 0 | 12 |
| 2 | 0 | 2 | 8 |
| 2 | 1 | 3 | 9 |
| 2 | 2 | 4 | 10 |
| 2 | 3 | 5 | 11 |
| 2 | 4 | 0 | 12 |
| 2 | 5 | 1 | 13 |
| 3 | 0 | 3 | 9 |
| 3 | 1 | 4 | 10 |
| 3 | 2 | 5 | 11 |
| 3 | 3 | 0 | 12 |
| 3 | 4 | 1 | 13 |
| 3 | 5 | 2 | 14 |
| 4 | 0 | 4 | 10 |
| 4 | 1 | 5 | 11 |
| 4 | 2 | 0 | 12 |
| 4 | 3 | 1 | 13 |
| 4 | 4 | 2 | 14 |
| 4 | 5 | 3 | 15 |
| 5 | 0 | 5 | 11 |
| 5 | 1 | 0 | 12 |
| 5 | 2 | 1 | 13 |
| 5 | 3 | 2 | 14 |
| 5 | 4 | 3 | 15 |
| 5 | 5 | 4 | 16 |

It should be mentioned that neither of the learning rules (3.92), (3.94)-(3.96) can be distinguished as the "best". Each of them can be good for solving different problems. It is also not possible to distinguish the best among Learning Strategies 1 and 2. For example, the "Iris" problem can be learned with no errors only using the Strategy 2, while for some other problems Strategy 1 gives better results.

A deeper study of advantages and disadvantages of the developed learning strategies and rules will be an interesting direction for the further research.

## 5.4  Concluding Remarks to Chapter 5

In this Chapter, we have considered the multi-valued neuron with a periodic activation function (MVN-P). This is a discrete-valued neuron (whose inputs can be discrete or continuous), which can learn input/output mappings that are non-linearly separable in the real domain, but become linearly separable in the complex domain.

First, we have considered the universal binary neuron (UBN). This neuron with a binary output was a prototype of MVN-P. We have shown that UBN with its periodic activation function projects a binary input/output mapping, which is non-linearly separable, into an $2l = m$-valued partially defined threshold function, which can be learned using a single neuron. We have considered the UBN learning algorithm, which is based on the MVN learning algorithm employing also self-adaptivity. We have shown that a single UBN may easily learn such problems as XOR and Parity.

Then we have introduced MVN-P. The MVN-P concept generalizes the two-valued UBN concept for the $k$-valued case. The MVN-P has a periodic activation function, which projects $k$-valued logic into $kl = m$-valued logic. Thus, this activation function is $k$-periodic and $l$-repetitive. The most wonderful property of MVN-P is its ability to learn $k$-valued input/output mappings, which are non-linearly separable in $k$-valued logic, but become linearly separable in $kl = m$-valued logic. This means that MVN-P may project a non-linearly separable $k$-valued function into a partially defined linearly separable $m$-valued function.

We have considered the MVN-P learning algorithm with the two learning strategies. This learning algorithm is semi-supervised and semi-self-adaptive. While a desired neuron output is known in advance, a periodicity of its activation function allows its self-adaptation to the input/output mapping. We have shown that the MVN-P learning can be based on the same error-correction learning rules that the regular MVN learning. The most important application of MVN-P is its ability to solve multi-class and multi-cluster classification problems, which are non-linearly separable in the real domain, without any extension of the initial space where a problem is defined.

MVN-P can also be used as the output neuron in MLMVN. This may help to solve highly nonlinear classification problems.

# Chapter 6
# Applications of MVN and MLMVN

"The scientist is not a person who gives the right answers, he is one who asks the right questions."

Claude Lévi-Strauss

In this Chapter, we will consider some applications of MVN and MLMVN. In Chapters 2-5 we have introduced MVN, we have deeply considered all aspects of its learning, we have also introduced MLMVN and its derivative-free backpropagation learning algorithm; finally we have introduced MVN-P, the multi-valued neuron with a periodic activation function and its learning algorithm. We have illustrated all fundamental considerations by a number of examples. Mostly we have considered so far how MVN, MVN-P, and MLMVN solve some popular benchmark problems. It is a time now to consider some other applications including some real-world applications. In Section 6.1, we will consider how MLMVN can be used for solving a problem of blur and its parameters identification, which is of crucial importance in image deblurring. In Section 6.2, we will show how MLMVN can be used for solving financial time series prediction problems. In Section 6.3, we will consider how MVN can successfully be used in associative memories. Some other MVN applications will be observed and some concluding remarks will be given in Section 6.4.

## 6.1  Identification of Blur and Its Parameters Using MLMVN

Identification of a blur mathematical model and its parameters is very important for restoration of blurred images. We would like to present here how MLMVN can be successfully used for solving this important problem. The results, which will be presented here, were obtained by the author of this book together with Dmitriy Paliy, Jacek Zurada, and Jaakko Astola and were published in their paper [77]. Thus, this Section will be mostly based on the paper [77], with some additional details and illustrations. Since there are different blur models, which may have different parameters, their identification is a multi-class classification problem, and it is very interesting to compare how this problem is solved by MLMVN and other machine learning techniques.

I. Aizenberg: Complex-Valued Neural Networks with Multi-Valued Neurons, SCI 353, pp. 207–248.
springerlink.com                                    © Springer-Verlag Berlin Heidelberg 2011

### 6.1.1  Importance of Blur and Its Parameters Identification

Let us first clarify why it is so important for image deblurring to be able to identify a mathematical model of blur and its parameters.

Usually blur is treated as the low-pass distortions introduced into an image.
It can be caused by many different reasons: by the relative motion between the
camera and the original scene, by the optical system which is out of focus, by at-
mospheric turbulence (optical satellite imaging), aberrations in the optical system,
etc. [96]. Any type of blur, which is spatially invariant, can be expressed by the
convolution kernel in the integral equation [96]. Hence, deblurring (restoration) of
a blurred image is an ill-posed inverse problem, and regularization is commonly
used when solving this problem [97].

Mathematically, a variety of image capturing principles can be modelled by the

Fredholm integral of the first kind in $\mathbb{R}^2$ space $z(t) = \int_X v(t,l)q(l)dl$ where

$t, l \in X \subset \mathbb{R}^2$, $v$ is a point-spread function (PSF) of a system, $q$ is an image
intensity function, and $z(t)$ is an observed image [98]. PSF determines what happens with a point after an image is captured by an optical system. In other words,
PSF describes how the point source of light is spread over the image plane. It is
one of the main characteristics of the optical system. The more spread is a point,
the more blurred is an image. PSF determines a specific type of this spreading, and
determines a type of blur, respectively.

A natural simplification is that the PSF $v$ is shift-invariant which leads to a
convolution operation in the observation model. We assume that the convolution
is discrete and noise is present. Hence, the observed image $z$ is given in the following form:

$$z(t) = (v * q)(t) + \varepsilon(t), \qquad (6.160)$$

where $"*"$ denotes the convolution, $t$ is defined on the regular $L_x \times L_y$ lattice,

$t \in T = \{(x,y); x = 0,1,...,L_x - 1; y = 0,1,...,L_y - 1\}$, and $\varepsilon(t)$ is the noise.

It is assumed that the noise is white Gaussian with zero-mean and variance $\sigma^2$,

$\varepsilon(t) \sim N(0,\sigma^2)$. In the 2D frequency domain the model (6.160) takes the form:

$$Z(\omega) = V(\omega)Q(\omega) + \varepsilon(\omega), \qquad (6.161)$$

where $Z(\omega) = F\{z(t)\}$ is a representation of a signal $z$ in a Fourier domain
and $F\{\cdot\}$ is a discrete Fourier transform, $V(\omega) = F\{v(t)\}$,
$Q(\omega) = F\{q(t)\}$, $\varepsilon(\omega) = F\{\varepsilon(t)\}$, and $\omega \in W$,

$W = \left\{ (\omega_x, \omega_y); \ \omega_i = 2\pi k_i / L_i, \quad k_i = 0, 1, ..., L_i - 1, \quad i = x, y \right\}$ is the normalized 2D frequency.

Image restoration is a process of the removal of the degradation caused by PSF. Mathematically this is an inverse problem. It is usually referred to as a deconvolution [98]. Usually this problem is ill-posed and it results in the instability of a solution. It is also sensitive to noise. The stability can be provided by constraints imposed on the solution. A general approach to such problems refers to the methods of Lagrange multipliers and the Tikhonov regularization [97]. The regularized inverse filter can be obtained as a solution of the least square problem with a penalty term:

$$ J = \left\| Z - VQ \right\|_2^2 + \alpha \left\| Q \right\|_2^2, \tag{6.162} $$

where $\alpha \geq 0$ is a regularization parameter and $\left\| \cdot \right\|_2$ denotes $l^2 -$ norm. Here, the first term $\left\| Z - VQ \right\|_2^2$ gives the fidelity to the available data $Z$ and the second term bounds the power of this estimate by means of the regularization parameter $\alpha$. In (6.162), and further, we omit the argument $\omega$ in the Fourier transform variables. The solution is usually obtained in the following form by minimizing (6.162):

$$ \hat{Q} = \frac{\overline{V}}{|V|^2 + \alpha} Z, \quad \hat{q}_\alpha(x) = F^{-1}\{\hat{Q}\}, \tag{6.163} $$

where $\hat{Q}$ is an estimate of $Q$, $\hat{q}_\alpha$ is an estimate of the true image $q$, and $\overline{V}$ is a complex-conjugate value of $V$.

There exist a variety of sophisticated and efficient deblurring techniques such as deconvolution based on the Wiener filter [96, 99], nonparametric image deblurring using local polynomial approximation with spatially-adaptive scale selection based on the intersection of confidence intervals rule [99], Fourier-wavelet regularized deconvolution [100], etc. A common and very important property of all these techniques is that they assume a prior knowledge of a blurring kernel, which is completely determined by a point spread function (PSF), and its parameter (parameters). If PSF is unknown, regularization based on (6.162) and (6.163) cannot be used to solve the deblurring problem. Hence, to apply any regularization technique, it is very important to reconstruct PSF, to create its mathematical model, which is as close as it is possible to the reality. This reconstruction can be based on the recognition (identification) of the mathematical model of PSF and its parameter (parameters). When a PSF model and its parameters are recognized, PSF can be easily reconstructed and then a regularization technique can be used for image restoration. To solve this recognition problem, some machine learning technique should be used. We would like to show, how this problem can be solved using MLMVN.

It should also be mentioned that when PSF is unknown, the image deblurring becomes a blind deconvolution problem [101]. Most of the methods to solve it are iterative, and, therefore, computationally costly. Due to the presence of noise, they suffer from instability and convergence problems. So it should be more efficient to use regularization. In turn, to use it, PSF must be identified first.

## 6.1.2  Specification of the PSF Recognition Problem

We will use MLMVN to recognize perhaps the most popular PSFs: Gaussian, motion, and rectangular (boxcar) blurs. We aim to identify simultaneously both the blur, which is characterized by PSF, and its parameter (parameters). Let us consider all these PSF models and show how they depend on the corresponding parameters.

For a variety of devices, like photo or video camera, microscope, telescope, etc., and for turbulence of the atmosphere, PSFs are often approximated by the Gaussian function:

$$v(t) = \frac{1}{2\pi\tau^2} \exp\left(-\frac{x^2 + y^2}{\tau^2}\right) \tag{6.164}$$

where the variance $\tau^2$ is a parameter of the PSF (Fig. 6.54a). Its Fourier transform $V$ is also a Gaussian function and its absolute values $|V|$ are shown in Fig. 6.54d.

Another source of blur is a uniform linear motion which occurs while taking a picture of a moving object relatively to the camera:

$$v(t) = \begin{cases} \dfrac{1}{h}, & \sqrt{x^2 + y^2} < h/2, \ x\cos\phi = y\sin\phi, \\ 0, & \text{otherwise,} \end{cases} \tag{6.165}$$

where $h$ is a parameter, which depends on the velocity of the moving object and describes the length of motion in pixels, and $\phi$ is the angle between the motion orientation and the horizontal axis. Any uniform function like (6.165) (Fig. 6.54b) is characterized by the number of slopes in the frequency domain (Fig. 6.54e).

The uniform rectangular blur is described by the following function (Fig. 6.54c):

$$v(t) = \begin{cases} \dfrac{1}{h^2}, & |x| < \dfrac{h}{2}, \ |y| < \dfrac{h}{2}, \\ 0, & \text{otherwise,} \end{cases} \tag{6.166}$$

where parameter $h$ defines the size of smoothing area. The frequency characteristic of (6.166) is shown in Fig. 6.54f.

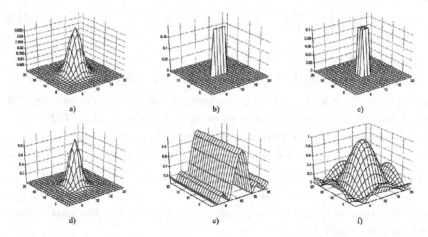

**Fig. 6.54** Types of PSF : a) Gaussian PSF with $\tau = 2$ and size $21 \times 21$ ; b) Linear uniform motion blur of the length 5; c) Boxcar blur of the size $3 \times 3$ ; d) frequency characteristics of a); e) frequency characteristics of b); f) frequency characteristics of c).

It is practically impossible to recognize PSF in spatial domain. It follows from Fig. 6.54 that it is much strongly marked in the frequency domain. Thus, our task is to recognize PSF definitely in the frequency domain using magnitude of the Fourier transform as a feature space. Then, after PSF and its parameters are recognized, the PSF $v$ can be reconstructed from (6.164)-(6.166) (depending on which model of blur was identified), then its Fourier transform $V$ can be found, and then

$\hat{Q}$ , which is the approximation of the Fourier transform of the original image $q$ (see (6.160) ) can be obtained from (6.163).

### 6.1.3   Choice of the Features

So, the observed image $z(t)$ is modeled as the output of a linear shift-invariant system (6.160) which is characterized by PSF. Those PSFs, which are considered here, are determined by (6.164)-(6.166). As we have realized, each of them has its own very specific characteristics in the frequency domain.

Hence it is natural to use magnitudes of their Fourier spectral coefficients as features for classification. Since originally the observation is not $v$ ($v$ is not known) but $z$, we use spectral coefficients of $z$ to form input training (and testing) vectors in order to identify the PSF $v$. Since this model in the frequency domain is the product of spectra of the true object $Q$ and the PSF $V$, we state the problem as recognition of the shape of $V$ and its parameters from the power-spectral density (PSD) of the observation $Z$, i.e. from $|Z|^2 = Z \cdot \overline{Z}$ , which in terms of statistical expectation can be rewritten as:

$$E\left\{|Z|^2\right\} = E\left\{|QV + \varepsilon|^2\right\} = |Q|^2 |V|^2 + \sigma^2 \tag{6.167}$$

where $\sigma^2$ is the variance of noise in (6.161).

Thus we will use here $\log|Z|$ to form pattern vectors $X$ whose membership we have to recognize. Our classes are determined not only by the corresponding PSF model, but also by certain values of its parameters ($\tau^2$ for the Gaussian PSF (6.164), and $h$ for the motion PSF (6.165) and the rectangular PSF (6.166), respectively).

Examples of $\log|Z|$ values are shown in Fig. 6.55. It is just necessary to take into account that distortions in the Fourier transform magnitude caused by blur are so clear just if there is no noise on an image. For noisy images (even when there is some very slight noise added) this picture becomes hidden, and blur model cannot be identified visually.

The distortions of PSD for the test image Cameraman (Fig. 6.55a) that are typical for each type of blur (Fig. 6.55b,c) are clearly visible in Fig. 6.55e,f (just take into account that DC (zero-frequency) in Fig. 6.55d,e,f is located in the top-left corner, not in the center. High frequencies are located closer to the center, respectively.

For simplicity (but without loss of generality) we may consider just square images $z(t)$, i.e. $L = L_x = L_y$ in (6.160), and (6.161). In order to obtain the pattern vector $X = (x_1,...,x_n)$ where $x_i; i = 1,...,n$ is located on the unit circle (to be able to use these pattern vectors as MVN and MLMVN inputs), we perform the following transformation. Taking into account that the PSF $v$ is symmetrical, PSD of $z(t)$ (6.167) is used as follows to obtain $x_j; j = 1,...,n$:

$$x_j = \exp\left(2\pi i \frac{\log\left(\left|Z\left(\omega_{k_1,k_2}\right)\right|\right) - \log\left(|Z_{\min}|\right)}{\log\left(|Z_{\max}|\right) - \log\left(|Z_{\min}|\right)}\right), \tag{6.168}$$

where

$$\begin{cases} j = 1,...,L/2-1, & \text{for } k_1 = k_2, \ k_2 = 1,...,L/2-1, \\ j = L/2,...,L-2, & \text{for } k_1 = 1, \ k_2 = 1,...,L/2-1, \\ j = L-1,...,3L/2-3, & \text{for } k_2 = 1, \ k_1 = 1,...,L/2-1, \end{cases} \tag{6.169}$$

$i$ is an imaginary unity, and $Z_{\max} = \max_{k_1,k_2}\left(Z\left(\omega_{k_1,k_2}\right)\right)$, $Z_{\min} = \min_{k_1,k_2}\left(Z\left(\omega_{k_1,k_2}\right)\right)$.

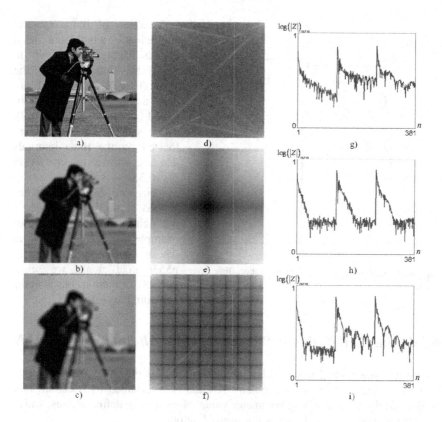

**Fig. 6.55** True test *Cameraman* image (a) blurred by: b) Gaussian blur with $\tau = 2$ ;
c) rectangular blur of the size $9 \times 9$ ;
log |Z| of the true test *Cameraman* image (d) blurred by: e) Gaussian blur with $\tau = 2$ ;
f) rectangular blur of the size $9 \times 9$ ;
The normalized log |Z| values used as arguments to generate training vectors in (6.168) and
(6.169) obtained from the true test *Cameraman* image (g) blurred by:
h) Gaussian blur with $\tau = 2$ ; i) rectangular blur of the size $9 \times 9$

Eventually, the number of the features and the length of the pattern vector, respectively was chosen as $n = 3L / 2 - 3$ .

Some examples of the values $\log|Z|$ normalized to be in the range $[0,1]$ as follows

$$\log\left(|Z|\right)_{norm} = \frac{\log\left(\left|Z\left(\omega_{k_1,k_2}\right)\right|\right) - \log\left(|Z_{\min}|\right)}{\log\left(|Z_{\max}|\right) - \log\left(|Z_{\min}|\right)},$$

are shown in Fig. 6.55g,h,i. They are used in (6.168) and (6.169) to produce the pattern vectors $X$.

$\omega(0,0)$

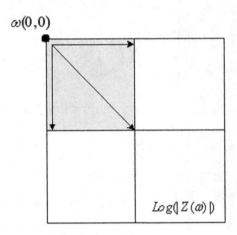

$Log(|Z(\omega)|)$

**Fig. 6.56** Selection of the features according to (6.169)

Selection of those spectral coefficients, which are used as the features, is shown in Fig. 6.56. Coordinates of the spectral coefficients, which are used as the features, are obtained according to (6.169). These spectral coefficients are located on the three intervals that are shown by arrows in Fig. 6.56.

Thus, we use just a very limited number of the spectral coefficients as the features. For example, for 256 x 256 images whose Fourier spectra contain 65536 coefficients, we select just 381 of them as our features according to (6.169).

### 6.1.4   Output Definition: "Winner Takes It All Rule" for MLMVN

As we have mentioned from the beginning, blur and its parameters identification is a multi-class classification problem. A class is determined by a certain blur model and the corresponding parameter value. Now we can define classes, which we want to recognize, and specify our desired outputs.

We are going to recognize the following six types of blur with the following parameters:

1) the      Gaussian      blur      is      considered      with
   $\tau \in \{1,\ 1.33,\ 1.66,\ 2,\ 2.33,\ 2.66,\ 3\}$ in (6.164);

2) the linear uniform horizontal motion blur of the lengths 3, 5, 7, 9 ($\phi = 0$) in (6.165);

3) the linear uniform vertical motion blur of the length 3, 5, 7, 9 ($\phi = 90$ degrees) in (6.165);

4) the linear uniform diagonal motion from South-West to North-East blur of the lengths 3, 5, 7, 9 ($\phi = 45$ degrees) in (6.165);

5) the linear uniform diagonal motion from South-East to North-West blur of the lengths 3, 5, 7, 9($\phi = 135$ degrees), in (6.165);

6) rectangular has sizes 3x3, 5x5, 7x7, and 9x9 in (6.166).

Hence, for the Gaussian blur we have 7 classes (determined by 7 parameter values), for the linear uniform horizontal, vertical, and two diagonal motion blurs we have 4 classes for each of them, and for the rectangular blur we also have 4 classes. Thus, there are 27 "blurred" classes in total. One more class has to be re-served for unblurred images. Hence, finally we have 28 classes.

Taking into account that we have six different blur models, for their identifica-tion we use six neurons in the MLMVN output layer. Each neuron is associated with a certain blur model. Since we want to recognize not only a blur model, but also its parameter, each of these six output neurons has to solve multi-class classi-fication problem. For each blur model we consider different parameter values, and each of these values determines a separate class.

Thus, on the one hand, each MVN in the output layer has to classify a parame-ter of the corresponding blur model, and, on the other hand, it has to reject other blurs, as well as unblurred images. Hence, if the $i$th type of blur is characterized by $p$ parameters, then the $m$th output neuron $N_m; m = 1,...,6$ (that one associated with this type of blur) has to work in $p+1=k$-valued logic. For instance, a neuron, which is asso-ciated with the Gaus-sian blur, has to work in 8-valued logic. There are 7 values of the Gaussian blur pa-rameter $\tau$, which de-termine 7 classes, and

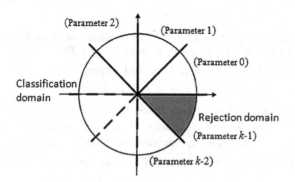

Fig. 6.57 Reservation of the $k$ domains in the $k$-valued acti-vation function of the output MVN

the 8<sup>th</sup> class label is reserved for the rejection and unblurred images. Since for other blurs we consider 4 possible parameter values, other 5 output layer MVNs have to work in 5-valued logic (4 values are reserved for the class labels corre-sponding to the parameter values and the 5<sup>th</sup> value is reserved for the rejection and unblurred images). This is illustrated in Fig. 6.57. Each of the six output neurons is the discrete MVN with the $k$-valued activation function (2.50), which divides the complex plane into $k$ equal sectors. The first $k$-1 sectors (0, 1, ..., $k$-2) form the Classification domain, where each sector corresponds to the certain value of the parameter of the corresponding blur model. The last sector ($k$-1<sup>st</sup>) is reserved for the Rejection domain. The desired output for the neuron $N_m; m = 1,...,6$ is formed in the following way. If the input pattern $X$ corresponds to the $m$th type of blur and $j$th value of the parameter ($j$=0, 1, ..., $k$-2), then the desired output is $e^{i2\pi j/k}$ (the desired domain is the $j$th sector). If the input pattern $X$ corresponds to any different type of blur, not associated with the neuron $N_m$, or to the absence of any blur, then the desired output is $e^{i2\pi(k-1)/k}$ (the desired domain is the last ($k$-1<sup>st</sup>) sector).

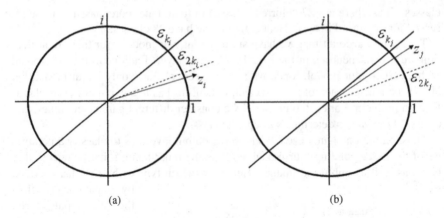

**Fig. 6.58** The weighted sum of the $i$th output neuron (fig. (a) ) is located closer to the bisector of the sector than the weighted sum of the $j$th output neuron (fig. (b) ). This means that the $i$th neuron $N_i$ is the winner, and that its output is the network output

To determine the network output, the "winner takes it all" technique is used. It is implemented for MLMVN in the following way. Ideally, only one out of $N$ output neurons should recognize a test sample. Suppose, this is the $i$th output neuron $N_i$. This means that only this neuron's weighted sum appears in one of the sectors $0,1,...,k_i-2$ ( $k_i$ is the value of logic, in which the neuron $N_i$ works), while all the other neurons' weighted sums are expected to appear in the sector $k_j-1; j=1,...,i-1,i+1,...,N$. The latter means that all the other neurons reject the corresponding test sample. This is a classical interpretation of the "winner takes it all" technique for MLMVN. However, this ideal situation may not be the case, because often more than one output neuron can classify a test sample as "its". This means that for more than one output neuron the weighted sum appears in one of the sectors $0,1,...,k_j-2; j=1,...,N$. The network output cannot be ambiguous, what can we do in this situation? We should use some additional criterion to determine the network output. Since the most reliable output should be that one, which is as distant as it is possible from the sector borders in terms of angular distance, that output neuron (out of those whose weighted sum appears in one of the sectors $0,1,...,k_j-2; j=1,...,N$ ) will be a winner whose weighted sum is as close as possible to the bisector of that sector where it is located. This means that this weighted sum is located as far as it possible from the sector's border. This is illustrated in Fig. 6.58. The weighted sum $z_i$ of the $i$th output neuron $N_i$ (Fig. 6.58a) is located closer to the bisector of the sector than the weighted sum $z_j$ of the $j$th output neuron $N_j$ (Fig. 6.58b). This means that the $i$th neuron $N_i$

is the winner, and that its output is the network output. If $k_i \neq k_j$, which means that the output neurons $N_i$ and $N_j$ work in different valued logic and sectors on the complex plane have different angular size for these neurons, respectively, a normalized angular distance from the current weighted sum to the bisector of the sector where it is located should be measured. To normalize this angular distance, its absolute value should be divided by the angular size of the sector $\dfrac{\left|\arg z - (2s+1)\arg \varepsilon_{2k}\right|}{2\pi / k}$ if the weighted sum is located in the $s$th sectror ($\arg \varepsilon_{2k}$ is the angle corresponding to the bisector of the $0^{th}$ sector).

It is important to mention that the "winner takes it all" technique should always be utilized for MLMVN with multiple output neurons in the same way, using the additional "angular" criterion for the network output formation, which we have just described. Of course, this approach works not only for solving the blur identification problem, but wherever MLMVN with multiple output neurons is used for solving any classification problem.

## 6.1.5  Simulation Results

To test the MLMVN's ability to identify a blur model and its parameters, we have created a database from 150 grayscale images with sizes 256x256. This database was split into the learning and testing sets. 100 images are used to generate the learning set and 50 other images are used to generate the testing set. Since we consider 28 classes (27 "blurred" classes formed by six types of blur, five of them with the four parameter values and one with the seven parameter values, along with the class of clean images - 5•4+7+1=28), the training set consists of 2800=28•100 pattern vectors, and the testing set consists of 1400=28•50 pattern vectors. The white Gaussian noise was added to all images. The level of noise in (6.160) is selected satisfying BSNR (the blurred signal-to-noise-ratio[1]) to be equal to 40 dB.

---

[1] The blurred signal-to-noise ratio is defined as $BSNR = 10\log_{10}\dfrac{\left(f - \tilde{f}\right)^2}{M\sigma^2}$, where $f$ is a blurred image without noise, $\tilde{f}$ is the expectation of the clean image, $M$ is the number of pixels, and $\sigma^2$ is the noise variance. To estimate the quality of the restoration, the improved signal-to-noise-ratio $ISNR = 10\log_{10}\dfrac{\left(f - g\right)^2}{\left(f - \hat{f}\right)^2}$ (where $f$ is an original image, $g$ is a degraded image, and $\hat{f}$ is a restored image) is used.

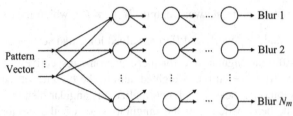

As we have calcu-
lated above, accord-
ing to (6.169) for
256x256 images, our
pattern vectors con-
tain 381 components.
Hence, our network
has 381 inputs. Ex-
perimentally, we
have found that the
best results are ob-

**Fig. 6.59** MLMVN used for blur model and its parameter
identification

tained with a network containing two hidden layers, 5 neurons in the first hidden
layer, and 35 neurons in the second hidden layer. As we agreed above, our network has to contain 6 neurons in the output layer (each neuron is responsible for a
certain blur model). Thus, the topology of MLMVN, which we used, is
381-5-35-6. A network with fewer hidden neurons needs more iterations for learning, while a network with more hidden neurons does not improve the results.

To evaluate the quality of classification, the classification rate

$$CR = \frac{N_{correct}}{N_{total}} \cdot 100\% ,$$

where $N_{total}$ is a total number of pattern vectors $X$ in the testing set, and $N_{correct}$
is a number pattern vectors correctly classified by the trained network, is used. It
expresses the number of correct classifications in terms of percentage (%).

For a comparison, we have chosen classical MLF with backpropagation learning algorithm and SVM. Fletcher-Reeves Conjugate Gradient [102] and Scaled
Conjugate Gradient [103] backpropagation training algorithms were used for MLF
training. They are implemented in MATLAB Neural Networks Toolbox. In order
to draw objective comparison, we used MLF with exactly the same topology as
MLMVN (381-5-35-6). The hidden layer neurons have tan-sigmoid activation
function, and the output layer neurons have linear activation function. For classification, the range of values [−1,1] of the output neurons activation function is
divided into intervals according to the number of classes associated with the corresponding neuron). Additionally, to avoid possible speculations on the complexity
of complex-valued and real-valued neural networks (a "skeptical" view is that a
complex-valued network employs twice more parameters[2]), we also used MLF
with a "doubled" number of hidden neurons 381-10-70-6.

For SVM modeling we used MATLAB code available in [104]. SVM-based
classification was performed following the well-developed strategy "one against
all" to use a combination of several binary SVM classifiers to solve a given multi-
class problem [105]. Since we have 27 "blurred" classes, the ensemble of the 27
binary decision SVMs is used, respectively. Each of these 27 SVMs recognizes a
certain blur with its parameter and rejects all other blurs along with unblurred images. Thus, a pattern rejected by all the SVMs considered unblurred.

---

[2] In fact, it is not possible to agree with this "skeptical" view, because mathematically a
complex number is a single number, not a pair of real numbers

The following results were obtained. MLF with both topologies and both learning algorithms failed to learn the problem with some acceptable accuracy. MLF could not drop the learning error below 5%, but it is even more important that its classification rate could not exceed 90%, which is significantly lower than the one for MLMVN and SVM. The results for MLMVN and SVM are shown in Table 6.25.

**Table 6.25** Classification results for blur identification

| BLUR | Classification Rate | |
| --- | --- | --- |
| | **MLMVN** | **SVM** |
| No blur | 96% | **100%** |
| Gaussian | 99% | **99.4%** |
| Rectangular | **98%** | 96.4% |
| Motion Horizontal | **98.5%** | 96.4% |
| Motion Vertical | **98.3%** | 96.4% |
| Motion Notrh-East Diagonal | **97.9%** | 96.5% |
| Motion North-West Diagonal | **97.2%** | 96.5% |

(a)  (b)

**Fig. 6.60** (a) Test noisy blurred Cameraman image with 9x9 rectangular blur (b) The blurred image reconstructed using the regularization technique [99] after the blur and its parameter have been identified using MLMVN (ISNR=3.88 dB)

The results are comparable with each other. Both tools, MLMVN and SVM, were trained with the zero error. Both demonstrate the excellent classification rate. Five out of seven classes are a little bit better classified by MLMVN. Clean images and images degraded by the Gaussian blur are a little better classified by SVM. However, the complexity of the SVM ensemble, which is used for solving this multi-class problem, is significantly higher than the one of the MLMVN! Each SVM uses approximately 2,500•381=952,500 support vectors, so there are about 952,500•28 = 25,717,500 support vectors in total, while the MLMVN uses (381+1)•5+35•(5+1)+6•(35+1)=2336 weights in total.

The results of using the MLMVN classification for image reconstruction are shown in Fig. 1.7 for the test Cameraman image. The adaptive deconvolution technique proposed in [99] has been used after the blur and its parameter identified[3]. The image was blurred by the rectangular blur (6.166) 9x9. Classified PSF coincides with the true PSF, and the value of improved signal-to-noise ratio (ISNR) criterion is 3.88 dB.

Thus, we see that MLMVN can be successfully used for solving a challenging and sophisticated real-world multi-class classification problem of blur and its parameters identification. It is interesting that the result shown by MLMVN is significantly better than the one shown by MLF. The result shown by MLMVN is also slightly better than the one shown by SVM. However, the complexity of the MLMVN in terms of the number of parameters employed is significantly smaller than of the one of SVM!

## 6.2  Financial Time Series Prediction Using MLMVN

### 6.2.1  Mathematical Model

We have already considered in Section 4.4 how MLMVN predicts time series. However, we have limited our considerations there just by the famous benchmark problem – prediction of the artificially generated Mackey-Glass chaotic time series. As we remember, MLMVN has shown very good results, outperforming other techniques.

However, time series prediction often appears in real-world applications. One of these applications is prediction of main stock indexes, stock prices, currency exchange rates, etc. We have to say from the beginning that we will not study here deeply time series, methods of their analysis and prediction. We should address the reader interested in this area, for example, to the fundamental monograph [106] by G. Box, G. Jenkins, and G. Reinsel. We also will not consider here examples of time series prediction from very different areas, which are usually offered to the participants of time series prediction competitions usually timed to neural networks conferences[4]. We just would like to show here how MLMVN can be successfully used for a long-term time series prediction.

According to the definition given, for example, in [106], a *time series* is a sequence of data points, measured typically at successive times spaced at uniform time intervals. Thus, mathematically, if $t_0, t_1, ..., t_n, ...$ are the time slots such that $\forall i = 0, 1, ... \ t_{i+1} - t_i = \Delta t = const$, and $x_0, x_1, ..., x_n, ...$ are data points measured at the time slots $t_0, t_1, ..., t_n, ...$, then we say that $x_0 = x(t_0), x_1 = x(t_1), ..., x_n = x(t_n), ...$ is a time series.

---

[3] MATLAB code for this technique is available following the link
    http://www.cs.tut.fi/~lasip/.

[4] See, for example http://www.neural-forecasting-competition.com/

Among many other time series there is a group of them, which is commonly referred to as financial time series. There are series of historical data of main stock
indexes, stock prices, and currency exchange rates usually measured at the moment when the corresponding market closes. A very popular problem is prediction
of such time series. *Time series prediction* in general is prediction of future members in the corresponding series based on other members known from historical
data. There are many different methods of time series analysis and their prediction, which are out of scope of this book. We just have to mention that any method
of prediction is based on some certain mathematical model, which is used for
modeling of the corresponding time series.

We will use here perhaps the simplest, but natural model, which is very suitable
for MLMVN. So let $x_0 = x(t_0), x_1 = x(t_1), ..., x_n = x(t_n),...$ be a time series.
Let us suppose that there exist some functional dependence among the series
members, according to which the $n+1^{st}$ member's value is a function of the preceding $n$ members' values

$$x_n = f\left(x_0, ..., x_{n-1}\right)$$
$$x_{n+1} = f\left(x_1, ..., x_n\right),$$
$$x_{n+2} = f\left(x_2, ..., x_{n+1}\right), \qquad (6.170)$$
$$...$$
$$x_{n+j} = f\left(x_j, ..., x_{n+j-1}\right).$$

This model can be used for prediction in the following way. Suppose we have
historical data for some time series $x_0, x_1, ..., x_{n-1}, x_n, x_{n+1}, ..., x_r$. Suppose that
there exist the functional dependence (6.170) among the time series members. Our
task is to predict the following members of the series, that is $x_{r+1}, x_{r+2}, ...$, which
are not known. How we can solve this problem? According to our assumption,
(6.170) holds for our time series, but $f$ is not known. However, we can approach
this function using some machine learning tool. This means that we have to from a
learning set from the known series members. Since the first $r$ members of the time
series are known, and according to (6.170) each following member is a function of
the preceding $n$ members, our learning set should contain the following learning
samples and desired outputs, respectively:

$$\left(x_0, ..., x_{n-1}\right) \to x_n,$$
$$\left(x_1, ..., x_n\right) \to x_{n+1},$$
$$\left(x_2, ..., x_{n+1}\right) \to x_{n+2},$$
$$...$$
$$\left(x_{r-n}, ..., x_{r-1}\right) \to x_r.$$

As soon as the learning process is completed, $f$ can be implemented as its approximation $\hat{f}$, which is resulted from the learning process, and future members of the time series can be predicted as follows

$$\hat{x}_{r+1} = \hat{f}\left(x_{r-n+1},...,x_r\right)$$

$$\hat{x}_{r+2} = \hat{f}\left(x_{r-n+2},...,x_r,\hat{x}_{r+1}\right),$$

$$\hat{x}_{r+3} = \hat{f}\left(x_{r-n+3},...,x_r,\hat{x}_{r+1},\hat{x}_{r+2}\right),$$ (6.171)

...

The heat sign in (6.171) above a series member means that the corresponding value is not a true value, but the predicted (estimated) value of the future member of the time series. It is important to mention that the predicting rule (6.171) differs from the one we used for Mackey-Glass time series prediction where only true values were used to predict future values. The latter is a commonly used approach in time series modeling, but it becomes not applicable, for example, to financial time series. Indeed, if we know just $r$ series members, and we want to predict the following members, only the $r+1^{st}$ member can be predicted from the true historically known members. To predict the $r+2^{nd}$ member, according to (6.170) and (6.171) we have to base this prediction on the values of $n-1$ true members $x_{r-n+2},...,x_r$, and on the just predicted value $\hat{x}_{r+1}$. In general, to predict each following member of the series, every time we have to base this prediction on a one more predicted member.

The question is: for how long it is possible to predict a time series in this way, particularly a financial time series? We would like to show here that using MLMVN it is possible to make realistic long-term predictions, up to one year ahead.

All financial time series are unsteady, they are usually characterized by high volatility. This means that a virtual function (6.170) is highly nonlinear and contains many small and possibly high jumps. To approximate such a function with a high accuracy using a feedforward neural network, we should use a network with two hidden layers. A network with just a single hidden layer will be able just to build a much smoothed approximation where all the mentioned jumps will be averaged. As a result, just very global long-term trends could be followed. Hence, we should use MLMVN with two hidden layers.

To make long-term predictions, we should also use a long array of historical data for the training purposes. Only learning many different trends during a long period of time, it is possible to make long-term predictions. Intuitively, we followed the next rule: to make prediction for a year ahead, it is necessary to learn the preceding two-year history.

Taking into account that the behavior of any financial market should depend on the preceding events that occurred not only yesterday, day before yesterday, or even a month ago, we have chosen $n=250$ in (6.170) (250 is the approximate number of business days during a year). This means that according to our model the value of each member of a time series depends on the values of the preceding 250 members.

Thus our task is to learn an input/output mapping described by a continuous function of 250 variables. This means that our network has 250 inputs. Since we want to learn a two year history and as we told, there are approximately 250 business days in a year, our learning set has to contain 500 samples, so $r=500$ in (6.171).

## 6.2.2 Implementation and Simulation Results

A simple empirical consideration shows that having 250 inputs, we should try first to reduce the space where our problem is considered, therefore the first hidden layer should contain a very small amount of neurons. But then, this space should be significantly extended, to ensure that we will follow all possible small jumps of our input/output mapping. Thus, we used MLMVN with the topology 250-2-32768-1 (250 inputs, 2 neurons in the first hidden layer, 32768 neurons in the second hidden layer and a single output neuron). The number 32768 is determined just by two reasons. The first reason is, as we told, the necessity to have as much as possible neurons in the second hidden layer, to follow all jumps of a function to be learned. That MLMVN software simulator, which we used, allows using up to 33000 neurons in a layer. The choice of the number 32768 was finally determined by its "beauty" as it is a power of 2.

To transform time series values into numbers located on the unit circle (to be able to use them as MLMVN inputs and outputs), transformation (2.53) (see p. 64) was used. We adapted this transformation as follows

$$y_j \in [a,b] \Rightarrow \varphi_j = \frac{y_j - a}{b-a} 2\pi \in [0, 2\pi[ \, ;$$

$$x_j = e^{i\varphi_j} \, ;$$

$$j = 0, \dots, n,$$

where $y_0, y_1, \dots, y_n$ is the original time series. If $y_{\min} = \min_{j=0,1,\dots,n} y_j$ and $y_{\max} = \max_{j=0,1,\dots,n} y_j$, then $a$ and $b$ were chosen in the following way $a = y_{\min} - 0.125(y_{\max} - y_{\min})$ and $b = y_{\max} + 0.125(y_{\max} - y_{\min})$. This extension of the range is important to avoid closeness to each other of the numbers on the unit circle corresponding to minimal and maximal values of a time series. Evidently, to return back to the original data scale, it is necessary to perform the inverse transformation

$$x \in O, \arg x = \varphi;$$

$$y = \frac{\varphi(b-a)}{2\pi} + a.$$

To control the learning process, the root mean square error (RMSE) stopping criterion (4.142) (see p. 156) was used. The local error threshold $\mu'$ (see (4.143) and footnote 3 in p. 156) was chosen $\mu' = \lambda / 2$, where $\lambda$ is the RMSE threshold from (4.143).

Let us now consider two examples of prediction. We predicted Dow Jones Industrial Average index and Euro/Dollar exchange rate.

## 1. Dow Jones Industrial Average

To train MLMVN, we created a learning set containing 500 samples corresponding to the period from December 29, 2005 to December 24, 2007. Then, after the learning process converged, we made prediction for the following year, from December 26, 2007 to December 19, 2008. The dataset was downloaded from Yahoo Finance. To control the learning process, we used RMSE threshold $\lambda = 110.0$ in (4.143) and $\mu' = 55.0$. The results are shown in Fig. 6.61.

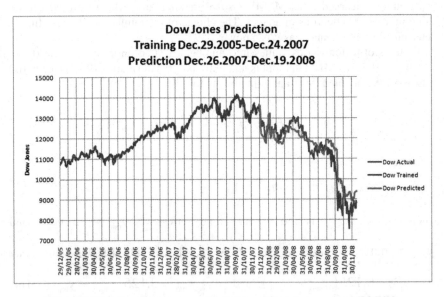

**Fig. 6.61** Results of learning and prediction of Dow Jones using MLMVN

The learning error (RMSE) on which the learning process was stopped is 104.6. The prediction error (RMSE) is 593.2. However, we did not expect that the predicted data should coincide with the actual data. Our goal was to check whether it is possible to predict all major trends. From this point of view the results are more than satisfactory. For example, dramatic fall of the market, which occurred during the first week of October 2008, was predicted with a very high accuracy. Thus, that global financial crisis, which spread over the world economy, could be predicted 9 months before it occurred.

## 2. Euro/Dollar Exchange Rate

The Euro/Dollar exchange rate was predicted for approximately the same period of time. To train MLMVN, we created a learning set containing 500 samples corresponding to the period from December 30, 2005 to December 18, 2007. Then, after the learning process converged, we made prediction for the following 10 months, from December 19, 2007 to October 31, 2008. The dataset was downloaded from the Bank of England data collection. To control the learning process, we used RMSE threshold $\lambda = 0.005$ in (4.143) and $\mu' = 0.0025$. The results are shown in Fig. 6.62.

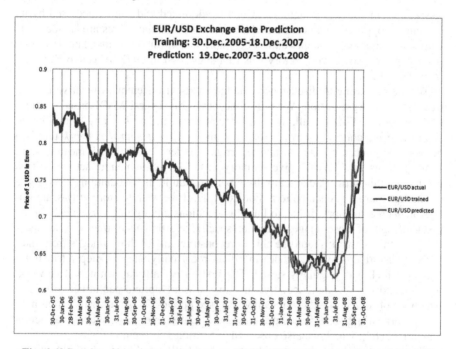

**Fig. 6.62** Results of learning and prediction of EUR/USD exchange rate using MLMVN

The learning error (RMSE) on which the learning process was stopped is 0.0042. The prediction error (RMSE) is 0.0213. As well as for Dow Jones time series, we did not expect that the predicted data should coincide with the actual data. Our goal was the same - to check whether it is possible to predict all major trends. As well as it was for the Down Jones time series, from this point of view the results are more than satisfactory. For example, dramatic fall of the Euro against the

US dollar, which occurred during September-October 2008 as a result of the global financial crisis, was predicted with a very high accuracy. Thus, again that global financial crisis, which spread over the world economy, could be predicted 9 months before it occurred.

Of course, we do not want to state that it is always possible to predict any financial time series with the same high accuracy. Such predictions are possible only if some similar trends are presented in the historical data used to train a network.

Actually, our goal was to show high potential of MLMVN in solving highly nonlinear problems. It was also very interesting to confirm in practice that the productivity of the MLMVN learning algorithm does not depend on the number of layers and neurons in layers.

## 6.3  MVN-Based Associative Memories

Associative memory is a memory, which is addressed and can be accessed by its content (as opposed to an explicit address). Thus, reference clues are "associated" with actual memory contents until a desirable match (or set of matches) is found. A concept of associative memory was first suggested in 1977 by Teuvo Kohonen in his book [22].

We are not going to discuss here fundamentals of different associative memories and their organization. We will concentrate on several MVN-based associative memories. First of all, it is important to mention that any neural-based associative memory is always based on some *recurrent neural network* that is a network, which can update its state recurrently. Indeed, what does it mean to store some patterns in an associative memory based on some neural network? This means that this network can learn these patterns, building associations among one or another "parts" of each pattern. For example, if we want to store gray-scale images in such a memory, it has to build associations among intensity values in the pixels of each of these images or among spectral coefficients of their Fourier spectra. Then this network should be able to reconstruct any of those images from their parts or distorted versions, processing them recurrently, step-by-step retrieving a pattern, which was originally stored. The first historically known neural network, which could work as an associative memory, was a Hopfield neural network [21], which we observed in detail in Section 1.3 (see pp. 36-37). Another recurrent neural network, which also can be used as an associative memory, is a cellular neural network (CNN), which we also observed in Section 1.3 (see pp. 37-38).

Two-state threshold neurons, which were initially used by J. Hopfield and his successors as basic network neurons determined a limitation of the corresponding associative memories: they could store only binary patterns. A classical binary associative memory is exactly a Hopfield network. In 1990, S. Tan, J. Hao, and J. Vandewalle suggested a CNN-based binary associative memory [107]. However, it was very attractive to develop multi-state associative memories, for example, to store there gray-scale images (but not only them – any multi-valued patterns). Of course, there are different multi-valued associative memories based on various ideas. We will observe here one of this ideas – the use of MVN as a basic tool for multi-valued neural associative memories. Since MVN is a multi-state neuron, its use in associative memories is very natural. Moreover, there are several original MVN-based associative memories developed by different authors. Let us observe them.

### 6.3.1 A CNN-MVN-Based Associative Memory

The first MVN-based associative memory was suggested by the author of this book and Naum Aizenberg in 1992 in their paper [38] (where, by the way, the term "multi-valued neuron" was used for the first time). It was a generalization of the CNN-based binary associative memory proposed in [107]. It was suggested to use discrete MVN as a basic neuron in a cellular neural network and to use this cellular neural network to store multi-valued patterns, for example, gray-scale images. To store $NxM$ patterns, a network of the same size, with 3x3 local connections should be used. 3x3 local connections mean that inputs of each neuron are connected to outputs of 8 neurons from its local 3x3 neighborhood and to its own output, and its output is connected to inputs of the same adjacent neurons. Thus, it is a network whose topology is illustrated in Fig. 1.13 (see p. 38). Taking into account that a basic neuron of the suggested CNN is MVN, the dynamics of the network is described by the following equation, which can easily be obtained from (1.39), which describes the CNN dynamics in general. The output $s_{mj}$ of the $mj$th neuron at cycle $t+1$ is

$$s_{mj}(t+1) = P\left( w_0^{mj} + \sum_{rp} w_{rp}^{mj} s_{rp}(t) \right);$$

$$m-1 \le r \le m+1, j-1 \le p \le j+1.$$

(6.172)

Here $P$ is the activation function (2.50) of the discrete MVN, $s_{rp}$ is the output of the $rp$th neuron (which is simultaneously the input of the $mj$th neuron), and $w_i^{mj}$ is the weight corresponding to the $i$th input of the $mj$th neuron.

Not being sure that time (in 1992) how this approach will work for large $k$ in (2.50) (for example, to store 256-valued gray-scale images, we should take $k=256$), the authors limited their considerations by $k=4$. Thus, let us suppose that we are going to store in this memory 4-valued $NxM$ gray-scale images. To encode these images in the form suitable for MVN, a simple transformation $f = e^{ig2\pi/k}; g \in \{0,1,...,k-1\}$ was used (in our particular case $k=4$, but in general this is the number of gray levels in images that we want to store). Two learning rules were suggested to train the network. The first is the Hebb rule (3.107) adapted to this particular case as follows

$$w_0^{mj} = \sum_{i=1}^{L} f_{mj}^i;$$

$$w_{rp}^{mj} = \sum_{i=1}^{L} f_{mj}^i \bar{f}_{rp}^i; \qquad\qquad (6.173)$$

$$m = 1,...,N; \ j = 1,...,M; \ m-1 \le r \le m+1; \ j-1 \le p \le j+1.$$

Here $mj$ are indexes of the current neuron, $rp$ are indexes of neurons from a 3x3 local neighborhood of the current neuron, $L$ is the number of patterns to be stored in the memory, $i = 1,...,L$ is the number of the current pattern, $f_{mj}^i$ is the intensity value in the $mj$th pixel of the $i$th pattern, bar stands for complex conjugation, and $w_0^{mj}$ is a free weight (bias) of the current neuron.

However, only a very limited amount of patterns could be retrieved from the memory without any error if the network was trained with the learning rule (6.173). It is important to understand that even a single error was very clearly visible (only 4-valued patterns were considered!), and, as a result, the second learning rule was suggested. Actually, it was the learning rule (3.80) adapted to this particular case. Whenever the desired output did not coincide with the actual output, the weights were updated in the following way.

$$\tilde{w}_0^{mj} = w_0^{mj} + \frac{1}{10} \omega f_{mj}^i;$$

$$\tilde{w}_{rp}^{mj} = w_{rp}^{mj} + \frac{1}{10} \omega f_{mj}^i \bar{f}_{rp}^i; \qquad\qquad (6.174)$$

$$m = 1,...,N; \ j = 1,...,M; \ m-1 \le r \le m+1; \ j-1 \le p \le j+1,$$

where the choice of $\omega$ depends on the angular distance between the desired and actual output and is described earlier (see p.104).

Evidently, the learning process based on the rule (6.173) is a single-stage process, while the one based on the rule (6.174) is iterative. The latter one continued until the zero error was reached. However, its convergence could be reached very quickly (maximum 10-20 iterations, depending on the number of stored patterns, and just a few iterations if the learning process based on (6.174) starts not from the random weights, but from the Hebbian ones obtained according to (6.173) ). After the network was trained, it could retrieve the stored patterns according to (6.172). Later, the same network was used as 256-valued associative memory [60]. To store 256-valued patterns, for example, gray-scale images with 256 gray levels, just a single change is required – it is necessary to take $k$=256 in the discrete MVN activation function (2.50). The learning time for 256-valued patterns and the learning process based on (6.174) is a little bit longer (100-200 iterations per a neuron for 20 images in the learning set) when the learning starts from the random weights, but still the learning process based on (6.174) requires just a few iterations when it starts from the Hebbian weights obtained according to (6.173). It should be mentioned that to avoid occasional jumps over the "$0/2\pi$ border" during the pattern retrieval procedure (if we store $\tilde{k}$-valued patterns in our associative memory), $k$ in (2.50) should be taken larger than $\tilde{k}$. Actually, for example, for 256-valued gray-scale images it was enough to choose $k=264$.

The retrieval capability of the network trained using the rule (6.174) was significantly higher than the ones of the network trained using the Hebb rule (6.173). It can easily retrieve corrupted patterns unless there are no more than 3 errors in each 3x3 local neighborhood.

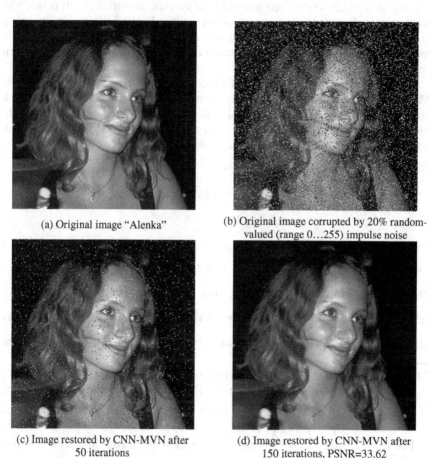

(a) Original image "Alenka"

(b) Original image corrupted by 20% random-valued (range 0...255) impulse noise

(c) Image restored by CNN-MVN after 50 iterations

(d) Image restored by CNN-MVN after 150 iterations, PSNR=33.62

**Fig. 6.63** Restoration of the image corrupted by impulse noise in the CNN-MVN associative memory

The example of image restoration in the CNN-MVN-based associative memory is shown in Fig. 6.63. The 256x256 CNN-MVN with 3x3 local connections was trained using a set of twenty 256x256 gray-scale images with 256 gray levels. One of these images in shown in Fig. 6.63a. To avoid occasional black-white inversion during the restoration process, $k=264$ was used in the discrete MVN activation function (2.50). After the learning process was completed, the original image from Fig. 6.63a corrupted by random-valued impulse noise (the corruption rate 20%, the range of noise is $0...255$) (Fig. 6.63b) was processed by CNN-MVN in order to retrieve it. The network retrieved the image iteratively, according to (6.172). The image retrieved after 50 iterations is shown in Fig. 6.63c. It is clearly visible that a significant part of corrupted pixels is already restored. The image retrieved after 150 iterations is shown in Fig. 6.63d. Visually it does not have any distinction with the originally stored image (Fig. 6.63a). In fact, PSNR[5] for image in Fig. 6.63d is 33.62, which corresponds to the standard deviation 5.3. It is worth to mention that the associative memory is not a filter. First of all, it cannot remove noise from any image, but just from prototypes of those images stored in the memory. Secondly, the associative memory should just retrieve a stored pattern with some acceptable accuracy.

A potential capacity (the number of patterns, which can be stored) of the CNN-MVN-based associative memory is huge. Since CNN is a network with a feedback connection, regardless of number of patterns to be stored, that input/output mapping, which has to be learned by each neuron in CNN-MVN with 3x3 local connections, is always described by the following function

$$f\begin{pmatrix} x_1 & x_2 & x_3 \\ x_4 & x_5 & x_6 \\ x_7 & x_8 & x_9 \end{pmatrix} = f\left(x_1,...,x_5,...,x_9\right) = x_5,$$

which is always a multiple-valued threshold function that can always be learned by a single MVN. Any variable $x_i$ is a threshold function because it always can be implemented by a trivial weighting vector $W = \left(0,0,...,0,w_i,0,...,0\right)$ with a single non-zero component $w_i$. Evidently, the mentioned function may take $k^9$ values. This means that the upper bound for the capacity of the

---

[5] Peak signal-to-noise ratio (PSNR), is the ratio between the maximum possible power of a signal and the power of corrupting noise. PSNR is usually expressed in terms of the logarithmic decibel scale. For example, for two $NxM$ images $A$ and $\hat{A}$ it is expressed as $\text{PSNR} = 20\log_{10}\left(\dfrac{MAX}{SD_{A\hat{A}}}\right)$ db, where MAX is the maximum possible intensity value (for example, for 256 gray-level images $MAX=255$), and $SD_{A\hat{A}}$ is the standard deviation between images $A$ and $\hat{A}$, that is $SD_{A\hat{A}} = \sqrt{\dfrac{\sum\limits_{i,j}\left(A_{ij} - \hat{A}_{ij}\right)^2}{M \times N}}$.

$NxM$ CNN-MVN-based associative memory with 3x3 local connections is $NMk^9$. Of course, it will be more and more difficult to obtain a non-trivial (with more than a single informally non-zero weight) weighting vector for each neuron when the number of the stored patterns increases. A problem is that the absolute value of the weight $w_5$ will become larger and larger in comparison to the one of the other weights. However, the number $NMk^9$ is so large, that it is difficult to say, where it is possible to find so many patterns to store.

CNN-MVN suffers more not from the possibility to get just near-trivial weights for its neurons, but from another disadvantage. Since CNN is a locally connected network, complete corruption (or even incomplete, but significant corruption) even of a single 3x3 local neighborhood means impossibility of retrieval of the information in that neighborhood. Nevertheless, this CNN-MVN-based associative memory was an important step to the future. It showed that MVN can be used as a basic neuron in recurrent neural networks and therefore it can be used as a basic neuron in associative memories. Let us consider other MVN-based associative memories, which were developed after that first and which employ not cellular, but other network architectures.

## 6.3.2  A Hopfield MVN-Based Associative Memory

In 1996, Stanislaw Jankowski, Andrzej Lozowski, and Jacek M. Zurada presented in their paper [108] a model of the Hopfield MVN-based associative memory. They considered a Hopfield network (see Fig. 1.12, p. 36) with discrete MVN as a basic neuron. On the one hand, this model of an associative memory follows traditional Hopfield's ideas. But one the other hand, being based on MVN, this model opens many new opportunities, first of all to store and retrieve multi-valued patterns. It also shows that MVN can be used as a basic neuron in a Hopfield network. It is important that this model became very popular and the paper [108] is one of the most cited papers devoted to MVN. That deep analysis of a Hopfield MVN-based network, which is done in [108] and which was later followed by a new design method developed by Mehmet Kerem Müezzinoğlu, Cüneyt Güzeliş, and Jacek M. Zurada in [109] is very important for better understanding of MVN properties and advantages. Let us consider an MVN-based Hopfield associative memory in detail, following [108, 109] and adapting some notations to those we use in this book.

Let us consider a Hopfield neural network composed from $N$ fully connected discrete MVNs with the activation function (2.50). Thus, the network state $S$ is an $N$-dimensional complex-valued vector $S = \left( s_1, s_2, ..., s_N \right)$ where $s_m \in E_k = \left\{ \varepsilon_k^0, \varepsilon_k, ..., \varepsilon_k^{k-1} \right\}$, $\varepsilon_k = e^{i2\pi/k}$. Each neuron in this network performs in the following way.

To train this network, the Hebb rule (3.107) with the normalization by the number $N$ of neurons in the network (and the number of inputs of each neuron, respectively) was used. Hence, to store $L$ $N$-dimensional patterns presented by vectors $\left( f_1^r,..., f_N^r \right); r = 1,..., L$, the following rule is used

$$w_{mj} = \frac{1}{L} \sum_{r=1}^{L} f_m^r \overline{f}_j^r; \quad m, j = 1,..., N, \quad (6.175)$$

where the bar sign stands for complex conjugation.

It easily follows from (6.175) that unlike in a classical Hopfiled network where always $w_{mj} = w_{jm}$, in MVN-based Hopfield network this property is modified and $w_{mj} = \overline{w}_{jm}$. This means that the synaptic $N$x$N$ matrix created from all the weights is a Hermitian matrix (a matrix, which is equal to its own conjugate-transpose)

$$\mathbf{W} = \begin{pmatrix} 0 & w_{12} & \cdots & w_{1N-1} & w_{1N} \\ \overline{w}_{12} & 0 & \cdots & w_{2N-1} & w_{2N} \\ \cdots & \cdots & \cdots & \cdots & \cdots \\ \overline{w}_{1N-1} & \overline{w}_{2N-1} & \cdots & 0 & w_{N-1N} \\ \overline{w}_{1N} & \overline{w}_{2N} & \cdots & \overline{w}_{N-1N} & 0 \end{pmatrix}.$$

The weighted sum of the $m$th neuron is

$$z_m = \sum_j w_{mj} s_j; \quad m = 1,..., N,$$

where $w_{mj}$ is the weight corresponding to the synaptic connection between the $j$th neuron and the $m$th neuron.

Network dynamics (that is the output of the $m$th neuron at cycle $t+1$) is determined by the following equation

$$s_m (t+1) = P\left( \varepsilon_{2k} z_m (t) \right); \quad m = 1,..., N, \quad (6.176)$$

where the multiplier $\varepsilon_{2k} = e^{i2\pi/2k} = e^{i\pi/k}$ is a phase shifter, which was suggested to adjust the weighted sum such that it should appear in that sector to whose lower border it is closer. In fact, it determines half a sector shift of the weighted sum in phase.

It is important to mention that it follows from (6.175) that this model does not employ a bias for its neurons. There is no "free weight" $w_0$.

As we have seen considering a classical Hopfield network (see Section 1.3, pp. 36-37), this network is characterized by the energy function (1.38). Updating its states during the retrieval process, the network converges to the local minimum of the energy function. To prove stability of the Hopfield associative memory based on (6.175)-(6.176), the following energy function was suggested in [108]

$$E(S) = -\frac{1}{2}\sum_m \sum_j w_{mj}\overline{s}_m s_j. \tag{6.177}$$

Since the synaptic matrix $\mathbf{W}$ is Hermitian, the energy function (6.177) is real-valued.

The network has asynchronous dynamics (that is its neurons update their states independently on each other). Network stability is proven in [108] in the following terms. The change of the energy function (6.177) of the Hopfield MVN-based neural network with the synaptic weights obtained according to (6.175) (which form a Hermitian matrix with nonnegative diagonal entries ($w_{mm} \geq 0$)), neuron outputs determined by (6.176), and asynchronous dynamics $S \rightarrow S'$, is nonpositive $\Delta E = E(S') - E(S) \leq 0$ only when $S' = S$.

Storage capacity estimation of a Hopfield MVN-based neural network was done in [108] in terms of probability of an error in the network response.

In 2003, in their paper [109], M. K. Müezzinoğlu, C. Güzeliş, and J. M. Zurada suggested a new design method for the Hopfield MVN-based associative memory, which improves its retrieval capability. Let us present it briefly, following [109] and adapting some notations. This method employs a set of inequalities to interpret each pattern stored in the memory as a strict local minimum of a quadratic energy function. This improvement makes it possible to improve retrieval capability of the network. The energy function (6.177) can be presented as a real-valued quadratic form

$$E(S) = -\frac{1}{2}S\mathbf{W}\overline{S}^T.$$

To ensure that this quadratic form reaches a local minimum at each element of the set $F = \left\{ \left( f_1^r, ..., f_N^r \right); r = 1, ..., L \right\}$ of patterns to be stored in the memory, the energy function has to satisfy the following condition for each $X \in F$

$$E(X) < E(Y), \forall Y \in \mathrm{B}_1(X) - \{X\}, \tag{6.178}$$

where $\mathrm{B}_1(Y)$ is the 1-neighborhood of $X$, which is defined as

$$\mathrm{B}_1(X) = \bigcup_{j=1}^{N}\left\{Y : y_j = x_j e^{i2\pi/k} \vee y_j = x_j e^{-i2\pi/k}, y_m = x_m, j \neq m\right\}\bigcup\{X\}.$$

Substituting (6.177) into (6.178), we obtain $2N$ inequalities that should be satisfied by the weights, that is by the elements of the matrix $\mathbf{W}$

$$\sum_m \sum_j w_{mj} \overline{x}_m x_j > \sum_m \sum_j w_{mj} \overline{y}_m y_j, \forall Y \in B_1(X) - \{X\},$$

which implies

$$\sum_{1 \le m < j \le N} w_{mj} \left( \overline{x}_m x_j - \overline{y}_m y_j \right) + \overline{w}_{mj} \left( x_m \overline{x}_j - y_m \overline{y}_j \right) > 0, \forall Y \in B_1(X) - \{X\}.$$

Taking into account that

$$w_{mj} \overline{x}_m x_j + \overline{w}_{mj} x_m \overline{x}_j = 2 \operatorname{Re}\{w_{mj}\} \operatorname{Re}\{\overline{x}_m x_j\} - 2 \operatorname{Im}\{w_{mj}\} \operatorname{Im}\{\overline{x}_m x_j\},$$

it is easy to obtain from the previous equation the following

$$\sum_{1 \le m < j \le N} \left\{ \operatorname{Re}\{w_{mj}\} \left( \operatorname{Re}\{\overline{x}_m x_j\} - \operatorname{Re}\{\overline{y}_m y_j\} \right) \right.$$

$$\left. + \operatorname{Im}\{w_{mj}\} \left( \operatorname{Im}\{\overline{y}_m y_j\} - \operatorname{Im}\{\overline{x}_m x_j\} \right) \right\} > 0, \forall Y \in B_1(X) - \{X\}.$$

Let us take into account that if $x_j = e^{i 2\pi \hat{x}_j / k} \in E_k$, and $x_m = e^{i 2\pi \hat{x}_m / k} \in E_k$, then

$$\operatorname{Re}\{\overline{x}_m x_j\} = \cos\left(2\pi / k(-\hat{x}_m + \hat{x}_j)\right); \operatorname{Im}\{\overline{x}_m x_j\} = \sin\left(2\pi / k(-\hat{x}_m + \hat{x}_j)\right),$$

where $\hat{x}_m, \hat{x}_j \in K = \{0, 1, ..., k-1\}$.

Hence (6.178), that is a necessary and sufficient condition for the energy function to reach its local minimum for each $X \in F$, can now be re-written as the following inequality

$$\sum_{1 \le m < j \le N} \left\{ \operatorname{Re}\{w_{mj}\} \left[ \cos\left( \frac{2\pi}{k} (\hat{x}_j - \hat{x}_m) \right) - \cos\left( \frac{2\pi}{k} (\hat{y}_j - \hat{y}_m) \right) \right] \right.$$

$$\left. + \operatorname{Im}\{w_{mj}\} \left[ \sin\left( \frac{2\pi}{k} \right)(\hat{y}_j - \hat{y}_m) - \sin\left( \frac{2\pi}{k} (\hat{x}_j - \hat{x}_m) \right) \right] \right\} > 0. \tag{6.179}$$

To obtain all the weights, it is enough to solve (6.179) with respect to $w_{mj}; m, j = 1, ..., N$. This approach guarantees that all patterns from the set $F$, which we want to store in the memory, are attractors of the network that is those fixed points where the energy function reaches its local minima.

However, this does not ensure that the energy function has no other local minima. If the latter is the case, the network may have other fixed points (attractors), which differ from the elements of the set $F$. This means that the network may occasionally "retrieve" some spurious patterns. For example, it follows from (6.178), (6.179), and the considerations above that (6.179) constructed for a pattern $\left( f_1^r, ..., f_N^r \right) \in F; f_j^r \in E_k; r = 1, ..., L$ would be exactly the same con-

structed for each vector $\left( f_1^r, ..., f_N^r \right) \otimes \left( \varepsilon_k^t, ..., \varepsilon_k^t \right); \varepsilon_k^t \in E_k$, where $\otimes$ stands

for component-wise multiplication of two vectors. This means that the weights

obtained from (6.179) may potentially produce at least $(k-1)|F| = (k-1)L$

spurious pattern vectors (there is a multiplier $k$-1, not $k$ in a front of $L$, because the

component-wise multiplication by the vector $\left( \varepsilon_k^0, ..., \varepsilon_k^0 \right) = \left( 1, ..., 1 \right)$ does not

produce any new pattern). As it was suggested in [109], at least such (trivial) spu-

rious pattern vectors can be eliminated by introducing a bias to a weighting vector

of each neuron. This is equivalent to adding a constant (for example, $\varepsilon_k = e^{i2\pi/k}$ )

as the $0^{\text{th}}$ component of each pattern vector, which become in this way

$\left( \varepsilon_k, f_1^r, ..., f_N^r \right); r = 1, ..., L$. Respectively, this is also equivalent to adding one

more neuron ($0^{\text{th}}$) to the network, and the output of this neuron is constantly equal

to $\varepsilon_k = e^{i2\pi/k}$ during the recurrent retrieval process. This means that the $m$th neu-

ron of the network $m = 1, ..., N$ gets a bias (a "free weight" or a "complex

threshold"), which is equal to $\tilde{w}_{m0} = \varepsilon_k w_{m0}$. Hence the weighted sum of the $m$th

neuron of the network is equal now to

$$ z_m = \tilde{w}_{m0} + \sum_j w_{mj} s_j; \ m = 1, ..., N , $$

and dynamics of this neuron is still described by (6.176).

It is important to mention that a little bit later we will see how this property of
the MVN-based Hopfield network to "memorize" spurious patterns along with the
true patterns can be successfully used to store in the memory rotated images along
with the original images. To get this effect, it will be necessary to represent images to be stored in the memory in the frequency domain. We will consider this
approach in the last subsection of Section 6.3.

To find the weights using this improved design technique, the following algorithm should be used.

1) For each element $X \in F$, which we want to store in the memory, and for
each $Y \in B_1(X) - \{X\}$, calculate the row vector

$$ \left( c_{12} s_{12} c_{13} d_{13} ... c_{10} d_{10} \mid c_{23} d_{23} c_{24} d_{24} ... c_{20} d_{20} \mid ... \mid c_{N0} d_{N0} \right), $$

where $c_{mj} = \cos \left( \dfrac{2\pi}{k} \left( \hat{x}_j - \hat{x}_m \right) \right)$ and $d_{mj} = \sin \left( \left( \dfrac{2\pi}{k} \right) \left( \hat{y}_j - \hat{y}_m \right) \right)$, and

append it to $N \times N(N+1)$ matrix $\mathbf{A}$ (which is initially empty).

2) Find a solution $\mathbf{q}' \in \mathbb{R}^{N(N+1)}$ for the inequality system $\mathbf{Aq}=0$ using any appropriate method.

3) Construct the Hermitian matrix

$$\mathbf{W}' = \begin{pmatrix} 0 & q_1'+iq_2' & \cdots & q_{2N-1}'+iq_{2N}' \\ q_1'-iq_2' & 0 & & q_{4N-3}'+iq_{4N-2}' \\ \cdots & \cdots & 0 & \cdots \\ q_{2N-1}'-iq_{2N}' & q_{4N-3}'+iq_{4N-2}' & \cdots & 0 \end{pmatrix}.$$

4) Extract the weights from $\mathbf{W}'$ as $w_{mj} = w_{mj}'$ for $m, j = 1, 2, ..., N$ and $w_{m0} = e^{i2\pi/k} w_{m,N+1}'$ for $m = 1, 2, ..., N$.

The Hopfield MVN-based associative memory shows wonderful retrieval capabilities. For example, it can successfully be applied to store gray-scale images and to retrieve them even when they are corrupted by heavy impulse noise. This network does not suffer from the disadvantage of local connectivity, which characterizes CNN-MVN-based associative memory. It is shown in [109] that the Hopfield MVN-based associative memory can retrieve gray-scale images corrupted by salt-and-pepper noise with even 60% corruption rate. It was also shown in [109] that the improved design method presented there leads to better retrieval capability than the Hebbian learning employed earlier in [108].

### 6.3.3  MVN-Based Associative Memory with Random Connections

We have just considered two associative memories with, let us say, "extreme" connecting topologies. One of them (CNN-MVN-based) has just a very limited amount of local connections where each neuron is connected only to its closest local neighbors. Another one (the Hopfield MVN network) is fully-connected. A disadvantage of the first network is its limited retrieval capability, while a disadvantage of the second network is its complexity. For example, to store $NxM$ gray-scale images in the Hopfield network, each neuron employs $NxM$ weights, thus, the total amount of weighting parameters is $(N \times M)^2$, which is a huge number. This complicates utilization of such a network. For example, in [109], where the Hopfiled MVN-based network was used to store 100x100 gray-scale images, they were split into 500 fragments; each of them contained just 20 pixels. Thus, in fact, 500 separate associative memories were integrated into an ensemble. Can such a structure be simplified? Is it possible to find some compromise between a full connectivity, which is less sensitive to distortions in retrieved patterns and a local connectivity, which is much easier to implement? In 1995, in the small conference paper [110] the author of this book together with Naum Aizenberg and Georgy Krivosheev suggested an MVN-based recurrent neural network with random connections. Later, in 2000, in [60], the author of this book together with

Naum Aizenberg and Joos Vandewalle developed this idea. To keep the integrity of our presentation, let us consider it here.

Let us again consider a problem of storing $NxM$ gray-scale images in an associative memory. Let us, as it is usual for such a task, associate each pixel of our images with a neuron in a network. Thus, to solve our problem, we should use a network containing exactly $NxM$ neurons. Evidently, if we need to store not 2D, but 1D patterns, nothing will change – to store $N$-dimensional vector-patterns, we should use a network containing $N$ neurons. In CNN, each neuron is connected only to its closest neighbors, which means that each component of a pattern (each pixel of an image) is associated only with its closest neighbors. In the Hopfield network, each neuron is connected to all other neurons, which means that each component (pixel) of each pattern is associated with all other components (pixels). While in the first case local associations can be insufficient to retrieve a corrupted pattern, in the second case associations could be redundant. Let us consider a network topology, which is somewhere in the "middle" between local cellular connectivity and full Hopfield connectivity.

So let us consider an MVN-based neural network containing $NxM$ neurons. We will use this network as an associative memory to store $NxM$ gray-scale images. Let us connect each neuron with just a limited amount of other neurons (like in a cellular network), but let us choose these "other neurons" to be connected with a given neuron, randomly. This means that they should be chosen not from a local neighborhood, but from the entire network. This is illustrated in Fig. 6.64.

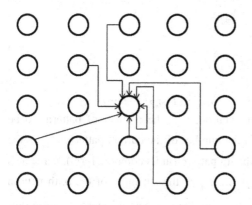

There are 25 neurons in the network. The inputs of the central neuron are connected to the outputs of 6 (out of 25) randomly selected neurons and to its own output.

The upper bound for the storage capacity of the $NxM$ MVN-based associative memory where each neuron has $n$ inputs (one of which is a feedback input connected to its own output) is $NMk^n$. This follows from the following. Each neuron in this network has $n$ inputs and implements an input/output mapping described by a multiple-valued threshold function of $n$

**Fig. 6.64** MVN-based neural network with random connections. The inputs of the central neuron are connected to the outputs of 6 (out of 25) randomly selected neurons and with its own output

variables $f\left(x_1,....,x_n\right)$. This function is threshold because the neuron has a feedback connection. If (without loss of generality) a feedback input is $x_n$, then $f\left(x_1,....,x_n\right) = x_n$. But such a function is always threshold, which means that it can be learned using MVN. This function may take $k^n$ values. Since there are $N \times M$ neurons in the network, and each of them implements a function, which may take $k^n$ values, then up to $NMk^n$ patterns can be stored in the network used as an associative memory. Similarly to CNN-MVN, it will be more and more difficult to obtain a non-trivial (with more than a single informally non-zero weight) weighting vector for each neuron when the number of the stored patterns increases. A problem is that the absolute value of the weight $w_n$ corresponding to the feedback input will become larger and larger in comparison to the one of the other weights. However, even for relatively small values of $n$, $NMk^n$ is so large, that it will not be possible to find so many patterns to store. To choose $n$, we should take into account the number of neurons in the network (which is the number of components (pixels) in our patterns to be stored). Experimentally it was shown that good results are shown with $n \sim 0.01 \times NM$. After $n$ is fixed, connections for each neuron should be generated randomly: inputs of each neuron have to be connected to the outputs of randomly selected other $n-1$ neurons and with its own output.

To train this network, the MVN learning algorithm based on the error-correction learning rule (3.92) adapted to this particular case was used. Evidently, all the neurons should be trained separately. Whenever the desired output of the $mj$th neuron does not coincide with the actual output, the weights are updated in the following way.

$$\tilde{w}_0^{mj} = w_0^{mj} + \frac{1}{(n+1)}\left(f_{mj}^i - Y_{mj}\right);$$

$$\tilde{w}_r^{mj} = w_r^{mj} + \frac{1}{(n+1)}\left(f_{mj}^i - Y_{mj}\right)\bar{f}_r^i; \tag{6.180}$$

$$m = 1, ..., N;\ j = 1, ..., M;\ r = p_{mj_1}, ..., p_{mj_n};\ i = 1, ..., L$$

where $mj$ are indexes of the current neuron, $L$ is the number of patterns to be stored in the memory, $i = 1, ..., L$ is the number of the current pattern, $f_{mj}^i$ is the intensity value in the $mj$th pixel of the $i$th pattern (in its complex form), bar stands for complex conjugation, $f_r^i;\ r = p_1, ..., p_n$ is the $r$th input of the $mj$th neuron taken from the $i$th pattern, $p_{mj_1}, ..., p_{mj_n}$ are the numbers of neurons whose outputs are connected to the inputs of the $mj$th neuron, $w_0^{mj}$ is a free weight (bias) of the $mj$th neuron, and $w_r^{mj};\ r = p_{mj_1}, ..., p_{mj_n}$ are its other $n$ weights.

The learning process based on the rule (6.180) is iterative. It is continued until the zero error is reached. However, its convergence could be reached very quickly (for example, it takes maximum 100-200 iterations for the 256x256 network where each neuron has 600 inputs and 20 training patterns). The learning process should converge faster when it starts from the Hebbian weights. After the network was trained, it could retrieve the stored patterns according to the following rule

$$s_{mj}(t+1) = P\left( w_0^{mj} + \sum_{r=p_{mj_1}}^{p_{mj_n}} w_r^{mj} s_r(t) \right);$$

$$m = 1,...,N; \ j = 1,...,M; \ r = p_{mj_1},...,p_{mj_n},$$

(6.181)

which describes the recurrent dynamics of the network. Here $P$ is the discrete MVN activation function (2.50). In other words, (6.181) determines the output $s_{mj}$ of the $mj$th neuron at cycle $t+1$.

To store 256-valued gray-scale images in our associative memory, and to avoid occasional jumps over the "$0/2\pi$ border" during the pattern retrieval procedure, $k$ in (2.50) should be taken larger than 256. Actually, for example, for 256-valued gray-scale images it was enough to choose $k=264$.

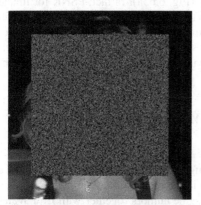

(a) Original image "Alenka" with 75 % of pixels contained just Gaussian noise

(b) Iterative Retrieval : 40 iterations

(c) Iterative retrieval: 80 iterations

(d) Iterative retrieval: 150 iterations, PSNR=33.74

**Fig. 6.65** Retrieval of the image corrupted by Gaussian noise, which replaces 75% of the original information, in the MVN-based associative memory with random connections

The retrieval capability of the network trained using the rule (6.180) is impressive. The example of image retrieval in the MVN-based associative memory with random connection is shown in Fig. 6.65.

The 256x256 MVN-based neural network with random connections (each neuron is connected to 59 other neurons and to itself) was trained using a set of twenty 256x256 gray-scale images with 256 gray levels. One of these images in shown in Fig. 6.63a. After the learning process was completed, the original image from Fig. 6.63a was corrupted by zero-mean Gaussian noise in such a way that a squared area, which is exactly 75% of the whole image area, was completely replaced by noise (Fig. 6.65a). Then the network retrieved the image iteratively, according to (6.181). The image retrieved after 40 iterations is shown in Fig. 6.65b. It is clearly visible that a significant part of corrupted pixels is already restored. The image retrieved after 80 iterations is shown in Fig. 6.65c. The image retrieved after 150 iterations is shown in Fig. 6.65d. Visually it does not have any distinction from the originally stored image (Fig. 6.63a). In fact, PSNR for image in Fig. 6.65d is 33.74, which corresponds to the standard deviation 5.24. It is again important to mention that the associative memory is not a filter. It does not remove noise, it restores a pattern, which was stored. In fact, there is no such a filter, which can remove Gaussian noise that replaces a signal completely in some connected domain, like it is in Fig. 6.65a.

### 6.3.4  An MVN-Based Associative Memory with Rotation Invariant Association

In 2000, Hiroyuki Aoki and Yukio Kosugi suggested in their paper [111] a method, which makes an MVN-based associative memory rotation invariant. This idea was further developed in 2001 by the same authors together with Eiju Watanabe and Atsushi Nagata in their paper [112] where they applied such a memory to store medical images. Let us present this idea here, following [111, 112] and adapting some notations to the style, which we use in this book.

As we have seen above, the Hopfield MVN-based associative memory, when it memorizes patterns $\left( f_1^r, ..., f_N^r \right) \in F; f_j^r \in E_k; r = 1, ..., L$, simultaneously memorizes patterns $\left( f_1^r, ..., f_N^r \right) \otimes \left( \varepsilon_k^t, ..., \varepsilon_k^t \right); \varepsilon_k^t \in E_k$, where $\otimes$ stands for component-wise multiplication of two vectors. Hence, while we want to store $L$ patterns in the memory and create exactly $L$ fixed points (attractors), the memory simultaneously creates $(k-1)L$ more fixed points, which are spurious patterns. In fact, they are really spurious, but just if images to be stored in the memory are represented in the spatial domain. Really, in the spatial domain, the operation $\left( f_1^r, ..., f_N^r \right) \otimes \left( \varepsilon_k^t, ..., \varepsilon_k^t \right); \varepsilon_k^t \in E_k$ determines shift, which means that all the components (pixels) of a pattern to be stored, are shifted, creating a spurious pattern. An elegant solution, which was suggested in [111, 112], is based on the frequency domain representation of patterns that we want to store in the memory and on the fundamental properties of phase (we considered them in detail in Section 1.4 (see pp. 43-44).

So let us consider again the same memory design problem. We need to store $L$ $N \times M$ gray-scale images with $\tilde{k}$ gray levels $\left( \tilde{f}_1^r, ..., \tilde{f}_N^r \right) \in \tilde{F}; \tilde{f}_j^r \in \tilde{K} = \{0,1,...,\tilde{k}-1\}; r = 1,...,L$ in the MVN-based associative memory. Instead of spatial domain representation of images and their recoding in the MVN-suitable form as $f = e^{ig2\pi/k}; g \in \tilde{K} = \{0,1,...,\tilde{k}-1\}; k > \tilde{k}$, we will use their frequency domain representation. Let

$$\hat{f} = \left\{ \hat{f} \left( n', m' \right) \right\} = DFT \left( \tilde{f} \right); n' = 0,1,...,N-1; m' = 0,1,...,M-1$$

be the Fourier transform of $\tilde{f}$ ($n', m'$ are the corresponding spatial frequencies).

Phases $\arg \hat{f} \left( n', m' \right)$ are the most suitable inputs for MVN. As we have seen (Fig. 1.15 and Fig. 1.16), phase completely contains all information about edges, boundaries, their orientation, etc. However, since we want not only to recognize a stored image, but to retrieve it, we should also memorize the magnitude information. It was suggested in [111, 112] to include this information in phase as follows.

Let $\left| f_{max} \right| = \max \left\{ \left| \hat{f} \left( n', m' \right) \right| \right\}; n' = 0,1,...,N-1; m' = 0,1,...,M-1$. Then

$$0 < \frac{\left| \hat{f} \left( n', m' \right) \right|}{\left| f_{max} \right|} \le 1.$$ Let us transform this ratio into the angular value

$$\gamma \left( n', m' \right) = \arccos \left( \frac{\left| \hat{f} \left( n', m' \right) \right|}{\left| f_{max} \right|} \right); n' = 0,1,...,N-1; m' = 0,1,...,M-1.$$

Let us now add $\gamma \left( n', m' \right)$ to phases $\arg \hat{f} \left( n', m' \right)$:

$$\varphi \left( n', m' \right) = \arg \hat{f} \left( n', m' \right) + \gamma \left( n', m' \right); \tag{6.182}$$
$$n' = 0,1,...,N-1; m' = 0,1,...,M-1.$$

Now we can form patterns, which will be stored in the network

$$f \left( n', m' \right) = e^{i\varphi(n',m')}; n' = 0,1,...,N-1; m' = 0,1,...,M-1. \tag{6.183}$$

Evidently, all $f(n', m')$ determined by (6.183) are located on the unit circle, and they are natural inputs for MVN. It is wonderful that transformation (6.182), which "incorporates" magnitude into phase, is easily invertible. An initial image $\tilde{f}$ can easily be restored as follows

$$\tilde{f} = |f_{\max}| \operatorname{Re}\{ IDFT(f) \}, \tag{6.184}$$

where $IDFT(f)$ is the inverse Fourier transform of $f$.

The most wonderful property of the representation (6.182)-(6.183) is that storing original images in the associative memory, we simultaneously store their rotated versions, without any additional effort. Let us show this.

It was suggested in [111, 112] to use $M$ separate Hopfield-like MVN-based associative memories to store $M$ $N$-dimensional vector-columns $f_r(m'); m' = 0, 1, ..., M-1; r = 1, ..., L$ of the $N$x$M$- dimensional pattern matrix $f_r(n', m'); r = 1, ..., L$ for each of our $L$ $N$x$M$ patterns. We call these networks Hopfield-like because they are fully connected as a Hopfield network, but each neuron has also a feedback connection – one of its inputs is connected to its own output.

Let us add to each $N$-dimensional vector- pattern $f_r(m'); m' = 0, 1, ..., M-1; r = 1, ..., L$, which we want to store in our memory, the $N + 1^{\text{st}}$ component, which is constantly equal to $\varepsilon_k^0 = 1$. Hence, we are going to store in the memory patterns

$$\vec{f}_r(m') = \left( f_r(m'), \varepsilon_k^0 \right) = \left( f_0^r, f_1^r, ..., f_{N-1}^r; e^{i2\pi l/k} \right);$$
$$m' = 0, 1, ..., M-1; r = 1, ..., L. \tag{6.185}$$

Let us consider an arbitrary memory out of our $M$ Hopfield memories. Its learning can be based on the normalized Hebb rule:

$$w_{tj} = \frac{1}{L} \sum_{r=1}^{L} \vec{f}_t^r(m') \overline{\vec{f}_j^r}(m'); \quad t, j = 0, 1, ..., N. \tag{6.186}$$

A wonderful property is that memorizing in this way a pattern $f(m')$, this memory also memorizes $k-1$ phase-shifted patterns $e^{i2\pi l/k} \vec{f}(m'); l = 1, ..., k-1$. This means that not only patterns $f(m'); m' = 0, 1, ..., M-1$ are the fixed point attractors for this associative memory, but also $k-1$ phase-shifted patterns

$$e^{i2\pi l/k} \vec{\hat{f}}(m') = \left(e^{i2\pi l/k} f_0, e^{i2\pi l/k} f_1, ..., e^{i2\pi l/k} f_{N-1}; e^{i2\pi l/k}\right);$$
$$l = 1, ..., k-1 \tag{6.187}$$

are the fixed point attractors for this associative memory either.

As any associative memory, this one, when shown any of patterns, which participated in the learning process (see (6.185) ) or their noisy versions (we do not discuss here a level of noise to which this network is tolerant), can retrieve them. The retrieval procedure is determined by the network dynamics, which determines the output of each neuron at cycle $t+1$

$$s_l(t+1) = P\left(\sum_{j=0}^{N} w_{lj} s_j\right); \; l,j = 0,1,...,N,$$

where $s_j; \; j = 0,1,...,N$ is a state of the $j$th neuron.

If now we are going to retrieve any of phase-shifted patterns (6.187), evidently, it will also be retrieved as a fixed point attractor.

Let us try now to retrieve a phase-shifted pattern $g(m') = (\alpha f_0, \alpha f_1, ..., \alpha f_{N-1}; 1)$ whose shift $\alpha$ is unknown (this means that the last (additional) component of the pattern vector $g(m')$ is equal to $1 = \varepsilon_k^0$) or its noisy version.

It is wonderful that this retrieval process converges to

$$\hat{g}(m') = (\beta f_0, \beta f_1, ..., \beta f_{N-1}; \beta),$$

where $\beta$ is a complex number $e^{i \arg \beta}$ located on the unit circle. Thus, it is enough to multiply all the components of the vector $\hat{g}(m')$ by $\beta^{-1} = e^{-i \arg \beta}$ to obtain $\vec{\hat{f}}(m') = (f_0, f_1, ..., f_{N-1}; 1)$, from which we obtain a pattern $\hat{f}(m') = (f_0, f_1, ..., f_{N-1})$, which must be a close estimation (or the exact copy) of one of our initially stored patterns $f_r(m'); r = 1, ..., L$. The example of this retrieval process is considered below.

Since we have considered an arbitrary of our $M$ associative memories, the same considerations are true for all of them. Thus, after we run the retrieval procedure in all of $M$ associative memories, we restore an estimation $\hat{f}(n', m'); n' = 0,1, ..., N-1; m' = 0,1, ...M-1$ of one of our initially stored Fourier "magnitude-rotated" patterns (6.183).

Then a spatial domain image $\tilde{\tilde{f}}(n,m); n = 0,1,...,N-1; m = 0,1,...M-1$ can be obtained using the procedure (6.184) based on the inverse Fourier transform. This image is a closer estimation or exactly one of the original images $\tilde{f}_r = (\tilde{f}_1^r,...,\tilde{f}_N^r) \in \tilde{F}; r = 1,...,L$.

This means that the described associative memory can retrieve the phase-shifted (rotated) patterns and therefore it is rotation invariant.

Finally, let us demonstrate one simple example, how a phase-shifted pattern can be restored using the described approach.

Let we want to store a pattern $(\tilde{f}_0, \tilde{f}_1, \tilde{f}_2); \tilde{f}_j \in \tilde{K} = \{0,1,2,3\}$. Since $\tilde{k} = 4$, let us choose $k = 6$ in the discrete MVN activation function (2.50) (see Fig. 6.66). Let without loss of generality representation (6.183) of our pattern is $f = (\varepsilon_6, \varepsilon_6^2, \varepsilon_6^3)$. Then $\hat{f} = (\varepsilon_6, \varepsilon_6^2, \varepsilon_6^3, 1)$. Since we have to learn 4-dimensional vector, our Hebbian network contains 4 neurons. Their weights we have to find according to (6.186). We can store them in the 4x4 matrix (each row corresponds to the weights of a particular neuron)

$$W = \begin{pmatrix} 1 & \varepsilon_6^5 & \varepsilon_6^4 & \varepsilon_6 \\ \varepsilon_6 & 1 & \varepsilon_6^5 & \varepsilon_6^2 \\ \varepsilon_6^2 & \varepsilon_6 & 1 & \varepsilon_6^3 \\ \varepsilon_6^5 & \varepsilon_6^4 & \varepsilon_6^3 & 1 \end{pmatrix}.$$

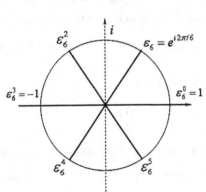

Let us now consider the following phase- shifted pattern

$$g = \varepsilon_6^2 f = (\varepsilon_6^2 \varepsilon_6, \varepsilon_6^2 \varepsilon_6^2, \varepsilon_6^2 \varepsilon_6^3) =$$
$$= (\varepsilon_6^3, \varepsilon_6^4, \varepsilon_6^5).$$

Let us process it using our associative memory. First, we have to add one more component to our

**Fig. 6.66** The discrete MVN activation function (2.50) for $k=6$

pattern $\tilde{g} = (g,1) = (\varepsilon_6^3, \varepsilon_6^4, \varepsilon_6^5, 1)$. Exactly this pattern should be processed in the network.

The results of this processing are as follows ($P$ is the discrete MVN activation function (2.50), $k=6$, see Fig. 6.66).

Neuron 0. $z_0 = 1 \cdot \varepsilon_6^3 + \varepsilon_6^5 \varepsilon_6^4 + \varepsilon_6^4 \varepsilon_6^5 + \varepsilon_6 \cdot 1 = \varepsilon_6^3 + \varepsilon_6^3 + \varepsilon_6^3 + \varepsilon_6; P(z_0) = \varepsilon_6^2$.

Neuron 1. $z_1 = \varepsilon_6 \varepsilon_6^3 + 1 \cdot \varepsilon_6^4 + \varepsilon_6^5 \varepsilon_6^5 + \varepsilon_6^2 \cdot 1 = \varepsilon_6^4 + \varepsilon_6^4 + \varepsilon_6^4 + \varepsilon_6^2; P(z_0) = \varepsilon_6^3$.

Neuron 2. $z_2 = \varepsilon_6^2 \varepsilon_6^3 + \varepsilon_6 \varepsilon_6^4 + 1 \cdot \varepsilon_6^5 + \varepsilon_6^3 \cdot 1 = \varepsilon_6^5 + \varepsilon_6^5 + \varepsilon_6^5 + \varepsilon_6^3; P(z_0) = \varepsilon_6^4$.

Neuron 3. $z_3 = \varepsilon_6^5 \varepsilon_6^3 + \varepsilon_6^4 \varepsilon_6^4 + \varepsilon_6^3 \varepsilon_6^5 + 1 = \varepsilon_6^2 + \varepsilon_6^2 + \varepsilon_6^2 + 1; P(z_0) = \varepsilon_6$.

Thus, the retrieved pattern is $\tilde{g} = \left( \varepsilon_6^2, \varepsilon_6^3, \varepsilon_6^4, \varepsilon_6 \right)$. The last component of this

retrieved pattern is $\varepsilon_6^2$. Let us find $\left( \varepsilon_6^2 \right)^{-1} \tilde{g}$ according to the described procedure:

$$\left( \varepsilon_6 \right)^{-1} \tilde{g} = \varepsilon_6^5 \tilde{g} = \varepsilon_6^5 \left( \varepsilon_6^2, \varepsilon_6^3, \varepsilon_6^4, \varepsilon_6 \right) = \left( \varepsilon_6, \varepsilon_6^2, \varepsilon_6^3, 1 \right) = \tilde{g}.$$

Extracting now the first three components of the last vector, we obtain nothing else than the original non-shifted pattern $f = \left( \varepsilon_6, \varepsilon_6^2, \varepsilon_6^3 \right)$, which we have stored in the memory.

This interesting approach presented in [111, 112] demonstrates not only a high potential of MVN and MVN-based neural networks, but also demonstrates how important it is to treat the phase information properly. This gives a wonderful possibility to develop new applications.

## 6.4 Some Other Applications of MVN-Based Neural Networks and Concluding Remarks to Chapter 6

In this Chapter, we have considered just some applications of MVN-based neural networks. There are perhaps the most remarkable earlier applications (associative memories) and recent applications (solving some real-world classification and prediction problems using MLMVN).

Those applications, which we have considered in this Chapter, confirm high efficiency of MVN and MVN-based neural networks, as well as they definitely confirm important advantages of complex-valued neural networks over traditional real-valued machine learning techniques. Indeed, we have considered, for example, the challenging multi-class classification problem of blur and its parameters identification. While classical MLF fails to solve this problem and SVM employs over 25 million parameters to solve it, MLMVN solves it employing more than 1000 (!) times fewer parameters. It is important that a high accuracy demonstrated by MLMVN when solving this problem shows a reliable "margin of safety". It is clearly visible that 28 classes considered in that problem is absolutely not a limit. If it would be necessary, MLMVN could learn and solve a classification problem with more classes. What just probably would be necessary – to increase amount of output neurons if new groups of classes will be considered (for example, more types of blur with their parameters). As we have seen, the MLMVN learning algorithm works efficiently regardless of how many neurons and layers are in the network.

This property was also confirmed by the next application, which we have considered in this Chapter, financial time series prediction using MLMVN. We wanted to show that MLMVN can learn highly nonlinear real-world problems and can show a high generalization capability. Financial time series are highly nonlinear, they are very unsteady, contain many local and global jumps of different size. To learn them using a neural network with an aim to make long-term predictions, it is necessary to learn from a longer interval of historical data and to use a network with as many hidden neurons as it is possible. Otherwise, it is impossible to follow all jumps in data corresponding to different trends. While, for example, it is very difficult to train MLF with thousands of neurons, it is not a problem to train MLMVN with even tens thousands of neurons. As a result, MLMVN can be efficiently used to make long-term predictions of time series.

Other impressive applications of MVN-based neural networks, which we have considered in this Chapter, are MVN-based multistate associative memories. There are a Hopfield MVN-based associative memory, a cellular MVN-based associative memory, an MVN-based associative memory with random connections, and an ensemble of Hopfield-like memories. We have considered learning rules for these memories and analyzed their stability. We have also considered how these memories may retrieve patterns, which are stored there, even when highly corrupted patterns are shown to them. It is also important that we convinced one more time how essential it is to treat phase properly and how helpful and suitable is MVN to work with phase.

We believe that all these applications along with those applications and benchmark tests, which we have considered earlier, clearly demonstrate advantages of MVN and MVN-based neural networks, their efficiency in solving pattern recognition, classification, and prediction problems.

However, the size of this book is limited and we have not considered some other applications of MVN-based neural networks. Our goal was first of all to consider in this book a fundamental character of complex-valued neural networks in general and MVN-based neural networks in particular. We have considered in detail some recently developed efficient applications (based on MLMVN) and some earlier fundamental applications (associative memories). Nevertheless, we would like to at least point out some other important applications of MVN-based neural networks, which have been developed earlier.

We have mentioned many times that MVN is perhaps the most suitable machine learning tool, which may work with phase. We have also distinguished that Fourier phase spectrum can be used as a feature space for solving pattern recognition and classification problems. In 2002, in the paper [113], the author of this book together with E. Myasnikova, M. Samsonova, and J. Reinitz employed low frequency phases for solving the problem of gene expression patterns classification. Gene expression patterns were taken from a confocal microscope, and a problem was to perform a classification corresponding to eight temporal stages of gene expression. Patterns from adjacent temporal classes were quite similar to each other and it was very difficult to find a feature space where they could be separated. It was suggested in [113] to form a feature space from phases corresponding to the lowest frequencies. What is interesting, in 2002, when this work was done, MLMVN was not yet developed. To solve a problem, it was suggested to use an MVN-based neural network, which is possible to call an MVN-ensemble, where each neuron was assigned to solve its part of the problem. This

approach was probably too artificial (first, two neurons classified patterns as belonging to classes 1-4 or 5-8, then other four neurons classified patterns as belonging to classes 1-2, 3-4, 5-6, and 7-8, and finally eight neurons classified patterns as
belonging to certain classes. The "wiener takes it all" technique was used on each
stage). Nevertheless, the results were quite good; the average classification accuracy (over all classes) was 78.4% (it ranged from 70% to 91% for different
classes). It is interesting that all other machine learning tools that were tested for
solving the same problem, failed to approach this accuracy. The best result shown
by an alternative technique (linear discriminant analysis) was 64.2%. We believe
that these results were very important for showing that phases can be used as features for image recognition and classification and for showing that since MVN
preserves the phase information and treats it properly, it is a very suitable tool for
working with phases. After MLMVN was developed in 2004-2007 and presented
in [62], we believe that such classification problems as that one considered in
[113] should be solved using MLMVN, and better results should be expected. But
that basic approach to image recognition using phases as features and MVN-based
neural network as a tool, which was presented in [113], merits to be mentioned
here.

There is another interesting line of MVN applications, which was not presented
in this book. There are applications in image processing. This book was mostly
devoted to MVN fundamentals. We also concentrated on recently developed solutions (like MLMVN, its learning algorithm, and applications) and considered applications of MVN-based neural networks in pattern recognition, classification,
and prediction.

Application of MVN (mostly of MVN-based cellular neural network, CNN-
MVN) in image processing is not directly connected with its ability to learn. However, they extend an area where MVN can be used, and they are quite efficient.

The first such an application was considered by the author of this book in his
paper [114]. It was so-called precise edge detection, which is based on the split-
ting of an image into binary planes, their separate processing using a Boolean
function and integration of the resulting binary planes into the resulting gray-scale
image. Actually, it is clear that this kind of processing can be implemented without any neural network. But what is interesting, those Boolean functions, which
were used for edge detection corresponding to upward and downward brightness
jumps and for edge detection by narrow direction in [114] and then in [60] and
[115], are non-linearly separable. Hence, it was very attractive to learn them using
the universal binary neuron (which is, as we remember, MVN-P with a periodic
two-valued activation function). Learning of some of these Boolean functions
was considered in [114] and [60]. Learning of Boolean functions for edge detection by narrow direction was considered later in [88]. After any of these Boolean
functions is learned, it can be implemented using MVN-P. Then these neurons
integrated in a cellular neural network process an image.

Another interesting application is the use of specific nonlinearity of the discrete
MVN activation function (2.50) for nonlinear filtering. Actually, in this case
(2.50) applied to the weighted sum of intensity values in the local neighborhood of
a processed pixel is simply considered as a filter, which is in fact a nonlinear filter.

This filter (multi-valued nonlinear filter) was considered in detail in [60, 116]. All properties of this filter are determined by its weights and the activation function (2.50). What is interesting, that depending on the weighting templates this filter can be used for additive and multiplicative noise reduction and for image enhancement. Evidently, this filter can be implemented using a cellular neural network with multi-valued neurons. In this case, each neuron processes a single pixel of an image. In the paper [115], a special attention was paid to application of multi-valued nonlinear filtering and neural edge detection methods in medical image processing.

# Concluding Remarks

"If it be now,
'tis not to come; if it be not to come, it will be
now; if it be not now, yet it will come: the
readiness is all…"

William Shakespeare, Hamlet

Thus we are close to finish. Actually, what is left - following tradition, we have to make some concluding remarks.

However, we would not like to repeat here again the formal content of the book. In fact, the reader can easily take a look at it or Chapter-by-Chapter overview in the Preface.

What is the most important in this book and behind it in the author's opinion?

Our thinking is organized in such a way that often it is difficult to accept something, which does not look clear from the very beginning, which from the first point of view even contradicts to something we consider an axiom. A typical example is how difficult it was for a scientific world to understand and to accept complex numbers. We have already mentioned in the Preface that it took about 300 years for mathematicians to accept them and to understand their importance. Often we are too conservative to easily accept new ideas.

The author may now share his own experience. A significant part of this book is devoted to MLMVN. However, to develop it, it was necessary to overcome a classical stereotype that the learning of any feedforward neural network can be based only on some optimization technique, which immediately leads to the necessity of differentiability of a neuron activation function to design a learning algorithm. But neither discrete nor continuous MVN activation functions are differentiable! It took at least 2-3 years to understand that to develop MLMVN, it is possible to use the error-correction learning, which is simpler, reliable and does not require differentiability of an activation function.

What does it mean? This means only that we should easier accept new approaches and new solutions. We should not see them just through some traditional stereotypes, which may prevent to understand them and to develop something, which is significantly different form something traditional.

The main goal of this book is to show that complex-valued neural networks and particularly the ones with the multi-valued neuron are natural and powerful tools for solving challenging real-world problems in pattern recognition, classification, prediction, and simulation of complex systems.

I. Aizenberg: Complex-Valued Neural Networks with Multi-Valued Neurons, SCI 353, pp. 249–251.
springerlink.com                                    © Springer-Verlag Berlin Heidelberg 2011

We hope that presenting multi-valued neurons as explicitly as possible, we convinced the reader that complex-valued neural networks are as natural as their traditional real-valued counterparts. But we also hope that it was succeeded to show that often complex-valued neurons and neural networks are more powerful than real-valued neurons and neural networks and many other machine learning tools.

We have started from a simple example, which shows how a single two-valued neuron with a periodic activation function easily solves the XOR problem. While most of neural networks textbooks state that since a single (real-valued!) neuron cannot learn such non-linearly separable problems as XOR, neural networks must be developed, we say here: not, many non-linearly separable problems can be learned by a single multi-valued neuron with complex-valued weights and complex-valued activations function. Of course, neural networks from these neurons also can be developed, but they should be used for solving those challenging problems, which cannot be solved using real-valued neural networks with a desirable accuracy or which it is difficult to solve using other traditional machine learning techniques.

New challenging problems generate new approaches and vice versa. This is a way of progress. The necessity of solving pattern recognition problems favored the development of machine learning in general, and artificial neurons and neural networks in particular. Today's achievements in machine learning make it possible to solve those problems of pattern recognition that were even not considered yesterday. The necessity of solving multi-valued problems (especially, multi-class classification problems) was a very important stimulus for the development of multiple-valued logic over the field of complex numbers. In turn, this model of multiple-valued logic was followed by multi-valued neurons and neural networks based on them as utilizations of its basic ideas. The development of new solutions in this area makes it possible to consider more and more new applications. The first artificial neuron (McCulloch and Pitts) and the first learning rule (Hebb) were developed as simplified mathematical models of biological neurons and their learning. Today's much more sophisticated artificial neurons and learning techniques make it possible to develop simulations of biological neurons and biological neural networks that are much closer to the reality. It is possible to demonstrate more and more such examples.

In fact, complex-valued neural networks are now a rapidly developed area. They open new and very promising opportunities in solving challenging real-world problems. We have seen that many problems considered in this book can be easier solved using MVN and MVN-based neural networks than using other techniques. Some other problems cannot be solved using other techniques at all.

Thus, we should not be afraid of new solutions, which may look not traditional and which should be even difficult to accept immediately. But the sooner we accept them and understand their efficiency, the more significant achievements we should expect, and the more challenging problems we may solve.

So what is the most important, the reader may get from this book? There are the following fundamental facts, which we would like to distinguish.

- Complex-valued neural networks are absolutely natural as well as their real-valued counterparts.
- Multiple-valued threshold logic over the field of complex numbers, being a generalization of traditional threshold logic, is a background on which the multi-valued neuron is based.
- A single multi-valued neuron with complex-valued weights and complex-valued activation function is significantly more functional than any real-valued neuron. It can learn multi-valued input/output mappings, which cannot be learned using a single real-valued neuron.
- A single MVN with a periodic activation function (MVN-P) may learn those input/output mappings, which are non-linearly separable in the real domain. Such problems as XOR and Parity $n$, for example, are about the simplest that can be learned by a single MVN-P.
- The most efficient MVN learning rule is the error-correction rule, which generalizes the classical Rosenblatt-Novikoff error-correction rule for the multi-valued case. The learning algorithm based on this rule should not be considered as the optimization problem, which is its advantage; it does not suffer from the local minima phenomenon.
- The latter is true not only for a single MVN, but for MLMVN, MVN-based feedforward neural network. Its error-backpropagation and its learning algorithm are derivative-free. The MLMVN learning algorithm is based on the same error-correction learning rule as the learning algorithm for a single MVN.
- MLMVN can be used for solving those multi-class classification problems, time series prediction problems, which cannot be efficiently solved using other machine learning techniques. MLMVN can also be used for simulation of complex systems.
- MVN treats the phase information properly and therefore it can easily employ phases as a feature space for solving classification and pattern recognition problems; it can also use phases to store patterns in an MVN-based associative memory.
- MVN can be a better model for simulation of a biological neuron. In biological neurons, the information transmitted by them to each other is encoded as the frequency of spikes. Since the frequency can easily be associated with phase, which determines the MVN state, inputs, and output, MVN can be used for simulation of a biological neuron.

Finally, the author sincerely hopes that his main goal, the explicit presentation of MVN and MVN-based neural networks, is achieved by writing this book.

Hopefully, this book will help to involve more people in research in the filed of complex-valued neural networks. The author will be glad and consider his work successful if more researches will use MVN and MVN-based neural networks for solving new challenging real-world problems.

# References

1. Aleksander, I., Morton, H.: An Introduction to Neural Computing. Chapman & Hall, London (1990)
2. Haykin, S.: Neural Networks: A Comprehensive Foundation, 2nd edn. Prentice Hall, Englewood Cliffs (1998)
3. Zurada, J.M.: Introduction to Artificial Neural Systems. West Publishing Company, St. Paul (1992)
4. McCulloch, W.S., Pits, W.: A Logical Calculus of the Ideas Immanent in Nervous Activity. Bull. Math. Biophys. 5, 115–133 (1943)
5. Gupta, M.M., Jin, L., Homma, N.: Static and Dynamic Neural Networks. John Wiley & Sons, Hoboken (2003)
6. Dertouzos, M.L.: Threshold Logic: A Synthesis Approach. The MIT Press, Cambridge (1965)
7. Muroga, S.: Threshold Logic and Its Applications. John Wiley & Sons, New York (1971)
8. Hebb, D.O.: The Organization of Behavior. John Wiley & Sons, New York (1949)
9. Rosenblatt, F.: The Perceptron: A Probabilistic Model for Information Storage and Organization in the Brain, Cornell Aeronautical Laboratory. Psychological Review 65(6), 386–408 (1958)
10. Rosenblatt, F.: On the Convergence of Reinforcement Procedures in Simple Perceptron. Report VG 1196-G-4. Cornell Aeronautical Laboratory, Buffalo, NY (1960)
11. Rosenblatt, F.: Principles of Neurodynamics. Spartan Books, Washington DC (1962)
12. Novikoff, A.B.J.: On Convergence Proofs on Perceptrons. In: Symposium on the Mathematical Theory of Automata, vol. 12, pp. 615–622. Polytechnic Institute of Brooklyn (1963)
13. Minsky, M.L., Papert, S.A.: Perceptron: An introduction to Computational Geometry. MIT Press, Cambridge (1969)
14. Touretzky DS, Pomerleau, D.A.: What's hidden in the hidden layers? Byte, 227–233 (August 1989)
15. Rumelhart, D.E., Hilton, G.E., Williams: Learning Internal Representations by Error Propagation. In: Parallel Distributed Processing: Explorations in the Microstructure of Cognition, vol. 1, ch. 8. MIT Press, Cambridge (1986)
16. Werbos, P.J.: The Roots of Backpropagation: From Ordered Derivatives to Neural Networks and Political Forecasting (Adaptive and Learning Systems for Signal Processing, Communications and Control Series). Wiley, NY (1994)
17. Kolmogorov, A.N.: On the Representation of Continuous Functions of many Variables by Superposition of Continuous Functions and Addition. Doklady Akademii Nauk SSSR 114, 953–956 (1957) (in Russian)

18. Powell, M.J.D.: Radial Basis Functions for Multivariable Interpolation: A Review. In: Proc. of Conference on Algorithms for the Approximation of Functions and Data, pp. 143–167. RMCS, Shrivenham (1985)
19. Hornik, K., Stinchcombe, M., White, H.: Multilayer Feedforward Neural Networks are Universal Approximators. Neural Networks 2, 259–366 (1989)
20. Hecht-Nielsen, R.: Theory of the backpropagation neural network. In: Proceedings of 1989 IEEE International Joint Conference on Neural Networks, Washington DC, vol. 1, pp. 593–605 (1989)
21. Hopfield, J.J.: Neural Networks and Physical Systems with Emergent Collective Computational Abilities. Proceedings of the National Academy of Sciences of the USA 79, 2554–2558 (1982)
22. Kohonen, T.: Associative memory - a Systemtheoretical Approach. Springer, Berlin (1977)
23. Hopfield, J.J.: Neurons with graded response have collective computational properties like those of two-sate neurons. Proceedings of the National Academy of Sciences of the USA 81, 3088–3092 (1984)
24. Chua, L.O., Yang, L.: Cellular Neural Networks: Theory & Applications. IEEE Transactions on Circuits and Systems 35(10), 1257–1290 (1988)
25. Vapnik, V.N.: The nature of statistical learning theory. Springer, New York (1995)
26. Vapnik, V.N.: The nature of statistical learning theory, 2nd edn. Springer, New York (1998)
27. Redlapalli, S., Gupta, M.M., Song, K.-Y.: Development of Quadratic Neural Unit with Applications to Pattern Classification. In: Proceedings of 4th International Symposium on Uncertainty Modeling and Analysis, pp. 141–146. College Park, Maryland (2003)
28. Cover, T.M.: Geometrical and Statistical Properties of systems of Linear Inequalities with application in pattern recognition. IEEE Transactions on Electronic Computers, EC 14, 326–334 (1965)
29. Aizenberg, I.N.: Model of the Element With Complete Functionality. Izvestia Akademii Nauk SSSR, Technicheskaia Kibernetika (The News of the Academy of Sciences of the USSR, Technical Cybernetics) 2, 188–191 (1985) (in Russian)
30. Aizenberg, I.N.: A Universal Logic Element Over Complex Field. Cybernetics and Systems Analysis 27, 467–473 (1991)
31. Oppenheim, A.V., Lim, J.S.: The importance of phase in signals. IEEE Proceedings 69, 529–541 (1981)
32. Aizenberg, N.N., Ivaskiv, Y.L., Pospelov, D.A.: About one generalization of the threshold function. Doklady Akademii Nauk SSSR (The Reports of the Academy of Sciences of the USSR) 196(6), 1287–1290 (1971) (in Russian)
33. Aizenberg, N.N., Ivaskiv, Y.L., Pospelov, D.A., Hudiakov, G.F.: Multiple-Valued Threshold Functions. I. Boolean Complex-Threshold Functions and Their Generalization. Kibernetika (Cybernetics) 4, 44–51 (1971) (in Russian)
34. Aizenberg, N.N., Ivaskiv, Y.L., Pospelov, D.A., Hudiakov, G.F.: Multiple-Valued Threshold Functions. II. Synthesis of the Multi-Valued Threshold Elements. Kibernetika (Cybernetics), 1, 53–66 (1973) (in Russian)
35. Aizenberg, N.N., Ivaskiv, Y.L., Pospelov, D.A., Hudiakov, G.F.: Multiple-Valued Threshold Functions. I. Boolean Complex-Threshold Functions and Their Generalization. Cybernetics and Systems Analysis 7(4), 626–635 (1971)

36. Aizenberg, N.N., Ivaskiv, Y.L., Pospelov, D.A., Hudiakov, G.F.: Multivalued Threshold Functions. Synthesis of Multivalued Threshold Elements. Cybernetics and Systems Analysis 9(1), 61–77 (1973)

37. Aizenberg, N.N., Ivaskiv, Y.L.: Multiple-valued threshold logic. Naukova Dumka, Kiev (1977) (in Russian)

38. Aizenberg, N.N., Aizenberg, I.N.: CNN Based on Multi-Valued Neuron as a Model of Associative Memory for Gray-Scale Images. In: Proceedings of the 2-d International Workshop on Cellular Neural Networks and their Applications (CNNA 1992), Munich, October1992, pp. 36–41 (1992)

39. Noest, A.J.: Discrete-State Phasor Neural Networks. Physical Review A38, 2196–2199 (1988)

40. Leung, H., Haykin, S.: The Complex Backpropagation Algorithm. IEEE Transactions on Signal Processing 39(9), 2101–2104 (1991)

41. Georgiou, G.M., Koutsougeras, C.: Complex Domain Backpropagation. IEEE Transactions on Circuits and Systems CAS- II. Analog and Digital Signal Processing 39(5), 330–334 (1992)

42. Hirose, A.: Complex-Valued Neural Networks. Springer, Heidelberg (2006)

43. Hirose, A. (ed.): Complex-Valued Neural Networks: Theories and Applications. World Scientific, Singapore (2003)

44. Hirose, A.: Dynamics of Fully Complex-Valued Neural Networks. Electronics Letters 28(16), 1492–1494 (1992)

45. Hirose, A.: Continuous Complex-Valued Back-propagation Learning. Electronics Letters 28(20), 1854–1855 (1992)

46. Nitta, T. (ed.): Complex-Valued Neural Networks: Utilizing High-Dimensional Parameters. Information Science Reference, Pennsylvania, USA (2009)

47. Nitta, T.: An Extension of the Back-Propagation Algorithm to Complex Numbers. Neural Networks 10(8), 1391–1415 (1997)

48. Nitta, T.: A Solution to the 4-bit Parity Problem with a Single Quaternary Neuron. Neural Information Processing. Letters and reviews 5(2), 33–39 (2004)

49. Mandic, D.P., Goh, S.L.: Complex Valued Nonlinear Adaptive Filters: Noncircularity, Widely Linear and Neural Models. Wiley, Chichester (2009)

50. Goh, S.L., Mandic, D.P.: An Augmented Extended Kalman Filter Algorithm for Complex-Valued Recurrent Neural Networks. Neural Computation 19(4), 1–17 (2007)

51. Goh, S.L., Mandic, D.P.: Nonlinear Adaptive Prediction of Complex Valued Nonstationary Signals. IEEE Transactions on Signal Processing 53(5), 1827–1836 (2005)

52. Fiori, S.: Extended Hebbian Learning for Blind Separation of Complex-Valued Sources. IEEE Transactions on Circuits and Systems – Part II 50(4), 195–202 (2003)

53. Fiori, S.: Non-Linear Complex-Valued Extensions of Hebbian Learning: An Essay. Neural Computation 17(4), 779–838 (2005)

54. Fiori, S.: Learning by Criterion Optimization on a Unitary Unimodular Matrix Group. Journal of Neural Systems 18(2), 87–103 (2008)

55. Amin, M.d.F., Murase, K.: Single-Layered Complex-Valued Neural Network for Real-Valued Classification Problems. Neurocomputing 72(4-6), 945–955 (2009)

56. Amin, M.d.F., Islam, M.d.M., Islam, K.: Ensemble of single-layered complex-valued neural networks for classification tasks. Neurocomputing 7(10-12), 2227–2234 (2009)

57. Buchholz, S., Sommer, G.: On Clifford Neurons and Clifford Multi-Layer Perceptrons. Neural Networks 21(7), 925–935 (2008)

58. Buchholz, S., Bihan, N.L.: Polarized Signal Classification by Complex and Quaternionic Multi-Layer Perceptrons. International Journal of Neural Systems 18(2), 75–85 (2008)
59. Łukasiewicz, J.: O logice trójwartościowej (in Polish). Ruch filozoficzny 5, 170–171 (1920); English translation: On three-valued logic. In: Borkowski, L. (ed.) Selected works by Jan Łukasiewicz, pp. 87–88. North–Holland, Amsterdam (1970)
60. Aizenberg, I., Aizenberg, N., Vandewalle, J.: Multi-Valued and Universal Binary Neurons: Theory, Learning and Applications. Kluwer, Boston (2000)
61. Aizenberg, I.: A Periodic Activation Function and a Modified Learning Algorithm for the Multi-Valued Neuron. IEEE Transactions on Neural Networks 21(12), 1939–1949 (2010)
62. Aizenberg, I., Moraga, C.: Multilayer Feedforward Neural Network Based on Multi-Valued Neurons (MLMVN) and a Backpropagation Learning Algorithm. Soft Computing 11(2), 169–183 (2007)
63. Si, J., Michel, A.N.: Analysis and Synthesis of a Class of Discrete-Time Neural Networks with Multilevel Threshold Neurons. IEEE Transactions on Neural Networks 6(1), 105–116 (1995)
64. Schwartz, L.: Analyse Mathématicue, Cours Professe a L'ecole Polytechnique I. Hermann, Paris (1967)
65. Schwartz, L.: Analyse Mathématicue, Cours Professe a L'ecole Polytechnique II. Hermann, Paris (1967)
66. Müller, B., Reinhardt, J.: Neural Networks: An Introduction. Springer, Heidelberg (1990)
67. Kandel, E.C., Schwartz, J.H.: Principles of Neural Science. Elsevier, Amsterdam (1985)
68. Gerstner, W., Kistler, W.: Spiking Neuron Models. Single Neurons, Populations, Plasticity. Cambridge University Press, Cambridge (2002)
69. Izhikevich, E.M.: Simple Model of Spiking Neurons. IEEE Transactions on Neural Networks 14(6), 1569–1572 (2003)
70. Mikula, S., Niebur, E.: Correlated Inhibitory and Excitatory Inputs to the Coincidence Detector: Analytical Solution. IEEE Transactions on Neural Networks 15(5), 957–962 (2004)
71. Izhikevich, E.M.: Dynamical Systems in Neuroscience: The Geometry of Excitability and Bursting. MIT Press, Cambridge (2007)
72. Valova, I., Gueorguieva, N., Troesche, F., Lapteva, O.: Modeling of inhibition/excitation firing in olfactory bulb through spiking neurons. Neural Computing and Applications 16(4), 355–372 (2007)
73. Izhikevich, E.M.: Which Model to use for Cortical Spiking Neurons. IEEE Transactions on Neural Networks 15(5), 1063–1070 (2004)
74. Schwarz, H.A.: Über ein die Flächen kleinsten Flächeninhalts betreffendes Problem der Variationsrechnung. Acta Soc. Scient. Fen. 15, 315–362 (1885); Reprinted in Gesammelte Mathematische Abhandlungen, vol. 1, pp. 224-269. Chelsea, New York (1972)
75. Conway, J.B.: Functions of One Complex Variable, 2nd edn. Springer, Heidelberg (1978)
76. Aizenberg, I., Moraga, C., Paliy, D.: A Feedforward Neural Network based on Multi-Valued Neurons. In: Reusch, B. (ed.) Computational Intelligence, Theory and Applications. Advances in Soft Computing, XIV, pp. 599–612. Springer, Heidelberg (2005)

77. Aizenberg, I., Paliy, D.V., Zurada, J.M., Astola, J.T.: Blur Identification by Multi-layer Neural Network based on Multi-Valued Neurons. IEEE Transactions on Neural Networks 19(5), 883–898 (2008)

78. Mackey, M.C., Glass, L.: Oscillation and chaos in physiological control systems. Science 197, 287–289 (1977)

79. Fahlman, J.D., Lebiere, C.: Predicting the Mackey-Glass time series. Physical Review Letters 59, 845–847 (1987)

80. Paul, S., Kumar, S.: Subsethood-Product Fuzzy Neural Inference System (SuPFu-NIS). IEEE Transactions on Neural Networks 13, 578–599 (2002)

81. Islam, M.M., Yao, X., Murase, K.: A Constructive Algorithm for Training Cooperative Neural Networks Ensembles. IEEE Transactions on Neural Networks 14, 820–834 (2003)

82. Kim, D., Kim, C.: Forecasting Time Series with Genetic Fuzzy Predictor Ensemble. IEEE Transactions on Neural Networks 5, 523–535 (1997)

83. Jang, J.S.R.: ANFIS: Adaptive-Network-Based Fuzzy Inference System. IEEE Transactions on Systems, Man and Cybernetics 23, 665–685 (1993)

84. Russo, M.: Genetic Fuzzy Learning. IEEE Transactions on Evolutionary Computation 4, 259–273 (2000)

85. Yao, X., Liu, Y.: A new Evolutionary System for Evolving Artificial Neural Networks. IEEE Transactions on Neural Networks 8, 694–713 (1997)

86. Lee, S.-H., Kim, I.: Time Series Analysis using Fuzzy Learning. In: Proceedings of the International Conference on Neural Information Processing, Seoul, Korea, vol. 6, pp. 1577–1582 (1994)

87. Aizenberg, N.N., Aizenberg, I.N.: Quickly Converging Learning Algorithms for Multi-Level and Universal Binary Neurons and Solving of the Some Image Processing Problems. In: Mira, J., Cabestany, J., Prieto, A.G. (eds.) IWANN 1993. LNCS, vol. 686, pp. 230–236. Springer, Heidelberg (1993)

88. Aizenberg, I.: Solving the XOR and Parity n Problems Using a Single Universal Binary Neuron. Soft Computing 12(3), 215–222 (2008)

89. Asuncion, A., Newman, D.J.: UCI Machine Learning Repository. University of California, School of Information and Computer Science, Irvine, CA (2007), http://www.ics.uci.edu/~mlearn/MLRepository.html

90. Benabdeslem, K., Bennani, Y.: Dendogram-Based SVM for Multi-Class Classification. Journal of Computing and Information Technology 14(4), 283–289 (2006)

91. Mizutani, E., Dreyfus, S.E.: MLP's hidden-node saturations and insensitivity to initial weights in two classification benchmark problems: parity and two-spirals. In: Proc. of the 2002 International Joint Conference on Neural Networks (IJCNN 2002), pp. 2831–2836 (2002)

92. Tian, M., Chen, S.C., Zhuang, Y., Liu, J.: Using statistical analysis and support vector machine classification to detect complicated attacks. In: Proc. of the Third International Conference on Machine Learning and Cybernetics, Shanghai, pp. 2747–2752 (2004)

93. Chen, J.H., Chen, C.S.: Fuzzy Kernel Perceptron. IEEE Transactions on Neural Networks 13, 1364–1373 (2002)

94. Sewak, M., Vaidya, P., Chan, C.-C., Duan, Z.-H.: SVM Approach to Breast Cancer Classification. In: Proc. of the 2nd International Multisymposium on Computer and Computational Science, pp. 32–37. IEEE Computer Society Press, Los Alamitos (2007)

95. Aizenberg, I., Caudill, M., Jackson, J., Alexander, S.: Learning Nonlinearly Separable mod k Addition Problem Using a Single Multi-Valued Neuron With a Periodic Activation Function. In: Proceedings of the 2010 IEEE World Congress on Computational Intelligence – 2010 IEEE International Joint Conference on Neural Networks, Barcelona, Spain, pp. 2577–2584 (2010)

96. Gonzales, R.C., Woods, R.E.: Digital Image Processing, 3rd edn. Prentice-Hall, Englewood Cliffs (2008)

97. Tikhonov, A.N., Arsenin, V.Y.: Solutions of ill-Posed Problems. Wiley, N.Y (1977)

98. Rushforth, C.: Signal Restoration, functional analysis, and Fredholm integral equations of the first kind. In: Image Recovery: Theory and Application, Academic Press, London (1987)

99. Katkovnik, V., Egiazarian, K., Astola, J.: A spatially Adaptive Nonparametric Image Deblurring. IEEE Trans. on Image Processing 14(10), 1469–1478 (2005)

100. Neelamani, R., Choi, H., Baraniuk, R.G.: Forward: Fourier-Wavelet Regularized Deconvolution for ill-Conditioned Systems. IEEE Trans. on Signal Processing 52(2), 418–433 (2003)

101. Lagendijk, R.L., Biemond, J., Boekee, D.E.: Identification and Restoration of Noisy Blurred Images Using the Expectation-Maximization Algorithm. IEEE Trans. on Acoustics, Speech and Signal Processing 38, 1180–1191 (1990)

102. Fletcher, R., Reeves, C.M.: Function minimization by Conjugate Gradients. Computer Journal 7, 149–154 (1964)

103. Moller, M.F.: A scaled conjugate Gradient Algorithm for Fast Supervised Learning. Neural Networks 6, 525–533 (1993)

104. Chang, C.C., Lin, C.J.: LIBSVM: a library for support vector machines. Software (2001), http://www.csie.ntu.edu.tw/~cjlin/libsvm

105. Duan, K.-B., Keerthi, S.S.: Which Is the Best Multiclass SVM Method? An Empirical Study. In: Oza, N.C., Polikar, R., Kittler, J., Roli, F. (eds.) MCS 2005. LNCS, vol. 3541, pp. 278–285. Springer, Heidelberg (2005)

106. Box, G., Jenkins, G., Reinsel, G.: Time series analysis: forecasting and control, 4th edn. Wiley, Chichester (2008)

107. Tan, S., Hao, J., Vandewalle, J.: Cellular Neural Networks as a Model of Associative Memories. In: Proceedings of the 1990 IEEE International Workshop on CNN and their applications (CNNA 1990), Budapest, pp. 23–26 (1990)

108. Jankowski, S., Lozowski, A., Zurada, M.: Complex-Valued Multistate Neural Associative Memory. IEEE Transactions on Neural Networks 7(6), 1491–1496 (1996)

109. Muezzinoglu, M.K., Guzelis, C., Zurada, J.M.: A New Design Method for the Complex-Valued Multistate Hopfield Associative Memory. IEEE Transactions on Neural Networks 14(4), 891–899 (2003)

110. Aizenberg, N.N., Aizenberg, I.N., Krivosheev, G.A.: Multi-Valued Neurons: Learning, Networks, Application to Image Recognition and Extrapolation of Temporal Series. In: Sandoval, F., Mira, J. (eds.) IWANN 1995. LNCS, vol. 930, pp. 389–395. Springer, Heidelberg (1995)

111. Aoki, H., Kosugi, Y.: An Image Storage System Using Complex-Valued Associative Memory. In: Proc. of the 15th International Conference on Pattern Recognition, vol. 2, pp. 626–629. IEEE Computer Society Press, Barcelona (2000)

112. Aoki, H., Watanabe, E., Nagata, A., Kosugi, Y.: Rotation-Invariant Image Association for Endoscopic Positional Identification Using Complex-Valued Associative Memories. In: Mira, J., Prieto, A.G. (eds.) IWANN 2001. LNCS, vol. 2085, pp. 369–374. Springer, Heidelberg (2001)

113. Aizenberg, I., Myasnikova, E., Samsonova, M., Reinitz, J.: Temporal Classification of Drosophila Segmentation Gene Expression Patterns by the Multi-Valued Neural Recognition Method. Mathematical Biosciences 176(1), 145–159 (2002)
114. Aizenberg, I.: Processing of Noisy and Small-Detailed Gray-Scale Images using Cellular Neural Networks. Journal of Electronic Imaging 6(3), 272–285 (1997)
115. Aizenberg, I., Aizenberg, N., Hiltner, J., Moraga, C., Meyer zu Bexten, E.: Cellular Neural Networks and Computational Intelligence in Medical Image Processing. Image and Vision Computing 19(3), 177–183 (2001)
116. Aizenberg, I., Butakoff, C.: Image Processing Using Cellular Neural Networks Based on Multi-Valued and Universal Binary Neurons. Journal of VLSI Signal Processing Systems for Signal, Image and Video Technology 32, 169–188 (2002)

# Index

Printed in the United States
by Bookmasters

Printed in the United States
By Bookmasters